Galactic Encounters

Frontispiece. Comet Lovejoy, projected against the southern Milky Way. Image by John Drummond at Gisborne, New Zealand, with a 20mm f2.8 Canon lens. *Courtesy: John Drummond.*

Galactic Encounters

Our Majestic and Evolving Star-System, From the Big Bang to Time's End

William Sheehan and
Christopher J. Conselice

Springer

William Sheehan
Willmar, MN, USA

Christopher J. Conselice
School of Physics and Astronomy
University of Nottingham
Nottingham, UK

ISBN 978-0-387-85346-8 ISBN 978-0-387-85347-5 (eBook)
DOI 10.1007/978-0-387-85347-5
Springer New York Heidelberg Dordrecht London

Library of Congress Control Number: 2014949181

Printed on acid-free paper

Springer is part of Springer Science+Business Media (www.springer.com)

to the brightest stars in our firmament,
Debb and Estelle.

* * *

Contents

Acknowledgments

This book has been an effort spanning more than a decade. At the beginning, the authors never guessed that it would take that long. However, the fact that it did bears out the truth of something Dr. Johnson said in his "Life of Pope":

> "The distance is commonly very great between actual performances and speculative possibility. It is natural to suppose that as much as has been done to-day may be done to-morrow; but on the morrow some difficulty emerges, or some external impediment obstructs. Indolence, interruption, business, and pleasure, all take their turns of retardation; and every long work is lengthened by a thousand causes that can, and ten thousand that cannot, be recounted. Perhaps no extensive and multifarious performance was ever effected within the term originally fixed in the undertaker's mind. He that runs against Time has an antagonist not subject to casualties."

Many people have contributed significantly to the book, and without their counsel and assistance it would have taken much longer to complete than it did—if indeed it would have been written at all.

When W.S. embarked on his biography of E. E. Barnard, *The Immortal Fire Within,* he found a tremendous mentor in the late Donald E. Osterbrock, and much of what is best in the following pages are thanks to the assistance and counsel, freely and generously rendered, of "DEO." Given that Barnard was the great pioneer of the wide-angle photography of the Milky Way and discoverer of the dark nebulae, work on Barnard began to lead to plans for further investigations of the Galaxy, and a Guggenheim Fellowship in 2001 for the "structure and evolution of the Galaxy" allowed time (and travel) away from professional commitments and a chance to begin work in earnest. Among those whose strong support at this stage was much appreciated were Michael Crowe, Steven J. Dick, and Owen Gingerich. Also, look-

ing back across decades, mention should be made of Robert Burnham, Jr. and Helen Horstman of the Lowell Observatory, who encouraged a fledgling interest that could as easily have been quashed as encouraged.

In 2001, W.S. spent a productive six weeks at the Cosmic Dust Laboratory at the University of the Witwatersrand with David Block, whose work on infrared imaging of galaxies is internationally known. He also learned much at the Conference on Galactic Bars, especially from David Block and Ken Freeman, held at the Pilanesberg Game Reserve in South Africa during the transit of Venus in 2004. At that time he began conversations with Chris Conselice, whom he had first met when he was at Yerkes emulating E.E. Barnard's feat of observing an eclipse of Iapetus by Saturn's rings in 1993, and who was now a professional astronomer interested in galaxies in the early universe. It was as a result of that contact that Chris first became involved in the project.

W.S. thanks Robert W. Smith, Michael Hoskin and Tony Simcock—and above all the late Peter Hingley, of the Library of the Royal Astronomical Society—for help and advice with the Herschels; also the Herschel Society, for permission to use pictures taken at the William Herschel Museum at 19 New King Street, Bath; Andrew Stephens and Brendan, the seventh Earl of Rosse, for Lord Rosse; C. Robert O'Dell on Barnard and the Orion Nebula; Michael Sims, who helped with the Barnard papers at the Joseph Heard Library at Vanderbilt University; Alison Doane and Owen Gingerich with G.P. Bond and archives the Harvard College Observatory; Irene Osterbrock and Dorothy Schaumberg for helping to access manuscripts in the Mary Lea Shane archives of the Lick Observatory; Dimitri Mihalas, Bob Garrison, Richard Kron, and Lewis Hobbs on W.W. Morgan, and members of the Morgan family for granting me access to private Morgan letters and notebooks; Eric Hillemann of Carleton College on Edward Fath; William Lowell Putnam, III, Kevin Schindler, Antoinette Beiser, Joe Tenn, and Deidre Hunter on V.M. Slipher and dwarf galaxies; the late Gale E. Christianson for his views on Edwin Hubble. An old physics lab partner (at University of Minnesota) Curtis Struck offered excellent advice on galaxy collisions. Andy Young offered astute insights on astronomical spectroscopy and galaxies. For insights into interstellar matter and French astronomy, thanks are due to James Lequeux and Françoise Launay.

For help with illustrations, Laurie Hatch, Rem Stone, Kyle Cudworth, Richard Dreiser, Judy Bausch, Tony Misch, John Drummond, Marilyn Head, Tony Simcock, Mike Hatcher, Terry Dickinson, Klaus Brasch, and Stephen Leshin, John Grula of the Carnegie Observatories—Carnegie Institution for Science, Dan Lewis of the Huntington Library, San Marino, California, and Alar Toomre of the Massachusetts Institute of Technology. All performed yeoman service. A special thanks to a special high school friend, Mike Conley, who has remained actively involved for forty years in

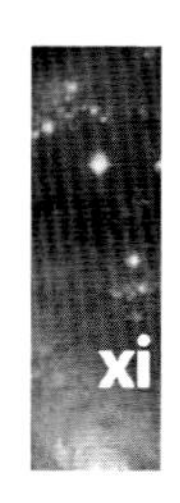

pursuit of astronomical "game" he helped get the small observatory at the North Fork of the Crow River up and running, and helped to obtain many of the images featured herein.

Drafts of the manuscript were improved immeasurably through the efforts of Mary Cain Buzzell, Jan Millsapps, Carl Pennypacker, Lee T. Macdonald, Parke Kunkle, and above all Bob O'Dell and David Sellers. Not all of them were professional astronomers, but those who were not were intelligent and perceptive lay readers. Long ago W.S. learned to heed Priscilla Bok's advice to her husband Bart, that it was wise to involve non-astronomers in reading manuscript materials, unless the book was meant for only experts (she mentioned Jan Oort). Heeding this advice, Bart and Priscilla produced the classic book *The Milky Way,* which set a standard all other books in the subject published since must heed and aspire to.

There were many others who contributed to the book over many years. The editors at Springer were efficient and professional, but one person stood out for special recognition, Joe Piliero, master book-designer, who produced the design and with patient and gentle prodding helped us fully realize our vision in what can be the most harrowing stage of the process for authors (there are good reasons Dickens never looked at any of his books after they were published!). Not least deserving acknowledgement are the members of W.S.'s family: wife Debb, sons Brendan and Ryan, and Cavalier King Charles spaniels Brady and Ruby, all of whom graciously granted hours of solitude, allowing the concentration that such a work demands.

CC would like to thank Bill Sheehan for inviting him to collaborate on this book, and for his patience over the years while it was being written. He also thanks colleagues, staff and students at the University of Nottingham, and especially Estelle Derclaye, for their support during the writing of this book.

* * *

— William Sheehan, Willmar, Minnesota

— Christopher Conselice, Nottingham, U.K.

February 17, 2014

Preface

Welcome to a masterful telling of a wonderful story. Man has a natural curiosity that extends to our place in the Universe. Satisfying that curiosity is the goal of this remarkable book by William Sheehan and Christopher Conselice.

Their historical approach to the tale is not only interesting in its own right, but also because it helps the reader to understand "why" we know what we do, not simply "what". The enterprising Italian Galileo Galilei's exploitation of the development of the telescope is an appropriate place to start, because the story is about the modern view of the Universe and the modern picture only began with the telescope.

There are heroes aplenty—the driven brother-sister collaborators William and Caroline Herschel; the pioneer of large telescopes, William Parsons Earl of Rosse, the self-educated Edward Emerson Barnard; and the morphologist W. W. Morgan, to mention only a few. There are entrepreneurs, con artists, and a considerable number of big egos. I'll leave it to the reader to identify them.

Lead author Sheehan draws on his multiple talents and areas of expertise. From his lifelong interest in astronomy, he shows his pleasure in observing things himself, not simply using the results obtained by others. As a historian of considerable stature, he has studied the lives and work of astronomers from the seventeenth through twentieth centuries. He is also a Doctor of Medicine, specializing in psychiatry, which informs his depictions of the personalities through the centuries in an insightful, informed, and wonderfully entertaining narrative.

Conselice is a practitioner of our shared craft, conducting original research on galaxy formation in the early universe. His description of the rapidly evolving picture of the large-scale Universe is told with remarkable clarity and an obvious depth of knowledge.

The last chapters delve into our totally unexpected present view, that we live in a Universe that defies expectation in that it expands ever faster through the action of mysterious dark energy and that our substance and everything that we see is but a few percent of all the material that is out there.

Solving today's problems continues, and it is a wise scientist indeed who recognizes that what he is saying and what his contemporaries believe may not be true. We know from studying the past that understanding comes in jumps, often by pursuing new directions, and that the term "the fog of war" also applies when we are trying to advance the frontiers of science.

* * *

— *Robert O'Dell*
Distinguished Research Professor of Astrophysics
Vanderbilt University
Nashville, TN

1. Setting the Scene

The lowing herd wind slowly o'er the lea,
The plowman homeward plods his weary way,
And leaves the world to darkness and to me.

—Thomas Gray, "Elegy written in a country churchyard"

Far from city lights, the view of the night sky is dominated by stars. They seem to be present in countless numbers. It comes as a surprise to learn that with the naked eye only about three thousand are visible at any given time. Most of these lie within a hundred, and all but a few within a thousand, light years of the Earth.

Despite our tendency to regard ourselves as the acme of creation, the human eye, at least as an astronomical instrument, performs rather poorly.

The human eye is the product of a long and complex evolutionary process that began billions of years ago with the eyespots of unicellular organisms. It has been molded by natural selection in response to the environments frequented by ancestral forms over thousands of generations. Thus, the human eye is a typical vertebrate eye, neither the most complex nor the most highly developed. (The most complex eye in the animal kingdom appears to belong to invertebrates, i.e., mantis shrimp, or stomatopods, have compound eyes with 16 different sorts of visual pigments, including 12 for color analysis in different wavelengths, four are sensitive to ultraviolet light and four for analyzing polarized light). The human eye is not so specialized; it is instead a generalist, like a Swiss army knife. It has a lens that accommodates quickly to keep objects near and far in focus; its retina contains a fovea, with close-packed cones providing detailed images and color information under high-intensity lighting conditions, and rods, sensitive to low-intensity light. Clearly, the human eye has adapted mostly for daylight conditions, and has a marvelous ability to detect motion; for

this, it uses short "exposure times" for integrating an image, but at the expense of not being able to see very faint objects. It was clearly not fashioned for pondering the night sky or for looking through telescopes.

Indeed, as a detector of low-intensity light, the human eye fares extremely poorly even compared to the eyes of other animals, like nocturnal animals or those that inhabit the gloomy regions of the deep ocean. The eye's light sensitivity is a function of its ability to capture photons. This is determined by: 1) the pupil diameter (in humans, ranging from 4 to 9mm), 2) the photoreceptor diameter, 3) the focal-length of the eye, where the shorter the focal-length, i.e., the "faster" the optics, just as in telescopes, the better, and 4) the fraction of light absorbed and converted into a signal, the quantum efficiency.

The photoreceptors of the human eyes have a quantum efficiency of about half, which means that about half of the photons received elicit a response. The superior sensitivity of nocturnal animals and benthic sea creatures does not have to do with the quantum efficiency of their photoreceptors but with the optimization of the other variables. They dispense with cones and have only rod receptors, have much larger pupils, and have eyes with very short focal lengths. In this way giant isopods, for instance, which live at depths of over a mile below the surface of the ocean and have huge reflective compound eyes, achieve a sensitivity of some 250 times that of the dark-adapted human eye. Of course, deep-sea creatures can never see the night-sky; among land animals, owls, tarsiers, and lemurs rate among the champions at seeing in the dark. They would presumably enjoy an awesome view of the stars and the Milky Way.

Incidentally, dung beetles (the species *Scarabaeus satyrus*), which have compound eyes but small brains, appear to orient themselves on moonless nights by the Milky Way as they roll their balls of muck along the ground. Perhaps astronomers on Earth appear that way to the condescending super-intelligent denizens of some vastly superior extraterrestrial civilization.

The point is that the human eye isn't very sensitive, and shows only a relatively small part of the contents of the night-sky. Obviously, during full moonlight, or sitting around a campfire, its space-penetrating power is even less. The naked (i.e., unassisted) eye can only nibble at the edge of the universe by revealing stars most of which are within a distance of 100 light years of us. On a clear dark night, we see only a little way out into the universe. Most of it is invisible to us.

Enter Galileo Galilei. He was the most distinguished member of the Academy of the Lynx-eyed (Accademia dei Lincei), founded in Rome in 1603, and named for a different kind of virtuouso of sight than owls or tarsiers:

the lynx. The lynx and other cats are known for the sharpness or *acuity* (as opposed to the *sensitivity*) of their eyesight. When Galileo turned a telescope to the sky over Padua in the winter of 1609-10, he enjoyed a marked enhancement both of acute (lynx-eyed) and sensitive (owl-eyed) vision. The first allowed him to make out details on the Moon, such as craters and mountains, that had never been seen before; both the first and second allowed him to make out the four largest satellites of Jupiter, while in the Milky Way itself—and in bits that appeared to be broken off from it, such as Praesepe (the "manger") in the constellation of Cancer—the increased sensitivity of his assisted vision revealed to him numerous stars "invisible to the unassisted sight."

He made sketches of two areas that seem rich in stars even with the naked eye. One was Pleiades, one of the most famous star-groups in the sky and named for the "seven sisters" of Atlas of Greek mythology. Curiously, the name is something of a misnomer, since the average person seems to see either six or eight stars (not seven). A person with an unusually sensitive eye (including young persons with normal vision, who are able to enlarge the pupil to the full aperture of 9mm or so, and perhaps some individuals possessed of unusually shortened, or hyperopic, eyeballs) can see 14 or more. Galileo's drawing made with the telescope shows 36 stars. His telescope had a lens with a clear diameter of 40 mm, making it four times the diameter of the most dilated human eyeball, thus giving it the power to penetrate some four times deeper into space and to comprehend sixteen times the volume. The universe known to man thus expanded by a factor of sixteen.

Larger and larger telescopes extended the empire of the eye even farther, until by the end of the 18th century William Herschel, by using the biggest telescopes built up to that time, could, briefly, imagine he had seen clear to the edge of the Milky Way star-system in which we reside. But in fact, as we now know, he was seeing only about 1% of stars in our galaxy. The contrast between his ambition and his achievement had to do partly with the human eye's poor performance as a detector of low-intensity light, but more so with the fact that large sections of the Galaxy were concealed from his view by clouds of interstellar dust. (as discussed in Chapter 3). In areas where he saw no stars, he did not suspect that they were being hidden as by a velvet curtain. In starless places, he thought he was seeing true voids—"holes" in the heavens. This was to be an oft-repeated mistake in the history of astronomy.

As noted above, the rods—the light-detectors in the retina—have a quantum efficiency of about 50%, which is actually very good. When photography, in the form of the gelatino-bromide dry plate, began to supplant the

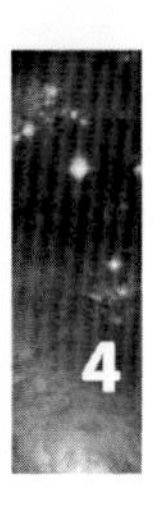

eye for the recording of faint stars and nebulae in the 1880s, it did so not by being more sensitive—in fact, it's not; the photographic emulsions used by pioneering figures in astronomical photography probably had a quantum efficiency of less than 1 percent of 1 percent. Photography's advantage was rather in its ability to accumulate light over long periods of time. In March 1882, the pioneer astrophotographer Henry Draper was able to expose a gelatino-bromide dry plate for 137 minutes on the Great Nebula of Orion. From this image he recorded more detail than had been painstakingly seen by visual observers who had worked for years drawing it with pencil and brush. Since the 1880s, technological improvements in photographic emulsions allowed films to reach a quantum efficiency of about 1 percent; though much poorer than the eye performs this means that only a few seconds' exposure is enough to produce an image equal to what the eye can see. Photographic films, in turn, have been rendered obsolete by solid-state detectors called Charge Coupled Devices (CCDs), arrays of tiny doped silicon pieces called pixels and which achieve a quantum efficiency of almost 100 percent. They have now become ubiquitous, as essential elements in our digital cameras. Film is now an anachronism. Most astronomers, amateur or professional, use CCDs to image stars and nebulae. With such instrumentation, an amateur observer with a quite small telescope can capture images that equal or exceed anything that was obtained in the photographic era at the great professional observatories like Lick, Mt. Wilson, and even Palomar. More of the universe is accessible to more people than ever before.

Personal Reflections on using a (relatively) small reflector

For the past several years one of the authors (W.S.) and his friend Mike Conley have operated a small observatory with such a CCD deployed on a 10-inch Ritchey-Chrétien telescope. The observatory is located in rural Minnesota, on the North Fork of the Crow River. The site is not the Canaries or Chile or Mauna Kea, but it still has impressively dark skies on clear moonless nights.

Both grew up in a city (Minneapolis) and from the mid-1960s onward, when dastardly mercury-vapor street lamps were put in, rarely caught more than a feeble glimpse of the Milky Way or the "little cloud" in Andromeda. For a long time we devoted ourselves to scrutinizing detail on the Moon and planets, since that was one domain where the eye's ability to use short "exposure times" for integrating the image favored it over film; so we could still do useful work (so we liked to imagine) even when equipped with nothing more than pencil and sketchpad. Moreover, dark skies weren't really necessary for the Moon and planets, since steady air, often to be had in cities owing to the "heat dome" effect and even when the sky was slightly hazy, was the most important factor in "seeing." With access to a dark site and the advent of CCD, we remain interested in the Moon and planets, but have also become inveterate "deep sky" enthusiasts.

On a typical night of observing, we would get together on a late Friday afternoon or early evening; a bit disheveled, feeling beaten up by the week's occupations, we would slide open the glistening white clamshell dome, to reveal the telescope, soon silhouetted against the rosy sunset, and wait for darkness to fall so we could begin to see.

Waiting, we drink in the beauty of the tranquil crepuscular time when the world seems to slide into a season of peace, and the last clouds in the west chase in vain after the sun. The line of trees along the river darkens, and with the gathering dusk, the swallows, half an hour before chasing swarms of gnats around the small bridge over the river, abandon their gustatory pursuits and retreat for the night. A bat or nighthawk makes an appearance, while in the distance, unseen, an owl readies itself for a night of hunting. We will hear its melancholy cry in the early morning hours, while imaging the Great Nebula of Orion.

We begin our own night-work. Powering up, red lights flash on within the dome; reawakened from its coma, the clock drive begins to purr. It slews the telescope around the polar axis at just the rate needed to compensate for the Earth's diurnal rotation. There! A lovely planet glistens against the sky. Then we see a star we think we know, Arcturus or Vega or Capella. We perform a three-star alignment to improve the accuracy of the guiding before beginning our night's imaging run.

We follow a set protocol. It is not something we could have ever worked out from any guide book, and when we first started trying to do this, we had more than our share of troubles and frustrations and were more than once on the verge of giving up and taking up an easier pastime. New technology always seems to bring in its wake a numerous train of applications of Murphy's Law. We have progressed as far as we have mostly on the basis of false starts and the solution of unanticipated difficulties. It's a sad fact: nothing works the way it's supposed to according to the book, and all technical writers are apparently incompetents. Experience alone is a reliable teacher.

We make sure that the finder scope is still aligned and expose a test image at one of our guide stars. Sometimes the guide star itself is so lovely that we can't resist getting an image, even though it is not in the way of our larger purposes for the night. Once we have the guide star in view, we know we are doing well enough for a rough pointing, but in this business accurate guiding is everything. We will need to use stars closer to the

Fig. 1.1. Small dome on the North Fork of the Crow River, Minnesota, housing a 10-inch Ritchey-Chrétien telescope. It awaits sunset so it can be unshuttered to the night skies. *Photograph by William Sheehan.*

Fig. 1.2. (below left) Mike Conley, biding time observing the Sun with an H-alpha filter telescope, before beginning a night of imaging. *Photograph by William Sheehan.*

Fig. 1.3. (below right) A mystical moment: sunset over Hollywood and Los Angeles, from the site where E.E. Barnard set up the Bruce telescope for imaging the Milky Way in 1905. *Photograph by William Sheehan.*

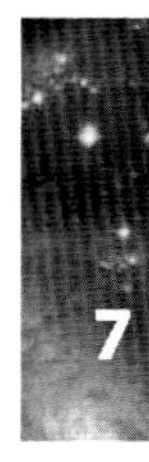

objects we will target, and "synch" the drive on them. In the process we have learned the Arabic names of many stars, each sounding like something out of Aladdin or the Thousand and One Nights: Almach (gamma Andromedae), Alphard (alpha Hydrae), Caph (Beta Cassiopeiae), Hamal (alpha Arietis), Mirfak (alpha Persei), reminding us that at one time Arabic astronomy was the envy of the western world, and Europe in the depths of its dark ages.

We have had to put up with our share of daunting clouds sweeping in just as we have been zeroing in on our prey. In November of our first imaging year, on three nights we were about to image the Great Nebula in Orion; each time—such was our poor luck—we obtained our first image just to watch the screen go blank under a veil of fog. We have often had to deal with mists rising from the nearby river and, while operating close to the dew point, have had to press into duty one of our wives' hair dryers to clear up fogged lenses. We have had to learn the intricacies of downloading images, to achieve mastery of protocols for stacking them, and figured out how to sort details of hue, saturation, and contrast. In the end, we have wasted many a night with nothing to show for it, giving up with the onset of clouds or the encroaching light of the dawn, cold, hungry, exhausted, defeated (but then even William Herschel, even Hubble must sometimes have felt that way). Each time the discouragement has been short-lived, and the enchantment has returned before long. It is in part the sheer beauty of the objects revealed to us, but in part also a human "fascination with what's difficult," that keeps us at our task. In the end, through perseverance, we learn our way about the sky, and everything seems worthwhile.

With our CCD attached to our smallish telescope, we have reached out to remote star-systems, some so far away that the light from them reaching us now—traveling at 300,000 kilometers per second (186,300 miles per second), year in and year out--left them during the Earth's Jurassic period. We owe it all to technology; but are unapologetic on that account. If it were not for technological advance, we should never have gathered just how splendid and dazzling the Earth appears looking down from space, nor guessed what worlds and suns and star systems exist above us far beyond the range of the unassisted sight but which are child's play for our telescopes, CCDs, and computers.

We remember one night when we were imaging from dusk to dawn. It was early November, but the weather was unseasonably warm. It was the weekend of the deer hunting opener in the Minnesota woods around us. We were not clad in orange blazers, we did not tramp the wood with guns; we would bring back no limp and bloodied trophy, no rack of antlers to display on our walls. Instead we were hunting a different kind of game. Our trophies were more cerebral, and consisted of a trove of images brought back from thousands, millions, and even tens of millions of light years away: images of celebrated objects, mostly in the Messier catalog, such as M1, M31, M33, NGC 891, M77, and—our hitherto nemesis—the Great Nebula in Orion, M42. Not one of our images is a six-pointer; we are under no illusions, and we do not overestimate what we have accomplished. Our successes are on a rather modest scale. We have not expected to rival the work of many an accomplished amateur, much less the professionals working at great observatories. We acknowledge that what we have accomplished is paltry compared to what can be downloaded off the web page of the Hubble Space Telescope or the other space observatories. It may not be much; but there is satisfaction, nevertheless, in that we have been participants rather than mere spectators. The results are our own, very own. The feeling of involvement is itself the source of satisfaction. It doesn't matter that our image of Orion doesn't show the detail of our friend Bob O'Dell's Hubble Space Telescope image.

And this gets to the heart of what drives us. It is the desire to be involved—to participate—in this vast (expanding) universe, in our own small way. We do not wish to sit back and let others do all the work; we refuse to be mere consumers of astronomical data, and to let the experts do everything for us.

Admittedly, we are living in an age in which the sheer quantity of knowledge—about every conceivable subject, but especially about science--is exploding. Once it was possible for a well-educated person to master, through the holdings of a small library of say a few thousand books, all the knowledge of the sages, saints, historians, poets and philosophers of the ages. But no more. The generalist—the jack of all trades—is today beleaguered as never before, and may no longer even be a viable species. The world increasingly belongs to the narrow specialist. This even applies within single scientific disciplines! George Rieke, University of Arizona infrared-astronomy researcher who worked to develop the infrared Spitzer Space Telescope—the infrared counterpart of Hubble—has written:

> "The various fields of space science—space physics, space biology, microgravity, astronomy, planetary studies—touch on nearly the entire range of science. Successful researchers need a high degree of focus and specialization. Few have any understanding or appreciation of the activities, aspirations, and potential scientific importance of fields far from their own. That is, the space science "community" is Balkanized into groups without a common language."[1]

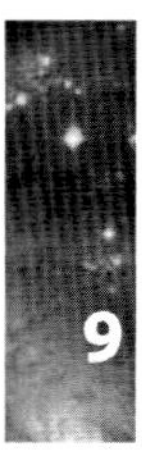

Fig. 1.4. A first success with the Orion Nebula, after many attempts., using an Orion Starshoot CCD on a 10-inch Ritchey-Chrétien telescope, November 26, 2009. This image stacks up well against those taken with the 200-inch Hale reflector in the classic film era. *Photograph by Michael Conley and William Sheehan.*

But it is not only the explosion of knowledge that puts polymaths at a disadvantage, but also the vast increase in the number of specialists and experts—monomaths--in every field. According to Edward Carr, "the very learning that creates would-be polymaths creates monomaths too and in overwhelming numbers. If you have a multitude who give their lives to a specialism, their combined knowledge will drown out even a gifted non-specialist. And while the polymath tries to take possession of second expertise in some distant discipline, his or her first expertise is being colonized by someone else."[2]

The people who are doing really first-class imaging work are technical monomaths, obsessive geniuses. For better or worse, we humble practitioners of the arts of CCD imaging on the North Fork of the Crow River are not in a position to do that, and will likely never be. We lack the technical aptitude. We lack, and are not afraid to admit it, the sheer obsessiveness. Our purposes are anyway different from theirs. We harvest diffuse nebulae, planetary nebulae, open clusters, globular clusters, and galaxies rather in the fashion that Henry David Thoreau did the wild cranberries of the New England woods. We have made the harvesting of the wonders of the night sky our business—though not by any means our sole or exclusive business. We realize that we are not one with the many of those around us (including those in the blaze orange that will be prowling these fields at dawn even as we are shuttering the dome after our night's labor, or the farmers who will be out in these fields with tractors and ploughs) who see us as eccentrics. Thoreau advised:

> If you would really take a position outside the street and daily life of men, you must have deliberately planned your course, you must have business which is not your neighbors' business, which they cannot understand. For only absorbing employment prevails, succeeds, takes up space, occupies territory, determines the future of individuals and states… I have always reaped unexpected and incalculable advantages from carrying out at last, however tardily, any little enterprise which my genius suggested to me long ago as a thing to be done, some step to be taken, however slight, out of the usual course.[3]

We imagers of the wonder-world of stars on a November night have neither corn to take to market nor antlers to hang on the walls. We have, in Thoreau's terms, been "cranberrying," and each of us can say with him:

> I enjoyed this cranberrying very much, notwithstanding the wet and cold, and the swamp seemed to be yielding its crop to me alone, for there are none else to pluck it or value it…. I am the only person in the township who regards them or knows of them, and I do not regard them in the light of their pecuniary value. I have no doubt I felt richer wading there with my two pockets full, treading on wonders at every step, than any farmer going to market with a hundred bushels which he has raked, or hired to be raked… I have not garnered any rye or oats, but I gathered the wild vine of the Assabet.[4]

What we have done has required deliberately planning of our course, setting aside time to do some business which was not our neighbors' business and which—alas—for the most part they cannot (and need not) understand. It has taken up space -occupied territory—in our minds, busied our hands and gotten us out under the open sky, beneath the starry traces of the Milky Way and the shining stars.

It has served as a needful respite from the worry and mental over-strain of mundane and mostly diurnal responsibilities. It has provided a needed change, a hobby and diversion, a flight from familiar toils into a realm of existence in which, for at least a brief while, we have been allowed a respite from melancholy and achieved a temporary calm.

Imaging these celestial scenes has been for us what painting was for Winston Churchill. He advised: "To rest our psychic equilibrium we

Fig. 1.5. Bob O'Dell's image of Orion nebula. A billion pixel image obtained with the Hubble Space Telescope's Advanced Camera for Surveys in 2006, showing the arcs, blobs, pillars, and rings of dust of the Orion Nebula, which appears to be a typical star-forming environment located 1500 light years from the Earth. *Courtesy: C. Robert O'Dell and Hubble Heritage Team (STScI/AURA).*

should call into use those parts of the mind which direct both eye and hand."[5] He took up painting at forty, when he left the Admiralty at the end of May 1915. We have taken up imaging at forty—and then some. Churchill had never handled a brush or fiddled with a pencil; we had never juggled a CCD camera or downloaded an image-processing program. Yet the results were much the same. We found ourselves belatedly "plunged in the middle of a new and intense form of interest and action with paints and palettes and canvases" and "not ... discouraged by our results."[6] In part we have done so because we have not been too ambitious. That would have spoiled it, made the result obtained more important than the process of obtaining it. Yet "we may content ourselves with a joy ride in a paint-box." An electronic paint-box, admittedly, but a paint-box nevertheless.

We have seen many things under those dark Minnesota skies as they have appeared, one by one, on the display monitor of our computer. The images displayed there have thrown our electronically enhanced eyes and minds a good distance out into the universe, allowing us to grasp things of which we would otherwise be largely unsuspecting and unaware.

Everything we see, we see because of the CCD detector's ability to register the arrival of photons, which are messengers from the depths of space. As a photon enters the doped silicon of the CCD array, it creates a charge deficit (hole) in that local area. During a time exposure these electronic holes build up in number in direct proportion to the number of photons coming in. The CCD array is wonderfully sensitive; its quantum efficiency approaches 1, which means that almost one charge hole is created for every incoming photon. Admittedly, the trick to the CCD is not the array but the readout. What we have so far is only a potential image; somehow the series of charge holes in a two-dimensional array has to be converted into an electronic image that can be displayed as an image. But these are details.

These photons are a form of electromagnetic radiation. They are radiated from objects consisting of ordinary matter, the familiar kind that is made up of protons and neutrons and electrons. The photons visible to the eye, and to the photographic plate and the CCD, all lie in the so-called visible range of the spectrum, which ranges from violet to red and is centered on the yellow-green, a reflection of the fact that our cones were shaped by natural selection to be most sensitive to the region of the Sun's radiation that is able to get through the Earth's atmosphere. (Obviously, the photographic plate and the CCD are not direct products of natural selection; they are products of human design. However, humans, their makers and their masters, chose their properties so that they would correspond to what the eye sees.)

The visible spectrum represents only a small segment of the entire electromagnetic spectrum. As was first appreciated at the beginning of the 19th century, when the invisible infrared was discovered beyond the red, and ultraviolet beyond the violet, the spectrum extends both ways far beyond what we can see. Beyond the infrared, in turn, lie the microwave and radio regions, beyond the ultraviolet the X-ray and gamma-ray regions. A large part of modern astronomy has concerned itself with developing detectors sensitive to and exploring the universe in these more exotic regions of the electromagnetic spectrum.

At the same time, the contents of the universe have come to be increasingly exotic. One of us remembers being an undergraduate in physics at a time when, after the introductory course in astrophysics, the only upper-level offerings in the department were: the astrophysics of diffuse matter, and the astrophysics of condensed matter. The first were masses of gas, the "diffuse" nebulae; the second were stars, planets, satellites, comets, and the like, as well as such exotic objects as neutron stars and black holes. Between them, diffuse matter and condensed matter seemed to cover everything. In the years since, we have discovered that matter accounts for about 5% of the total mass of the universe. The other 95% is something else—invisible and so far unaccounted for.

The history of astronomy is about how humans—the product of natural selection, with their so-so eyes compensated for by their rather complicated brains—have come to see (or at least be aware) of more and more of the universe. The universe is expanding, and so is the human mind trying to take it in. Still it is strange to think that most of the universe consists of what cannot be seen.

At least astronomy has made it visible to our minds if not to our senses.

* * *

1. George Rieke, *The Last of the Great Observatories: Spitzer and the era of faster, better, cheaper at NASA* (Tucson: University of Arizona Press, 2006), 28.
2. Edward Carr, "The Last Days of the Polymath," *Intelligent Life Magazine,* August 2009.
3. Henry David Thoreau, *Wild Fruits,* ed. Bradley P. Dean (New York and London: W. W. Norton, 2000), 165.
4. Ibid., 167-168.
5. Winston S. Churchill, *Painting as a Pastime* (New York: The McGraw-Hill Book Company, Inc., 1950), 12.
6. Ibid., 14.

2. Catchpole of the Nebulae

The glory and the nothing of a name.

—George Gordon, Lord Byron

Among Galileo's most momentous discoveries was his demonstration that Milky Way's river of light consists of a host of small stars. It is a discovery that can be repeated by anyone with a dark sky site and an ordinary pair of field glasses.

Though most have been lost, one of Galileo's lenses (though broken) has survived, and is preserved in the Museum of the History of Science in Florence. The diameter is 60 mm, though in order to improve its performance Galileo seems to have used an aperture-stop to reduce the diameter to 40 mm. This can be compared to the diameter of the pupil of the dark-adapted human eye, which is at most about 9mm. This means, as we noted earlier, that Galileo's lens was more than four times the diameter of the dilated human pupil, thus its area—or light-gathering power—was some sixteen times that of the eye. Instead of the 3,000 or so stars visible to the naked eye on a clear dark night, Galileo's telescope might have shown, on the basis of these proportions, 50,000 or more, a staggering advance.

To demonstrate the power of his telescope, Galileo in 1610 noted a number of telescopic stars in the "Sword" of Orion, which consists of three stars hanging down like a scabbard from the belt of the Giant. But though his drawing shows many faint stars, there is not a trace of the *nebula* (from the Latin word for cloud) that invests the middle star—the nebula which will be a leading character in this book, and which is now known as the Great Nebula of Orion.

Whatever the reason Galileo failed to make note of it—perhaps with the telescope he was using, he was so accustomed to seeing glare around all bright objects that he just assumed it was an optical effect, or perhaps, as astronomer Bob O'Dell has suggested,[1] he had no commercial

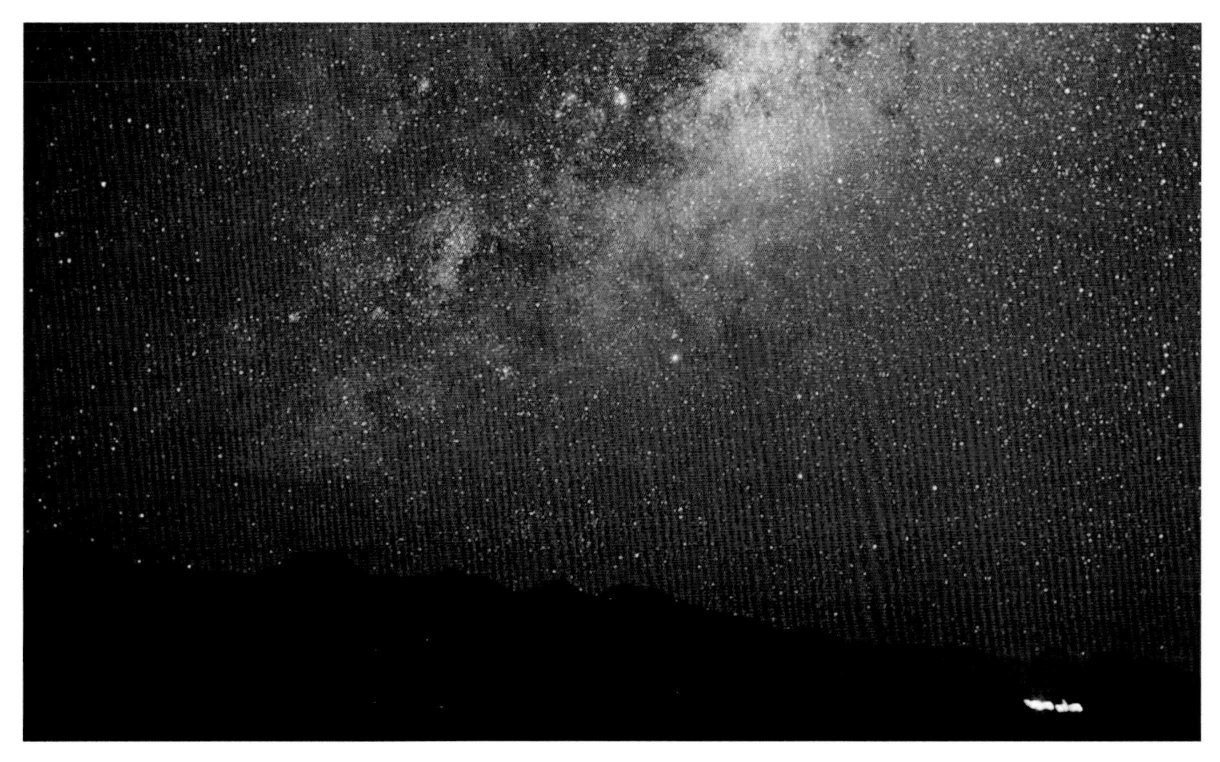

Fig. 2.1. The Milky Way, photographed with an 18mm wide angle lens from Chile by Terence Dickinson. *Courtesy: Terence Dickinson.*

Fig. 2.2. "A congeries of innumerable stars." In Cygnus, we are looking along the direction of the Sun's motion through the Galaxy, and seeing features in the so-called Orion arm: HII regions (including the North America Nebula) on the left, the bright star Deneb, the most distant of the 20 brightest naked-eye stars, at 1400 light years away, and the great dust cloud known as the Northern Coal Sack dimming the light from more distant fields of stars. Imaged by Klaus Brasch with a 50 mm f/3 lens. *Courtesy: Klaus Brasch.*

interest in broadcasting that there were some cloudy objects in the heavens he could not resolve into individual stars—the Great Nebula was recorded only a year later by Nicolas de Peiresc at Marseilles, who didn't publish, and then in 1618 by the Jesuit astronomer Johann Baptist Cysat, who described it as like a radiant white cloud. But it would be Christiaan Huygens of Holland, one of the greatest scientists who ever lived, who made the first really satisfactory observations of the Great Nebula, in 1655.

Unaware of either Peiresc's or Cysat's work, Huygens rediscovered it with a telescope whose lens was 50 mm in diameter and tube 12-feet long. His drawing shows the bright central part of the Orion Nebula (later called *Regio Huygeniana*—the Huygenian Region—in his honor).

The Great Nebula of Orion was a member of what seemed to be a new class of celestial objects. It was, in the phrase of the early 18th century clergyman-astronomer William Derham, who observed it with one of Huygens's telescopes bequeathed to the Royal Society of London on Huygens's death, like "a visible entrance to the empyrean." Another similar object was the Great Nebula in Andromeda, an elongated hazy patch northwest of the Great Square of Pegasus recorded as the "little cloud" by the 10th century Persian astronomer Abd-al-Rahman Al-Sufi.

In succeeding centuries, these two objects—the Great Nebula of Orion and that of Andromeda--would prove among the most studied and discussed objects in astronomy. Gradually, a few more (but lesser) objects of the same type were discovered. Giovanni Hodierna, a Sicilian astrono-

Fig. 2.3. Christiaan Huygens. *Courtesy: Yerkes Observatory.*

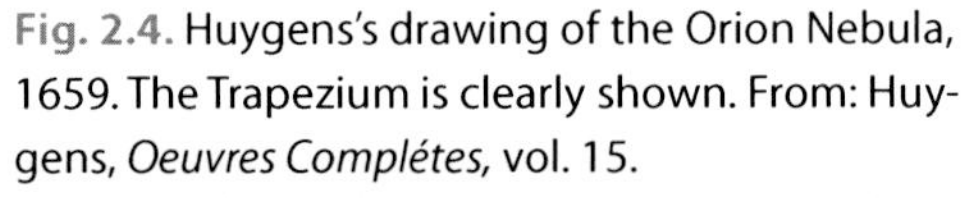

Fig. 2.4. Huygens's drawing of the Orion Nebula, 1659. The Trapezium is clearly shown. From: Huygens, *Oeuvres Complétes*, vol. 15.

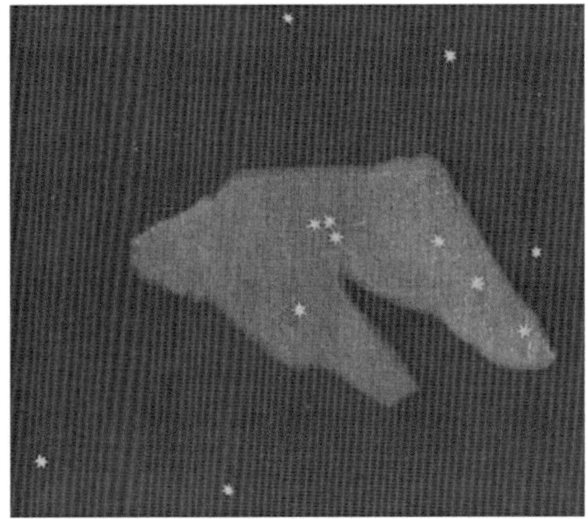

mer who died in 1660, added a few more, but his work remained unknown; in fact, it has been rediscovered only in recent times. Edmond Halley, famous for his comet, published a list of six in 1715.

The reason for the slow progress was partly owing to the limitations of 18th century telescopes. Chromatic aberration—the tendency of a simple lens like those used by Galileo, Peiresc, Cysat, and Huygens to act as a prism and to bring light of each color to a slightly different focus, so as to produce a haze of unfocused purple or magenta light around bright objects such as the Moon or brilliant planets—was a serious problem. By making telescopes of great length, this haze could be subdued, and so some telescopes of this era were to be made very long indeed; Hevelius, for instance, a brewer-astronomer at Danzig (now Gdansk) on the Baltic, made a telescope that consisted of a monstrous tube slung from a mast towering 27 meters from the ground and raised or lowered by means of cranks, ropes, and pulleys. Another vexing problem was the very small field of view enjoyed by such telescopes, which made them useful for studying the Moon and planets but atrocious for studying faint objects like nebulae. Finally, astronomers of the period were chiefly concerned with such down-to-Earth problems as creating tables of the Moon's motions to help navigators calculate their longitude at sea, which had been the main purpose for which the great national observatories at Paris (established in 1667 by King Louis XIV) and Greenwich (founded by his not-to-be outdone cousin King Charles II in 1675) were founded. Preoccupied with such problems, they could hardly be bothered with the remote, indistinct, and seemingly unimportant nebulae.

After the publication of the *Principia* of Isaac Newton, a professor of mathematics at Trinity College, Cambridge, in 1687, introducing the theory of gravitation and showing how it could be applied to the detailed calculation of the intricate motions of the Moon and planets, astronomers preoccupied themselves almost exclusively with the Solar System, and would do so until the end of the 18th century. In terms of Newton's world-view, the stars were little more than a remote backdrop against which the Moon and planets moved. The great man scarcely concerned himself with their distribution or their arrangement in space. Rarely did he allow his powerful mind to wander among the distant stars. Once he made a crude attempt to measure the distance to Sirius, on another occasion he hazarded a guess as to how a system of stars might form through gravity acting upon different kinds of "luminous" and "opaque" matter:

> It seems to me that if the matter of our sun and planets and all the matter of the universe were evenly scattered throughout all the heavens, and every particle had an innate gravity toward all the rest, and the whole space through which this matter was scattered was finite, the matter

would, by gravity, tend toward all the matter on the inside and, by consequence, fall down into the middle of the whole space and there compose one great spherical mass. But if the matter was evenly disposed throughout an infinite space, it could never convene into one mass; but some of it would convene into one mass and some into another, so as to make an infinite number of great masses.... And thus might the sun and fixed stars be formed.... But how ... matter should divide itself into two sorts ... part of which is fit to compose [shining bodies like stars and part fit to compose opaque bodies like the planets] ... I do not think explicable by mere natural causes, but am forced to ascribe it to the counsel and contrivance of a voluntary Agent.

—Isaac Newton to Richard Bentley;
quoted in Charles A. Whitney, The Discovery of our Galaxy
(Ames, Iowa: Iowa State University, 1988), 52.

Not for the first time, and certainly not for the last, did a thinker perplexed with the origins and structure of the cosmos invoke the "contrivance of a voluntary Agent," though by the end of the 18th century at least some astronomers, such as Laplace, would regard it as the ultimate "hypothesis non fingo."

Modest beginnings

The 18th century dominance of astronomy by Solar System studies partly explains how the first reasonably complete catalog of nebulae—the famous Messier catalog—was produced more or less accidentally by an astronomer who was not even interested in them for their own sake. He was Charles Messier, and his catalog, published in installments in 1771, 1780, and 1784, consists of a disparate collection of 103 objects, of which only about a third were discovered by Messier himself.[2] Messier was a celestial magpie who unapologetically appropriated the lists of earlier astronomers such as Halley, William Derham, Jean-Dominique Maraldi II, Chéseaux, Le Gentil, and especially Nicholas-Louis de Lacaille, a recently deceased French astronomer who had gone to the Cape of Good Hope to catalog southern stars and nebulae, to expand his list. By type, about two-thirds of the objects in Messier's catalogs are star clusters, open or globular, or diffuse or gaseous nebulae associated with our own Milky Way Galaxy; the rest consist of true "extra-galactic" systems (galaxies). Of course, neither Messier nor his contemporaries knew anything of their true nature. Other catalogs have appeared since, but his list will always be foundational. Because of his catalog, his name is well-known to every amateur astronomer smitten enough by the night-sky to acquire a small telescope and eager to see its most notable objects, even if he or she knows nothing about the man himself or the circumstances in which he came to draw it up.

Messier was born in 1730 at Badonviller (near Nancy), the tenth of twelve children. His father was a catchpole--an official charged with finding and arresting people who were in debt—and since there have always been a lot of people in debt in every age, the family seems to have been moderately well to do. When Messier was 11, his father died. As a result, his older brother Hyacinthe, employed in the administration of the now-forgotten Principality of Salm-Salm, assumed responsibility for him. This soon included educating the lad at home after he fell from a window while playing and broke his leg.

All the while Charles's interest in astronomy was taking shape. At the end of 1743, the magnificent six-tailed Chéseaux' Comet appeared (so-called, although at least two other observers had seen it before Phillipe Loys de Chéseaux). If there was any doubt after this comet's appearance as to his future direction in life, it was removed by his observation of an annular eclipse of the Sun, visible from Badonviller on July 25, 1748.

When Messier was 21, the Princes of Salm withdrew from Badonviller. Hyacinthe, deciding to remain in their service, moved to Senones. Charles went his separate way and hoped to find his fortune in Paris.

When he arrived in the "City of Light," its population was 800,000, of whom 100,000 were servants and 20,000 were beggars. Most lived in slums, a lucky few in magnificent palaces. In 1742, some of the roads were paved with rolled stones, but most of the streets were plain dirt, or laid with cobblestones. Street lighting, provided by lanterns, began to appear in 1745. No other city in Europe—not even Rome—had street lighting at the time. The lanterns were lit only when there was no moonlight, which meant it was still possible to see the fainter stars and the Milky Way from within the great city.

One can imagine that young Messier's first impression of the daunting city might well have resembled that of Montesquieu's Persian visitor: "The houses are so high that one would suppose they were inhabited only by astrologers." He would soon have a room in one of the highest.

Obviously, it would have been difficult for a young man without connections to make his way forward in such a place. But thanks to his brother, Messier had friends in high places. Hyacinthe, in consultation with a local abbé, found two possible positions for him: one was as curator of a palace, the other in service of an astronomer. One can guess what Messier's own inclinations must have been, but Hyacinthe advocated for the latter as the better opportunity. So it was that young Messier came to knock at the door of the naval astronomer Joseph-Nicholas Delisle, at the Hôtel de Clugny (now Cluny). Built on the site of old Roman baths, the hotel had served as the town house of the powerful Medieval abbots of Cluny, and had harbored within its spacious apartments such famous personages as the daughter of the English King Henry VII, Mary Tudor who was briefly the Queen of France, and Cardinal Mazarin.

Fig. 2.5. Charles Messier by Ansiaume. From: Wikipedia Commons.

Within one of the hotel's towers, Messier's new boss had erected a small wooden observatory, housing a 4 ½-foot focal length Newtonian reflector. This telescope used a mirror to collect light instead of a lens, as used in the refracting telescopes of Galileo and Huygens. Until the latter part of the eighteenth century, reflectors were very much out of fashion, partly due to the problem of speculum metal quickly tarnishing but also due to the fact that astronomy then was primarily a science of position, so there was little demand for the larger light-grasp of reflectors.

Messier was not, at first, an astronomer, but rather Delisle's clerk. With a calligraphic hand, and with habits of system and order that were

Fig. 2.6. The Tower at the Hôtel de Cluny in the West Bank area of Paris, where Messier had his small observatory and hunted for comets. *Photograph by William Sheehan.*

Fig. 2.7. Sundial at the Hôtel de Cluny. *Photograph by William Sheehan.*

extreme even by standards of the notoriously tidy French, Messier's first assignment was to copy a large map of China. Delisle himself was then preoccupied with the grand project of applying the method of Edmond Halley (whom Delisle had met in London in 1723) to measure the solar parallax from observations of the rare transits of Venus[3]—work that would lead, it was hoped, to a highly precise measurement of the all-important value of the astronomical unit, the distance from the Earth to the Sun. Messier began to assist him by making and recording observations with the telescope in the tower. As a warm-up for the forthcoming transit of Venus of 1761, Messier made his first documented observations--of the transit of Mercury on May 6, 1753. (For various reasons that need not concern us here, transits of Mercury are not as useful as those of Venus in calculating the value of the astronomical unit.)

In other work, building on Halley's foundations, Delisle had been immersed in trying to calculate the place in which the famous comet Halley had shown to be periodic would appear at its expected return in 1758-59. After four and a half years of calculating, he had his result, and assigned Messier to search for the comet. The eager young man set to work at once. Unfortunately, Delisle had made an error, so Messier spent many months looking in the wrong positions. Even so, his efforts weren't entirely in vain; he soon learned his way around the sky, and on August 14, 1758, he was rewarded with what he thought was the discovery of a new comet. He was disappointed, however. It turned out that the comet had already been seen at Bourbon Island, in the Indian Ocean, but as he tracked it across the sky, Messier encountered yet another comet-like patch, on August 28, 1758, "a nebulosity above the southern horn of Taurus… It contains no star; it is a whitish light, elongated like the flame of a taper." Returning to measure it again on September 12, he found it had not moved in the interim as a comet would have. This observation would lead to the first entry in what became his famous "Catalogue of Nebulae and Star Clusters":

> What caused me to undertake the catalog was the nebula I discovered above the southern horn of Taurus on September 12, 1758, whilst observing the comet of that year. This nebula had such a resemblance to a comet in its form and brightness that I endeavored to find others, so that astronomers would no more confuse these same nebulae with comets just beginning to appear. I observed further with suitable telescopes for the discovery of comets, and this is the purpose I had in mind in compiling the catalog.

This nebula, like many of the so-called Messier objects, had already been noted by an earlier astronomer, the now almost-forgotten English astronomer Dr. John Bevis, who discovered it around 1731. As the first object entered in Messier's catalog, it became M1. Messier, of course, had no idea as to its nature, and could never have guessed that it would turn out to be one of the most singular objects in the sky, the object known today as the Crab Nebula, the most spectacular of the 100 or so known supernova remnants in our Galaxy and observed as a "new" star by Chinese and Japanese astronomers in 1054 (but not, apparently, in Europe, where the Pope of Rome and the Patriarch of Constantinople were busy excommunicating one another, events which would led to the "Great Schism" between the Eastern and Western churches).

Resuming the quest for Halley's Comet at its predicted return, Messier finally succeeded in finding it on January 21, 1759 (although unbeknownst to him, he had been beaten to it by a German amateur, Johann Palitzsch, a farmer at Prohlis, near Dresden, who spotted it on Christmas Night). Though Messier and Delisle kept it under observation as it sank into the evening sky, and re-observed it as it emerged before the dawn into the morning sky in April, only then did the secretive Delisle finally announce Messier's independent discovery; other astronomers were skeptical about this late announcement, and Messier got little credit for his work.

Finally, in January 1760, Messier made an undisputed comet discovery. The first "comet Messier" became bright enough to be seen with the naked-eye, and eventually developed a tail 10 degrees long. Henceforth Messier was hopelessly addicted to the excitement of comet discovery. He channeled all his energy into their single-minded pursuit, in the process covering himself in the *gloire* attached to their discovery. (Famously, Louis XV referred to him as the "ferret of comets.") Between the first in 1760, and his last in 1801, Messier discovered thirteen comets on his own, and was co-discoverer of another seven. His was easily the most brilliant record of comet-discovery of the 18th century.

All the while, in order to improve his chances and to optimize the time he spent searching for comets, Messier adapted a method of "double-entry bookkeeping," tallying not only actual comets but comet-masqueraders he came across that cost him valuable time (their significance would emerge only later). Messier thus became a "catchpole" of nebulae. After the M1 nebula in Taurus, the next object entered into Messier's list was M2 in Aquarius, now known to be a globular cluster. It had been recorded already in 1746 by another French astronomer named Maraldi. Only in 1764 did he begin to make frequent entries to the catalog, beginning with M3, a fuzzy ball in Canes Venatici, which he spotted on May 3. This noble globular cluster is just visible to the naked eye from dark skies, but had never been recorded before.

By the end of 1764, Messier's list had grown to 40, many of which were true comet-masqueraders—apt, when seen in a small telescope, to be mistaken for comets.

Fig. 2.8. The periodic comet Hartley 2, an object of the class that Messier sought so eagerly, moves past NGC 869 and 884, the Double Cluster (OB association) in Perseus, on October 9, 2010. These clusters, located at a distance of 7600 light years, are among the finest objects in the northern skies, but for some reason Messier did not include them in his catalog. Imaged by Klaus Brasch with a TMB 92mm apochromatic refractor. *Courtesy: Klaus Brasch.*

Fig. 2.9. Messier no. 1: the Crab Nebula in Taurus. *Image by Michael Conley and William Sheehan with a 10-inch Ritchey-Chrétien telescope and an Orion Starshoot imaging system, October 2, 2010.*

Fig. 2.10. M3, a globular cluster in Canes Venatici, one of the objects Messier actually discovered himself, imaged with an 85mm f/7 apochromatic refractor by Michael Conley and William Sheehan.

The first installment of "A Catalogue of Nebulae and Star Clusters" was published in the *Histoire de l'Academie Royale des Sciences* in 1771. It included all the objects in the 1764 list, with a few others tacked on, such as the Orion Nebula, Praesepe, and the Pleiades, to give a total of 45.

In 1774, he met a younger astronomer, Pierre Méchain, who shared his enthusiasm for comet-seeking. Henceforth the two collaborated in their efforts to add to the growing catalogue. An additional 23 objects were added in 1780, while the third and last installment, bringing the total to 103 objects, appeared in 1784.

The Messier catalog now accepted contains 110 entries, where an additional seven were added by later astronomical historians based on their review of the original observing records of Messier and Méchain. Though some of the objects are not significant—i.e., M40 and M73 consist only of a few faint stars; a few others, M47, 48, 91, and 102, have gone "missing," due to Messier's errors in transcribing the positions of the objects he observed—as a whole the objects within are among the best known

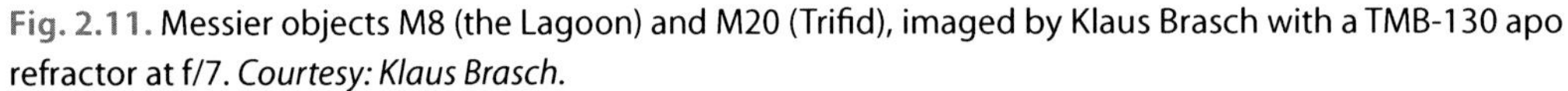

Fig. 2.11. Messier objects M8 (the Lagoon) and M20 (Trifid), imaged by Klaus Brasch with a TMB-130 apo refractor at f/7. *Courtesy: Klaus Brasch.*

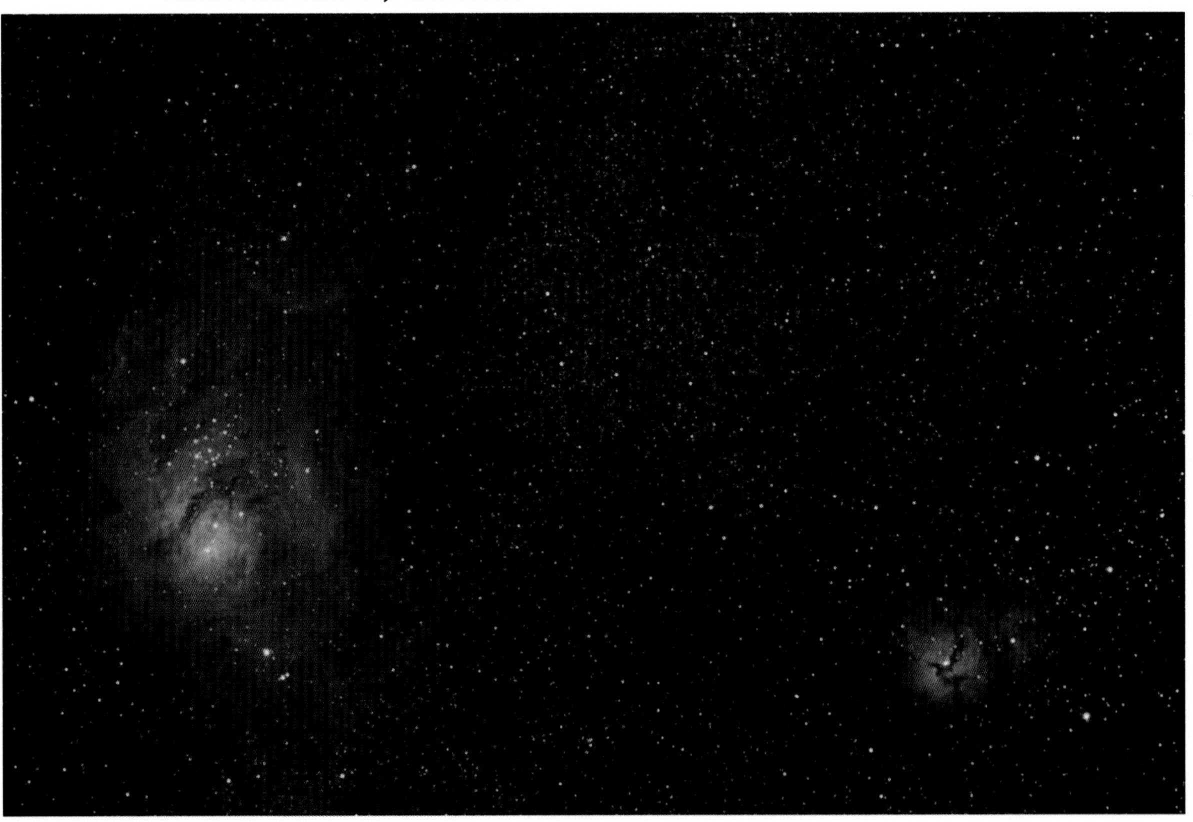

and most sublime objects in the Northern heavens. Among Messier's own discoveries are such standouts as M3, the globular cluster in Canes Venatici mentioned above; M27, the "Dumbbell" planetary nebula in Vulpecula; M33, the Great Spiral in Triangulum; M46, the fine open cluster in Puppis; M49, the first Virgo Cluster galaxy discovered; M51, the famous spiral in Canes Venatici.

The rest of Messier's career can be summarized briefly here. After his "breakout year" of observing, 1764, his boss Delisle retired. Messier remained in residence at the Hôtel de Cluny, and continued to patrol the heavens with the small telescope in his tower. In 1771 he was appointed astronomer of the French Navy. With the increased security of that position, he married Marie-Francoise de Vermauchampt, whom he had known for fifteen years. However, she died in March 1772, eleven days after giving birth to a son, who also died, and he never married again.

Fig. 2.12. HII region: The Swan Nebula, M17, in Sagittarius: 30 second exposure with a 10-inch Ritchey-Chrétien telescope and an Orion Starshoot imaging system, June 27, 2009. *Imaged by Michael Conley and William Sheehan.*

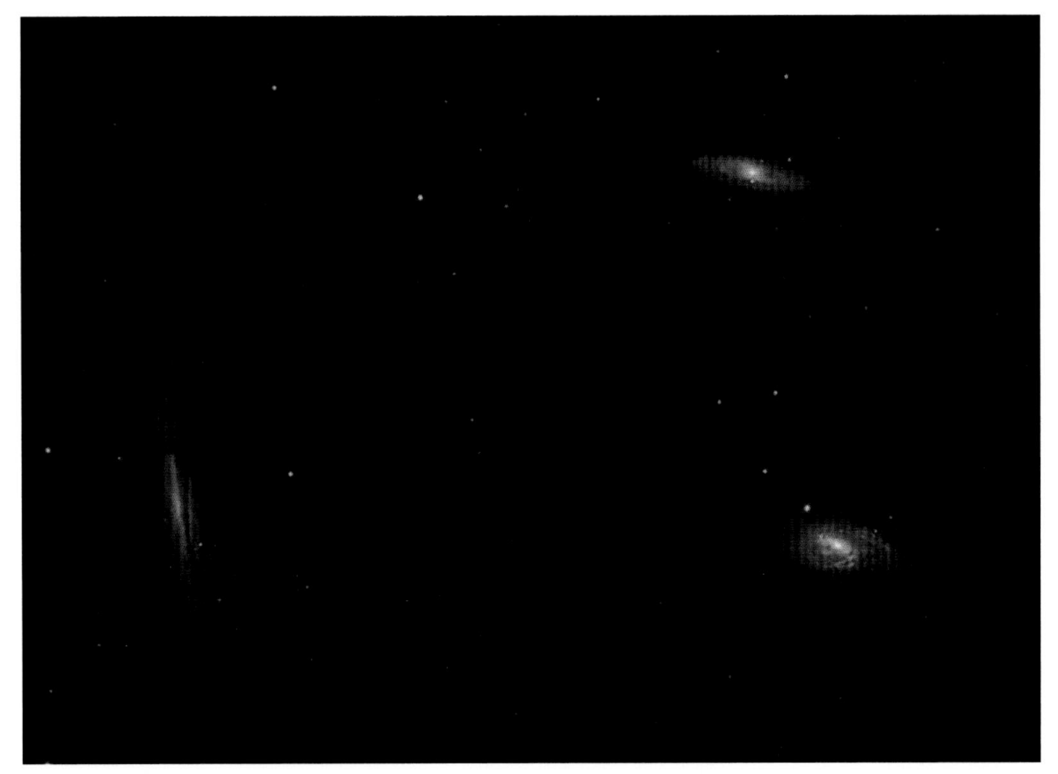

Fig. 2.13. The Leo Triplet of galaxies: M65 and M66 and, to the left, NGC 3628, which Messier missed. *Imaged with a 10-inch Ritchey-Chrétien by Michael Conley and William Sheehan*

He was becoming famous now, and was elected to various Academies, including the Academy of Brussels and the Royal Academy of Hungary in 1772 and the Academy of St. Petersburg in 1777. Celebrity cut into his observing time. Most of the objects in the latter installments of his catalog were discovered by Méchain.

An even more serious setback occurred in 1781. Messier was walking in a park with a friend, the astronomer Jean-Baptiste-Gaspard Bochart de Saron, when he tripped, fell into an ice cellar, and was badly injured. Although he returned to his work as soon as he could, he was distracted not only by poor health, but also by the political upheaval of the French Revolution of 1789. (During the "Terror" of 1793-94 his friend Saron, president of the Parliament of Paris before the Fall of the Bastille, would be guillotined on April 20, 1794, soon after calculating the orbit of Messier's latest comet.)

As late as 1800-01, when Messier discovered the last "comet Messier," a bookend to a career of comet discovery that had begun with the first "comet Messier" four decades earlier, he continued to ponder revisions to his catalog, although by then new nebulae were turning up by the bushelfull, thanks to the systematic surveys being conducted with large telescopes by William Herschel in England. When Herschel visited Paris in 1802, he sought out his great predecessor in the study of the nebulae, who was still in his lodgings at the Hôtel de Cluny. Herschel recorded in his journal: "He complained of having suffered much from his accident... He is still very assiduous in observing, and regretted that he had not interest enough to get the windows mended in [the] tower where his instruments are."[4] Messier observed the Great Comet of 1807. By then, his eyesight badly failing, he shuttered up the observatory for the last time. Suffering a paralytic stroke in 1815, he lingered on for two more years. On April 12, 1817, he expired in the Hôtel de Cluny; by then it had served as his residence for almost seventy years.

The irony of Messier's career is that the stone that was rejected became the cornerstone of his fame. Messier's interest was single-minded. As strange as it now must seem, he appears to have had no curiosity about his nebulae; he regarded them rather as orphan stepchildren, as nuisances to be avoided by the comet-seeker. In his defense, it is difficult to imagine how he could have made much progress toward understanding the nature of the nebulae, which as we now know, were a disparate set of objects, some inside the Milky Way galaxy, others consisting of remote star systems far beyond it. To understand the nature of the disparate objects classed together as "nebulae" and their role in the universe itself would take another two centuries, and is discussed in the remainder of this book.

If Messier could somehow be summoned in a *séance* from the Great Beyond, he would undoubtedly be shocked to find that posterity hardly remembers his comets. His immortality has come instead from the objects he so studiously and assiduously avoided in his relentless pursuit of them.

* * *

1. C. Robert O'Dell to William Sheehan, July 1, 2013, personal communication.
2. The most complete biography of Messier is Jean-Paul Philbert, *Charles Messier: Le furet des comètes*. Sarreguemines: Edition Pierron, 2000. Regarding Messier's attempt to discover Halley's Comet, see: Peter Broughton, "The First Predicted Return of Comet Halley," *Journal for the History of Astronomy, xvi* (1985), 123-133. The third and definitive edition of Messier's catalog published during his lifetime is: "Catalogue des Nebuleuses et des amas d'Etoiles," in *Connaissances des Temps, 1784* (Paris, 1781), 117-272. An English translation of the catalgo is found in Kenneth Glyn Jones, *Messier's Nebulae and Star Clusters* (Cambridge: Cambridge Univeristy Press, 2nd rev. ed., 1991).
3. For details about Delisle's career, see: William Sheehan and John Westfall, *Transits of Venus* (Amherst, New York: Prometheus, 2004), 136-138.
4. J.L.E. Dreyer, ed., *The Scientific Papers of Sir William Herschel* (London: the Royal Society and the Royal Astronomical Society, 1912), vol. 1, lxii.

3. "I Have Looked Farther...."

> I have looked farther into space than ever a human being did before me. I have observed stars of which the light, it can be proved, must take two million years to reach the earth.... Nay more ... if those distant bodies had ceased to exist millions of years ago, we should still see them, as the light did travel after the body was gone.
>
> *—William Herschel to the poet Thomas Campbell (1813)*

Friedrich Wilhelm Herschel—or William Herschel as he naturalized his name after moving to England—was the great pioneer of stellar astronomy and the nebulae in the 18th century. At the time he began, the nebulae—catalogued by Messier only to save time in eliminating them as comet suspects—were hieroglyphs of the universe, which no one had a clue how to decipher. Herschel left a field having at least begun to ask the right questions.

Herschel was born in the German city of Hanover in 1738, into a family of accomplished musicians. His own musical skills—already by age four, his father was setting him on a table with a fiddle to play a solo—led him to follow in his father's footsteps, and in his late teens he obtained a position as oboist in the Hanoverian regimental band. However, during a particularly unpleasant skirmish with the French during the Seven Years' War between France and England—Hanover, whose Grand Elector sat on the English throne, was allied with England—he and his father were forced to seek shelter beneath a tree. As the musket balls whizzed over their heads through the branches, his father shouted at him, "Go! Go!" waving his hand in the direction of Hamburg. Since, as William afterward reflected, "no body seemed to mind whether the Musicians were present or absent," he left for Hamburg. From there he made his way to England.

William arrived in England with "good linen and clothing," a serviceable smattering of French, Latin, and English, and skill on the oboe, violin, harpsichord, and organ. In addition, he had the enormous drive of an immigrant, and hated nothing more than wasting time. He started as a music copyist in London, but finding the city "overstocked with musicians," he set out for the provinces, and by 1759 used his military contacts to find employment, on a month-to-month retainer, in Yorkshire, as head of the Earl of Darlington's private militia band. He was now traveling by horseback all over Yorkshire, "crossing over the moors alone at night" and finding himself "studying the panoply of stars overhead as he had done as a boy."[1] By day, he read while he rode, and on at least one occasion became so absorbed in his reading that he somersaulted from his horse and landed, to his surprise, upright, still holding the book in his hand. He was drawn to composition, and began churning out symphonies and concertos, all the while directing and performing concerts in the great houses of the North—and even a castle or two.

Fig. 3.1. William Herschel, an engraving based on the portrait by John Russell, 1794. Russell, a portrait artist, had made sketches of the Moon with a Herschel telescope several years earlier (see Fig. 3.5). *Image from: William Sheehan Collection.*

Eventually, Herschel began to find the harassments of business wearing on him. Fortunately, he found a more stable situation in Leeds, as Director of Public Concerts, and his loneliness was assuaged by the company of a Mr. and Mrs. Bulman and their daughter, with whom he shared lodgings. He now had more leisure for private studies and schemes for self-improvement. "During all this time," he reflected afterwards," ... I had not forgot my former plan [to give] all my leisure hours to the study of languages. After I had improved myself sufficiently in English I soon acquired the Italian, which I looked upon as necessary to my business; I proceeded next to Latin, and having also made considerable progress in that language I made an attempt at Greek, but soon after dropped the pursuit... as leading me too far from my other favorite studies by taking up too much of my leisure. The theory of music being connected with mathematics had induced me very early to read in Germany all that had been written upon the subject of harmony; and when not long after my arrival in England the valuable book of Dr [Robert] Smith's *Harmonics* came into my hands, I perceived my ignorance and had recourse to other authors for information, by which I was drawn on from one branch of mathematics to another."[2]

After a couple years in Leeds, something better came up: he was appointed organist at the private Octagon Chapel then being organized in the resort town of Bath. With its healthful waters (later described by Dickens's Sam Weller as having "a very strong flavor o' warm flat irons") and its new Pump Room and Theatre Royal, Bath was becoming the most fashionable city outside London. It had already achieved distinction as a cultural center second only to London itself, and was then undergoing the major expansion of Georgian-era architecture for which it remains famous. The Octagon Chapel was a symptom of Bath's growing prosperity. It was designed to cater to Christians of the aristocracy, as a place where they could worship their Creator in style without mixing with the vulgar members of the working classes.

By obtaining a foothold in Bath, Herschel, at 28, watched the curtain fall on a decade of obscure and lonely struggle. Henceforth he would make his way among scenes of brilliant architecture, dazzling society, and lucrative possibilities. And in Bath, too, William Herschel the musician underwent a dramatic metamorphosis into William Herschel the astronomer. "He was sure to have emerged sooner or later," writes his nineteenth-century biographer E. S. Holden, "but every year spared to him as a struggling musician was a year saved to Astronomy."[3]

When Herschel arrived in Bath, the Octagon Chapel was still unfinished. He could not afford to sit idle, and with his usual precipitancy threw himself into the local music scene. In addition to public engagements, Herschel advertised for pupils, offering to give lessons in guitar, harpsichord, violin, and singing. Teaching became the most lucrative source of income. At his peak, he was meeting with four, five, six and even more "private scholars" a day, mostly ladies of the aristocracy striving to improve themselves (and their marriage prospects) by learning to sing or play the harpsichord. (This was the Bath Jane Austen knew and heartily disliked: the elegant, well-ordered, cut-throat world of *Northanger Abbey* and *Persuasion,* where a successful marriage was a woman's only way of securing her social status and where no advantage was ever neglected in the goal of "marrying up").

Though a composer by inclination, Herschel's time after 1770 was increasingly squeezed by demands for concert performances and the instruction of pupils. Nevertheless, by then he had already produced an impressive *oeuvre* consisting of countless symphonies, concertos, and oratorios. Much of his musical *oeuvre* is only now being rediscovered. As a composer, he was, as Patrick Moore has said, "not of the first rank, but perfectly acceptable."[4] To some, he may seem like Mozart gone stale, but at least one authority has called his compositions "arresting, innovative works, the product of a superb analytic mind driven by an obsession for order and coherence."[5] These manuscripts are of particular value to musicologists since, in contrast to most composers of his day, Herschel was meticulous in giving precise indications to his musician, and expected them to follow his instructions to the letter; they thus furnish insights into the practices of the great virtuosi of the period. They also foreshadow his career as an astronomer. He learned to handle astronomical instruments with the same dexterity as he handled the organ and the fiddle, and his precision about musical indications would continue in the detailed records he kept of his astronomical "scores."

Fig. 3.2. Caroline Herschel; silhouette. Photographed by William Sheehan, 2009, with permission of the William Herschel Museum, Bath.

Though Herschel was a successful professional musician, he was coming to regard music—and especially the grim routine of teaching (for him) —as "an intolerable waste of time." Over the next several years he was joined in Bath by his younger brother, Alexander, who was mechanically as well as musically gifted and by his younger sister Caroline.

When she arrived in August 1772, Caroline was 22, and her whole life so far had been spent in a Cinderella-like existence under the thumb of her mother who kept her around as a kind of all-purpose household drudge. Unlike her brothers, she had had no formal schooling, and though William later taught her to add and subtract, she continued till the end of her life to carry a set of multiplication tables around with her. During a visit to Hanover, William whisked her away to begin a new existence. There she hoped that, while serving as female head of her bachelor-brother's household, she might receive lessons from him and pursue a career as a soprano. Her prospects can hardly have seemed good. Though not without the family talent for music, she was only four-foot nine, her complexion was badly scarred by smallpox, her manners were coarse, and she spoke only German. She was hardly likely to command a stage, but against all odds she did for a time—until William's astronomical preoccupations forced her to choose between a career and a brother. She chose the brother.

After Easter, Bath emptied for the summer months, and though William kept busy doing concerts out of town, he again had time for other interests. When the winter season of 1772-73 ended, Caroline not unreasonably expected to see more of him and to receive more regular instruction. But for several years he had been working at mathematics, physics, and the theory of music, and that winter, after his fatiguing day of business working sometimes 14 or 16 hours a day, Caroline recalled finding him "retiring to bed with a basin of milk, sago or a glass of water and [Robert] Smith's *Harmonics* or *Opticks* and [James] Ferguson's *Astronomy*, etc., and so went to sleep under his favorite authors."[6]

Though he began with the *Harmonics*, it was the *Opticks* that William found all-absorbing. "When I read [in the *Opticks*] of the many charming discoveries that had been made by means of the Telescope," he later recalled, "I was so delighted with the subject that I wished to see the heavens and Planets with my eyes thro' one of those instruments."[7]

Indeed, that spring, as soon as the heavy music season ended, the memoranda Herschel had long been in the habit of keeping register a dramatic change. Hitherto almost all of his entries relate to music. Now he recorded the acquisition of a quadrant, of texts on astronomy and trigonometry, and of assorted lenses and tubes. With the lenses, he put together some small telescopes, including one with a tube of 12 feet, and "contrived a stand for it." It was good enough to show Jupiter and its satellites. He attempted another of 18 or 20 feet focal length, but the tube, being of pasteboard, was too flimsy, so that, as Caroline recalled, "no one beside my Brother could get a glimpse of Jupiter or Saturn, for the great length of the telescope could not be kept in a straight line."[8] Replacing the paper tube with a tin tube made it sturdier. Still, William was finding such long tubes "almost impossible to manage." These were all refractors. In despair, he turned to the reflector. Commercial instruments were at the time beyond his means, so he did as many an amateur has done since: he began to grind and polish his own mirrors.

With the manual dexterity of a musician, Herschel was not put off by the difficulty of acquiring, learning to use, and finally building his own instruments. It was just as well, since there were at the time only a few other amateur telescope-makers in England; how-to information was scarce, and Herschel had to work everything out for himself, by trial and error. The only abrasives available at the time were sand and emery. The mirrors were made of speculum-metal, a rather hard, brittle bronze alloy, which required a mirror-maker to be a metallurgist and foundry man as well as an optical worker. Herschel had to select and refine the often crude raw materials for their alloys, then melt suitable disks using a small iron furnace and special molds. He eventually found that the best molds were formed from pounded horse-dung. The castings had to be slowly cooled so they could be properly annealed. Because speculum metal is much harder than glass, working a mirror was physically demanding and very fatiguing. Moreover, since speculum metal tarnishes in the open air, any mirror made has to be periodically re-polished and re-figured—Herschel's mirrors often deteriorated noticeably within a month, and were unusable after three or four months. In the interests of being able to continue to use his telescopes

without interruption, Herschel eventually produced two or three specula for each of his telescopes, deploying them relay-style while he reworked the tarnished members.

As he began to work his mirrors, his rooms underwent a dramatic transformation. "To my sorrow," Caroline complained, "I saw almost every room turned into a workshop; a cabinet-maker making a tube and stands of all descriptions in a handsomely furnished drawing-room; Alex[ander] putting up a huge turning machine ... in a bedroom for turning patterns, grinding glasses and turning eye-pieces &c."[9] Naturally, Caroline wanted to continue her music lessons but now found herself "much hindered in my practice by my help being continually wanted" by her brothers in the execution of their telescope-making schemes. By January 1774, Herschel had managed to complete a 4½-inch mirror, and mounted it in a square wooden tube. On March 1, with a power of 40x, he made his first recorded observations: of Saturn, which appeared "like two slender arms," and of the "lucid spot in Orion's sword" (the Great Nebula). Comparing his view to Huygens's old sketch reproduced in Ferguson's *Astronomy,* he found them like but not identical, and began to suspect that the Great Nebula might be subject to change over time. He made this entry in his notebook: "perhaps from a careful observation of this Spot something might be concluded concerning the Nature of it."[10] He had identified an important problem of astronomical research. In later years he would not let go of it.[11]

Up to this time, he had been renting a house at 7 New King Street, on the same street as the house which he would make famous only a few years later. He now moved to a house on Walcot Parade near the outskirts of Bath. Despite its greater distance from the concert halls, it had sheds and stables that he turned into workshops for his mirror-making. There he built a 7-foot telescope and furnished it with "many different object mirrors, keeping always the best of them for use and working on the rest at leisure."

When William got a feel for a mirror's shape, he dared not let go. Recalling his work on one of these mirrors, Caroline recalled:

> ... my time was so much taken up with copying Music and practicing, besides attendance on my Brother when polishing, that by way of keeping him alife [sic] I was even obliged to feed him by putting the Vitals by bitts into his mouth; this was once the case when at the finishing of a 7 feet mirror he had not left his hands from it for 16 hours together.[12]

Though Herschel would later be remembered as the "father of stellar astronomy," at first he was "moonstruck." Like most men of the time, he was always fascinated by—and a firm believer in—the habitability of other worlds, including the Moon. In Ferguson's *Astronomy,* he had read: "From

Fig. 3.3. (left) William Herschel's workshop on the ground level at 19 New King Street, Bath; showing grinding equipment and damage to flagstones from a near-fatal casting mishap. Photographed by William Sheehan, 2009, with permission of the William Herschel Museum, Bath.

Fig. 3.4. (right) William Herschel's seven-foot reflector; the celebrated instrument used to discover Georgium Sidus—the Star of George. Photographed by William Sheehan, 2009, with the permission of the William Herschel Society, Bath.

what we know of our own System, it may be reasonably concluded that all the rest are with equal wisdom contrived, situated, and provided with accomodations for rational inhabitants." This is an important side of Herschel's interests too long neglected, or more precisely, airbrushed out as the unfortunate idiosyncrasy of an otherwise sober researcher. But the record has been set straight by the historian of the 18th century extraterrestrial life debate, Michael Crowe, who sees Herschel "less as an isolated empiricist than a speculatively inclined celestial naturalist, quixotically caught up in a quest for evidence of extraterrestrials."[13]

On May 28, 1776, he was testing a new 10-foot telescope on the Moon. With a power of 240, he scrutinized an area in Mare Humorum near the imposing crater Gassendi, when he found himself

> struck with the appearance of something I had never seen before, which I ascribed to the power and distinctness of my instrument [but which may perhaps be an optical fallacy]... I believed [myself] to perceive something which I immediately took to be *growing substances*. I will not call them trees, as from their size they can hardly come under that denomination.[14]

The region is merely one of rough terrain seen by low illumination, and Herschel himself seems to have given up on the "forests" of Mare Humorum soon afterward, but he had not, by any means, given up on the habitability of the Moon.

In 1778 William's records of his telescope-making efforts become very lengthy. Obsessed with aperture, he began work on a 20-foot telescope, for which he made three 12-inch mirrors in order to use them as usual relay-style. It was a monstrous instrument by the standards of the time, and tricky to put into working order. Caroline recalled the precariousness of Alexander's situation "standing at the top of the house supporting himself with his left arm on the chimney stack whilst with the right at full stretch he was guiding [a] plumbline."[15] When in use, the telescope was elevated by means of a long pole to the top of which was fastened a short arm holding a set of pulleys, and a movable ladder was employed on which the observer perched precariously and looked sideways into an eyepiece near the top of the tube. If the instrument was pointed at something near the zenith, the observer found himself standing some 20 feet above ground level—in the dark.

Given the difficulty and downright danger of using the "small 20-foot," Herschel did most of his routine work with a new 7-foot telescope, with a 6.2-inch mirror, which was conveniently mounted on a folding wooden frame. It was also finished in 1778, and proved to be one of his most productive instruments.

Herschel was still giving concerts in town, and despite its advantages as a workshop, the location at Walcot Parade was proving increasingly inconvenient. So Herschel moved again, to No. 19 New King Street (for the first time) where he had a workshop at ground level, then, two years later, to No. 27 Rivers Street, near the Circus and New Assembly Rooms where he had many of his concerts. By now, his astronomical passion was so all-consuming that he regularly dashed home from the theater between acts in order to observe, and was increasingly regarded, by the more charitably disposed, as an eccentric. The less charitably disposed presumably thought him quite mad. A revealing description of him during his time on Rivers Street is given in the *Retrospections of the Stage* of the Bath actor John Bernard, who says he called on Herschel twice a week for singing lessons, "at his own lodgings, which then resembled an astronomer's much more than a musician's, being heaped up with globes, maps, telescopes, reflectors, &c., under which his piano was hid, and the violoncello, like a discarded favorite, skulked away in a corner."[16] On one occasion, having difficulty with a song,

> I went as usual to my clever friend to rehearse it. It was cold and clear weather, but the sky that night was rather cloudy, and the Moon peeped out only now and then from her veil. Herschel had a fire in his back apartment, and placed the music stand near its window, [a circumstance] which I could not account for. He then procured his violin, and commenced the song, playing over the air twice or thrice to familiarize me with its general idea; and then leading me note by note to its thorough acquaintance. We got through about five bars pretty well, till of a sudden the sky began to clear up, and his eye was unavoidably attracted by the celestial bodies coming out, as it were, one by one from their hiding places: my eye, however, was fixed on the book: and when he exclaimed "Beautiful! beautiful!" squinting up at the stars, I thought he alluded to the music. At length, the whole host threw aside their drapery, and stood forth in naked loveliness:—the effect was sudden and subduing,—"Beautiful, beautiful," shouted Herschel, "there he is at last!" dropping the fiddle, snatching a telescope, throwing up the window, and (though it was a night in January) beginning to survey an absentee planet, which he had been looking for.[17]

Herschel Discovered

As Herschel had no yard or garden, he simply set up his telescopes on the cobblestone street in front of the house. In due course this was to lead to a fateful meeting. At the end of December 1779, he had the 7-foot reflector trained on the Moon just as William Watson, Jr., a physician and son of William Watson, Sr., the secretary to Sir Joseph Banks, the president of the Royal Society of London, happened to pass by. Watson was a founding member of Bath Philosophical Society, which had its rooms nearby, and asked if he could look through the homebuilt telescope. To his surprise, the view of the Moon was the best he had ever had. The two men then entered into a lively conversation that continued inside Herschel's rooms, and seems to have lasted till dawn. Watson invited Herschel to join the Bath Philosophical Society, and not only did Herschel begin to attend the meetings, he became its most enthusiastic member; over the two years of its existence, he submitted 31 papers, on topics ranging from the growth and measurement of corallines (lime-impregnated red algae), static electricity, for which he volunteered himself as a kind of human Leyden jar to determine whether the discharge of static electricity affected the heart rate, and the nature of matter and space. Though the Bath Philosophical Society was a local affair, because of Watson's connections Herschel was soon submitting, through Watson, some of his papers to the prestigious *Philosophical Transactions of the Royal Society of London,* including one on the variable star Mira in the constellation Cetus which—as was typical of Herschel's luck—had just had the brightest flare-up ever recorded, and "Astronomical Observations related to the Mountains of the Moon," a product of his lunar enthusiasm.

In March 1781, Herschel moved again, back to the house at 19 New King Street. (On the brink of being torn down only a few years ago, it was fortunately saved and restored, and is now open to the public as the William Herschel Museum). Probably he wished he had never left; the site had the advantage of a good workshop at ground level where he could cast and grind his telescope mirrors as well as a southward-facing garden, "behind the house and beyond its walls all open as far as the River Avon,"[18] in which he could set up his telescopes.

The superb optics of his telescopes and eyepieces—and his ensuing ability to employ magnifications that were regarded as insanely high by contemporary astronomers—now opened up a new field to him that rapidly eclipsed even his studies of possible life in the Moon. As long ago as August 1779, he had begun a survey—a "first review" of the heavens —with his 7-foot reflector. His purpose was to scrutinize all the naked-eye stars down to the fourth magnitude, in order to make out which of them were double stars.

On the very first night of searching, he discovered what previous astronomers had missed. Polaris, the Pole Star, was a double.

As always with Herschel, his preoccupation was not an idle one. He was trying to apply a method of working out the distances to the stars proposed as long ago as 1632 by Galileo. The principle of the technique is simple. Assume the stars of a pair are unrelated to each other, and only happen to lie along the same line of sight from the Earth. Assume, further, that wherever there is a difference, the brighter star must lie at a much closer distance than the fainter. By measuring the separation between the stars from opposite sides of the baseline of the Earth's orbit (say in January and in June), it ought to be possible to detect a slight shift—called a parallax—of the nearer star relative to the more distant one. This happens for the same reason that, in holding one's thumb at arm's length and looking at it alternately with each eye, the thumb appears to "jump" relative to a distant backdrop of trees. To continue the analogy, the brighter star is the thumb and the distant star the trees.

During his first "review," Herschel carefully listed all the stars that appeared double. He then decided to take the project a step farther, and in 1780 set about a much more ambitious "second review", in which he hoped to examine for "duplicity" all stars down to 8th magnitude in the great star

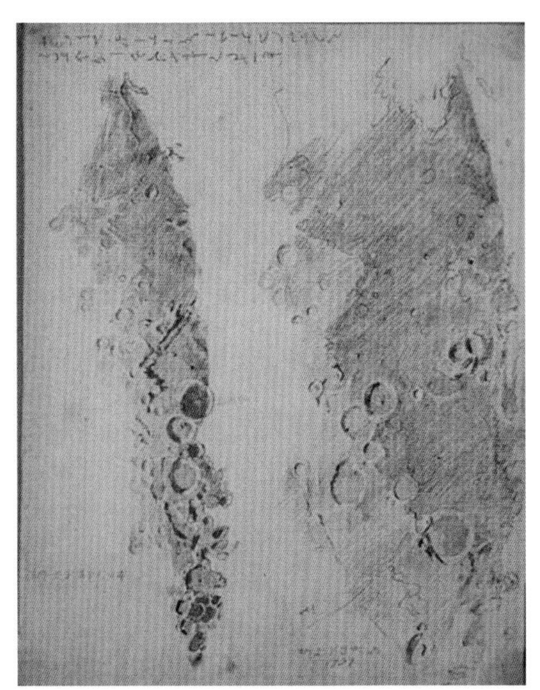

Fig. 3.5. (above) John Russell, the famous portrait artist who would paint Herschel in 1794, made these exquisite sketches of the Moon a decade earlier from London with a Herschel telescope. They recall Herschel's early infatuation with the Moon. *Photographed by Laurie Hatch, with permission of the Oxford Science Museum.*

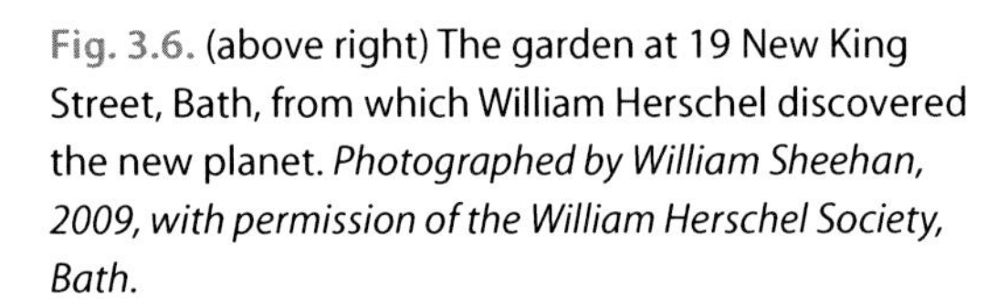

Fig. 3.6. (above right) The garden at 19 New King Street, Bath, from which William Herschel discovered the new planet. *Photographed by William Sheehan, 2009, with permission of the William Herschel Society, Bath.*

Fig. 3.7. (right) Albireo. This famous double, one of the most beautiful in the sky, consists of the brighter golden component (magnitude 3.1) and blue-green companion (magnitude 5.1). At a distance of 430 light years, it is still not certain whether they are a physical system; if so, the period of revolution around the barycenter must be on the order of 100,000 years. Imaged by Dietmar Hager, Linz, Austria. *Courtesy: Dietmar Hager.*

catalog drawn up by John Flamsteed at the Royal Observatory at Greenwich. It was a colossal undertaking, and would occupy him for two years, at the end of which he published a first catalog of 269 double stars; 227 were new.

The "Georgian Star"

With Caroline still tying up ends at the Rivers Street residence, William, in the throes of this vast project, was observing alone with the 7-foot reflector—the instrument he regarded, "for distinctness of vision ... the equal of any ever made"—on the nights of March 12 and 13, 1781. His observing log for March 12 records that he looked at Mars at 5:45 in the morning. "Mars," he wrote, "seems to be all over bright but the air is so frosty & undulating that it is possible there may be spots without my being able to distinguish them." At 5:53 he added: "I am pretty sure there is no spot on Mars." He next looked at Saturn and noted, "The shadow of Saturn lays [sic.] at the left upon the ring."

On March 13 his log book reads:

> "Pollux is followed by 3 small stars at about 2′ and 3′
> "Mars as usual.
> "In the quartile near Zeta Tauri the lowest of two is a curious either nebulous star or perhaps a comet.
> "A small star follows the Comet at 2/3rds of the field's distance."

It thus appears that Herschel observed Pollux and Mars in the morning of March 13, and then—in the evening, presumably after returning from his professional duties, such as giving concerts—he recommenced his close scrutiny of Flamsteed stars, looking for new doubles. As was his custom, he was using an ocular magnifying 227x. Suddenly, a new planet swam into his ken. He did not at first recognize it as such. He could only be certain that the object he had in his grasp was no ordinary fixed star; it had a disk. The excellence of his telescope and his customary use of higher powers allowed him to recognize it, and Herschel noted it down as a "curious either nebulous star or perhaps a comet."

It was "perhaps fortunate," John Louis Emil Dreyer would observe, that "at the time Herschel had never come across a planetary nebula ... as he might have taken the stranger for one of these bodies, noted its place, and not looked it up again for some time; in which case it might have figured for years as a lost nebula, like the lost stars of Flamsteed."[19] (Precisely this misapprehension was to deprive his son, Sir John Herschel, of the chance to discover the next planet out from the Sun, Neptune, in 1830.)

But in any case, the object's motion was soon apparent to Herschel, and this showed it to be an object held firmly by the reins of the Sun's gravitational force. It was not a nebula; it must, then, be a comet.

In "An Account of a Comet," dated March 22, 1781, presented to the Bath Philosophical Society and communicated by Watson to the Royal Society, Herschel gave additional details about the discovery. At "between ten and eleven" in the evening, that is after his concerts would have ended, he began scrutinizing stars in the interesting region between M1 (the Crab Nebula) in Taurus and M35, an open cluster in Gemini. Using an ocular magnifying 227x, which was his usual practice when looking for double stars, he examined two small stars near Zeta Tauri and found that one of them didn't quite look right:

> While examining the small stars in the neighborhood of H Geminorum [1 Geminorum], I perceived one that appeared larger than the rest; being struck with its uncommon magnitude, I compared it to H Geminorum and the small star in the quartile between Auriga and Gemini, and finding it so much larger than either of them, suspected it to be a comet.[20]

Though a Messier or a Méchain would have found their hearts skipping a beat at the discovery of a new comet, Herschel seems to have taken it in stride; its discovery was inevitable, given the quality of his telescope and the thoroughness of his methods. Messier, indeed, found what Herschel had done little less than incredible. In congratulating "Monsieur Hertsthel at Bath," he added: "I cannot conceive how you were able to return several times to this star—or comet ... since it had none of the characteristics of a comet."[21] Herschel turned over the routine work of tracking the comet and computing its orbit to the professionals, like Nevil Maskelyne, the Astronomer Royal at Greenwich, and Oxford astronomer Thomas Hornsby. Maskelyne wrote on April 4, 1781 that the comet's motion was "very different from any comet I ever read any description of or saw," and added on April 23 that it was "as likely to be a regular planet moving in an orbit very nearly circular around the sun as a comet moving in a very eccentric ellipsis."[22] This proved to be the case. By May, Messier's friend, Jean Baptiste Gaspard Bouchart de Saron, demonstrated to the Paris Academy of Sciences that the object was very remote—at least as far out as 14 astronomical units (14 A.U.) from the Sun. As better orbits were computed, its mean distance was increased to 19 A.U. The size of the Solar System had doubled.

Herschel was almost the very last to recognize or appreciate the importance of his own discovery. He was like Columbus, the man who having found an authentic New World believed he had simply reached the East Indies—conceiving, as do most of us, that the world holds fewer surprises than is actually the case. As late as November, when Sir Joseph Banks

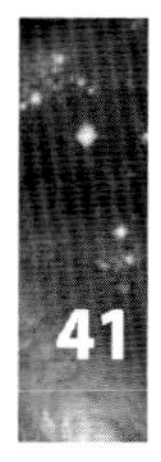

Fig. 3.8. Jupiter and the Galilean satellites, with Uranus in the same field (at far upper right) for comparison. *Imaged on September 5, 2010, with an 85 mm apochromat and a Canon 35mm digital camera by Michael Conley and William Sheehan.*

awarded him the prestigious Copley medal of the Royal Society, Herschel still hedged as to the object's significance, and said only: "With regard to the new star I may still observe that tho' we are not sufficiently acquainted with its nature, yet enough has been seen already to shew that it differs in many essential particulars from Comets and rather resembles the condition of Planets."[23] But this is the last note of diffidence Herschel expressed, for by then there could no longer be any doubt. Herschel had discovered nothing less than a new major planet of the Solar System, the first to be added in modern times.

Astronomy writer Richard Baum has conjured up the importance of what Herschel had achieved:

> Herschel had broken with immemorial tradition. His sighting of a body, identified as a massive world beyond the icy rim of the known, a world so remote as to be barely visible to the unaided eye, at once doubled the size of the known solar system. Importantly it sent a breath of fresh cosmic air coursing through the dusty corridors of thought.[24]

It was indeed the beginning of a revolution in astronomy, a revolution as great as that which took place in North America the following October with the surrender of Lord Cornwallis's army to George Washington at Yorktown—an event that has been regarded as the "dirge" of the British Empire in America, the first slippage of that imperial grasp that had led to Britain's subjugation of an empire on which the Sun never set. Historian Piers Brendon has written:

> The Old World did regard the New World's victory as an ominous inversion of the established order. It was an unbeaten revolt of children against parental authority—the first successful rebellion of colonial subjects against sovereign power in modern history.[25]

In astronomy, the new planet was the harbinger of a similar revolt against the starry status quo, but it was also, for Herschel, the harbinger of a potential sea-change in his life circumstances—at least if he played it right, he could expect the new planet to allow him to give up music and devote himself full-time to astronomy. The discovery of a new planet would have been enough for most men to sit contented on their laurels, but Herschel was always restless and eager to move on to new frontiers. A new planet was for him almost small change. He was a man who did not recognize any limits to his own mind, or—in Shakespeare's words—he was

> like one that stands upon a promontory,
> And spies a far-off shore where he would tread,
> Wishing his foot were equal to his eye.
>
> —*(Henry VI, Part III, ii)*

His promontory was the eyepiece of his splendid telescopes; the far-off shore where he would tread lay far beyond even the distant new planet. For his most important work—on the Galaxy and the nebulae—was just to begin.

Packing up the 7-foot telescope with its folding stand in a mahogany traveling case, Herschel and Watson took the stage to London, and went from thence to Greenwich (where his telescope trounced the performance of the professional telescopes of the Royal Observatory) and finally to the King's palace at Kew, near which (at Deer Park, Richmond) George had his personal observatory. Urged on by Watson and other friends, Herschel angled for an official position. As a step in that direction, he flattered the King by proposing that the new planet receive the name *Georgium Sidus,* George's Star, and spelled out the reasons for his choice:

> As a subject of the best of kings, who is the liberal protector of every art and science ... and ... as a person now more immediately under the protection of this excellent monarch, and owing everything to his bounty; I cannot but wish to take this opportunity of expressing my sense of gratitude, by giving the name *Georgium Sidus* (the Georgian Star) ... to a

> star, which (with respect to us) first began to shine under His Auspicious Reign.[26]

Fig. 3.9. George III; mezzotint by Johann Zoffany. *From: Wikipedia Commons.*

(Needless to say, the name didn't stick; it was rejected by Continental astronomers, who instead proposed a name from Greek mythology, Uranus, and Uranus it would be.)

George was genuinely interested in science, and could hardly resist. He offered to sponsor Herschel's work in astronomy, but since the position as astronomer at his private observatory at Kew had already been promised to another, George had to create a new post for Herschel. He would become the King's "Personal Astronomer," at a salary of £200 per annum (though a substantial sum, Herschel had been making at least £400 per annum as a professional musician). The disbursement was contingent on Herschel's moving his residence from Bath to the neighborhood of Windsor Castle so that he could (when summoned) present occasional astronomical entertainments to the royals and their guests. Even before he had received this grant, Herschel had written with keen anticipation to Caroline:

> Among Opticians and Astronomers nothing is now talked of but what they call my great discoveries. Alas! This shows how far they are behind, when such trifles as I have seen and done are called great. Let me but get at it again! I will make such telescopes & see such things...[27]

At the time he discovered the new planet, Herschel was only 43. Still possessed of tremendous energy and drive, his portraits show an eager, agile, cat-like face that only gradually, as he aged, became increasingly slabby and jowly, until it resembled the face of a prosperous German burgher. As much as he had accomplished in astronomy so far, he was on the verge of the most productive decade of his career, and would soon begin to fulfill the prophetic words that he had uttered to Caroline, "I will make such telescopes & see such things."

Even before he left the residence at 19 New King Street, Herschel—who always wanted larger and larger telescopes: more aperture, more light, more "space-penetrating power"—was working on a new giant telescope. Twice—in August 1781—he had attempted to cast a three-foot mirror in the basement where he did his own foundry work. As was his usual practice, the mold of loam was prepared from loam or horse dung, of which, says Caroline, "an immense quantity was to be pounded in a morter and sifted through a fine seaf; it was an endless piece of work and served me for many hours exercise and Alex frequently took his turn, for we were all eager to do something towards the great undertaking."[28] Even Watson took a turn at pounding horse dung. Unfortunately, two attempts to cast the large speculum failed, and on one of the two attempts the mirror cracked on cooling. Both Caroline and William gave accounts; Caroline's is the more dramatic. "As soon as the season for the Concerts was over," Caroline writes, "and the mould &c in readiness a day was set apart for casting and the metal was in the furnace; which unfortunately began to leak at the moment when ready for pooring, and both my Brothers, and the Caster with his men were obliged to run out at opposite doors; for the stone flooring (which ought to have been taken up) flew about in all directions as high as the ceiling. My poor Brother fell exhausted by heat and exertion on a heap of brick-batts."[29] Clearly, William had narrowly escaped with his life.

On the last day of July 1782, the three Herschel siblings—William, Alexander, and Caroline—left Bath forever, and began a new chapter of their lives at Datchet, a village between Slough and Windsor on the banks of the Thames. They moved to a large rambling property that had not been lived in for years; "the ruins of a place," Caroline groused. It no longer exists, though the grander adjoining house, the "Lawns," remains, much altered. William and Alexander saw more possibilities in it than Caroline did. They adapted the stables for a workshop for mirror-making and set up a library in the laundry. Alexander, searching the overgrown garden for a place to set up the 20-foot telescope, had a close call of his own—he nearly fell in a well. At first, Herschel was much put upon by the King, and regularly had to transport the seven-foot reflector to Windsor castle so the Royal Family could look through it. The "small" 20-foot telescope was too large to transport, but the King was eager to see it. On December 1 he came to Datchet expressly for the purpose. "The precarious observing ladder was no place for royalty,"[30] notes Herschel scholar Michael Hoskin, and the King settled for a view of the "Georgian" through Herschel's 10-foot telescope. Eventually the novelty of looking through Herschel's telescope wore off at Windsor. Herschel had fewer interruptions.

At Datchet, Herschel resumed the quest for double stars with the seven-foot telescope. This remained his main preoccupation until September 1783. Meanwhile, thinking to improve the knowledge of the heavens of his "assistant," Caroline, he set her to work with a small refractor, placed on an ingenious mounting of his own design that allowed her to make horizontal sweeps across the sky in the quest for interesting objects. He later fabricated an equally ingenious instrument for doing vertical sweeps. She complained as she always did of her hard lot, and found life hard to bear when William was away. Then she was

> left solely to confuse myself with my own thoughts, which were anything else but cheerful; for I found I was to be trained for an assistant Astronomer and by way of encouragement a Telescope adapted for sweeping consisting of a Tube with two glasses such as are commonly used in a finder [was furnished to me], I was to sweep for Comets, and by my Journal No. 1, I see that I began August 22, 1782 to write down and describe all remarkable appearances I saw in my Sweeps.... But it was not till the last two months of the same year before I felt the least encouragement for spending the starlight nights on a grass-plot covered by dew or hoar frost without a human being near enough to be within call.[31]

As her sweeps continued, she came across some of the nebulae listed in Messier's catalog, which William had in his possession but, being otherwise preoccupied, had hardly so much as glanced at. Then, in the summer of 1783, Caroline began finding new nebulae. It was this success which first opened her brother's eyes to the great number of them awaiting discovery. Thus, Caroline—long relegated to a footnote in her brother's triumphant story—played a key role in inspiring Herschel's most important work.

At the time, the biggest telescope Herschel had available was the "small" 20-foot telescope, with its mirror of 12 inches, but that fall, he successfully cast two 18 ¾-inch mirrors to be used relay-fashion in what would become his "large" 20-foot telescope. While the "small" 20-foot had been difficult and even dangerous to use with its precarious and primitive mounting, its successor would be furnished with a solid wooden mounting on wheels which could be rotated by a single workman. It was also equipped with a seat, later replaced by a platform extending from one side of the tube to the other that could be raised and lowered on a ladder made for the purpose.

Herschel began a series of "sweeps" with the "large" 20-foot on October 28, 1783. At first he tried to go it alone, and tried by candlelight to sketch the positions of stars in the finder and the field of the telescope whenever an object of interest passed by. This work was not only very fatiguing, but each time he wrote something down he found that he spoiled the sensitivity of his eye. By mid-December he discarded this inefficient method of operation. Henceforth he resorted to vertical sweeps in which a workman was employed to raise and lower the telescope. The tube was locked on a selected zone (or "gage") of sky, and William used a handle to rack it up and down in 2 degree steps. Bells were placed at the extreme positions as a way of alerting him of the need to change direction. The frequent sound of the bell of the telescope announced that he was happily, and productively, sweeping. Caroline, meanwhile, had been impressed into service as his amanuensis, carefully recording the observations so that he did not need to take his eye from the telescope or expose it to candlelight.

Though the work went on with rather mind-numbing routine, there was always danger of a serious accident. Caroline recalled one such occasion:

> My brother began his series of sweeps when the instrument was yet in a very unfinished state, and my feelings were not very comfortable when every moment I was alarmed by a crack or fall, knowing him to be elevated fifteen feet or more on a temporary cross-beam instead of safe gallery. The ladders had not even their braces at the bottom; and one night, in a very high wind, he had hardly touched the ground before the whole apparatus came down.[32]

Caroline herself sustained a serious injury soon after she began to assist her brother. On December 31, 1783, the clouds began to part and at about 10 o'clock a few stars were visible. According to Caroline's own rather breathless account:

> In the greatest hurry all was got ready for observing. My brother, at the front of the telescope, directed me to make some alteration in the later motion, which was done by machinery, on which the point of support of the tube and mirror rested. At each end of the machine or trough was an iron hook, such as butchers use for hanging their joints upon, and having to run in the dark on ground covered a foot deep with melting snow, I fell on one of these hooks, which entered my right leg above the knee. My brother's call, "Make haste!" I could only answer by a pitiful cry, "I am hooked!"[33]

Though Caroline was unhooked by her brother and the workmen, the hook tore away about two ounces of her flesh. Afterwards she was examined by Dr. James Lind, celebrated for his idea of using citrus to prevent scurvy on long sea voyages. He declared that a soldier meeting with such a wound he would have been entitled to six weeks' nursing in a hospital. Caroline concludes: "I could give a pretty long list of accidents which were near proving fatal to my brother as well as myself."

Bigger Things

In the 18th century, the spot in Datchet where the 20-foot telescope stood and Caroline was "hooked" was a very lonely and desolate place—and it still is. On a visit in 2009, one of us (W.S.) found nothing of the house still standing, but appreciated the fact that because of its low situation near the Thames it would have suffered frequent floods and been miserably damp during the winter. (Even today, the UK Environment Agency's maps show this area as being at risk of flooding.) Despite this insalubrious location, William and Caroline labored on tirelessly through all seasons, rubbing their faces and hands with onions to warm themselves and ward off the ague—though on at least one occasion, William succumbed to the ague at last.[34]

With the 7-foot used to discover the "Georgian," the "large" 20 foot, with its 18.7-inch mirror, was Herschel's most productive instrument. He would use it for what was nothing less than a survey of the dimensions and structure of the entire system of stars. Even today, the scope of his ambition astounds.

As with the double-star survey, he began with certain initial assumptions. Though hitherto the starry heavens had usually been regarded as if they were painted on the surface of a concave sphere, Herschel grasped that the objects therein were disposed along the far line of a third dimension—the heavens had not only breadth and height but depth. It had been in order to get an estimate of the measure of that depth that he had tried to use the double-star method to obtain distances to the stars, but so far the results were inconclusive, while by 1803 he was forced to admit that he had failed to obtain the parallax of a single star this way. Instead he perceived something different: one of the doubles he re-measured, Castor, showed a change in the relative positions of its components. Parenthetically, it wasn't Herschel's own re-measurements that revealed this. By a stroke of luck, when Herschel told Maskelyne what he was doing, the latter recalled that his predecessor as Astronomer Royal, James Bradley, had taken note of the orientation of the components of Castor in 1759, and when Herschel compared Bradley's observation with his own he concluded that the component stars were, in fact, revolving "either in circles or ellipses, round [a] common centre of gravity."[35] In other words, they were members of a gravitationally bound system, and the difference between the brighter and fainter member of a pair was a true physical difference.

At this time a Scottish farmer and naturalist, James Hutton, was examining geological strata in order to theorize about the processes at work on and within the Earth. Herschel, by analogy, proposed studying the "geological" structure of the star system, examining beds of stars and nebulae, "variously inclined and directed, as well as consisting of very different materials." The Milky Way itself, Herschel asserted, was "nothing but a stratum of fixed stars."

In the "large" 20 foot telescope, Herschel saw that the Milky Way about the hand and club of Orion was completely resolved into "small stars, which my former telescopes had not light[-grasp] enough to effect." In an initial foray, on January 18, 1784, he counted the number of stars visible in six fields chosen randomly and found them to contain 110, 60, 90, 70, and 74 stars each. This was Herschel's first attempt to apply what he called the method of "star-gages." He conceived of himself extending plumb-lines along different visual rays and reaching far into the depths of the starry stratum. By counting stars in fields along different directions, he could work out the dimensions of the starry stratum in each direction and so to arrive at "the interior construction of the heavens."

In his surveys with instruments of unprecedented power under the inky-black 18th- century skies, he encountered in his sweeps not only a glittering multitude of stars but countless nebulae—including one "nebulous bed" which was so rich that, in passing through a section of it in a time of only 36 minutes, he detected no less than 31 nebulae, "all distinctly visible upon a fine blue sky." As we now know, he was then sweeping in part of the elongated area of sky running up through Virgo and Coma Berenices into Ursa Major and containing what is now known as the Virgo Cluster, or Super-cluster, of galaxies.

In yet another "stratum"— perhaps, Herschel added, "a different branch" of the one just described—he delighted in the splendid variety of what passed before his eye as he called out each object, as it passed, for Caroline to write down:

> I have seen double and treble nebulae, variously arranged; large ones with small, seeming attendants; narrow but much extended, lucid nebulae or bright dashes; some of the shape of a fan, resembling an electric brush, issuing from a lucid point; others of the cometic shape, with a seeming nucleus in the center; or like cloudy stars, surrounded with a nebulous atmosphere; a different sort again contain a nebulosity of the milky kind, like that wonderful, inexplicable phenomenon about [theta] Orionis; while others shine with a fainter, mottled kind of light, which denotes their being resolvable into stars.[36]

Much of the work by which Herschel is still remembered was accomplished by the time he left Datchet in June 1785 (about the time Lemuel Francis Abbott painted his portrait; Fig. 3.1). To stellar astronomy, he contributed his first and second catalogs of double stars (containing 269 and 434 pairs, respectively). He also published an important paper in which he announced the important discovery of the motion of the Sun with respect to nearby stars. This last was a particularly vivid example of his genius. Though the first proper motions of stars had been discovered

by Halley in 1718, Herschel argued in 1783 that the dozen or so proper motions known by then were not random, but revealed a common component due to the motion of the Sun itself—it was, he concluded, hurtling through space and carrying the Earth and other objects of the Solar System in the direction of the constellation Hercules.

At Datchet, too, he carried out 683 star-gages. These were counts of stars of different brightness occurring in randomly selected fields. Under the assumption that they were all of the same intrinsic brightness and knowing that the apparent brightness of any one star falls of inversely with the square of the distance, there should be a well-defined relation of stars per magnitude interval versus magnitude, which would allow him to work out what he called the "interior construction" of the heavens.[37] He observed the expected relation for brighter stars, but found fewer stars at the faint end. He interpreted this as running out of stars in certain directions. He demonstrated this result in a sketch, which was a cross-sectional map like one of those that the pioneering geologist James Hutton was producing at the same time of geological formations. (It was not, nor was it intended to be, a 3-dimensional map; this is sometimes a source of confusion.). That he should have found the star system to be elongated was hardly a surprise; that result was implied by the very existence of the "stratum" of the Milky Way. As we are situated within the disk, we see many more stars than when we look along it length-wise than when we look upwards or downwards through it. The form that Herschel's plot of his star- gages traced was irregular, with marked prolongations in certain directions and indentations in others. Herschel himself referred to his model as a "Grindstone." We think it looks rather like a snapping alligator with open jaws about to bite down. The open jaws mark the bifurcation of the Milky Way produced by the dark marking known as the "Great Rift" of Cygnus.

At first blush, the resemblance of Herschel's diagram to a modern Galaxy map seems uncanny. This resemblance is, however, misleading. Lacking an exact distance to a star, Herschel had had to make assumptions. With uncanny intuition, he had supposed that, on average, the stars might be of about the same brightness as the one star best known—the Sun. Here, as so often, he was helped by analogies drawn from natural history. As early as 1782 he had written to the Astronomer Royal Nevil Maskelyne:

> When I say "Let the stars be supposed one with another to be about the size of the Sun," I only mean this in the same extensive signification in which we affirm that one with another Men are of such or such a particular height. This does neither exclude the Dwarf, nor the Giant. An Oak-tree also is of a certain general size tho' it admits of very great variety....

> If we see such conformity in the whole animal and vegetable kingdoms, that we can without injury to truth affix a certain general Idea to the sizes of the species, it appears to me highly probable and analogous to Nature, that the same regularity will hold with respect to the fixt stars.[38]

Assuming, then, that the stars were more or less like oaks, a count of the number of stars visible along different routes or lines of sight through his telescope would allow him to map the length and breadth of

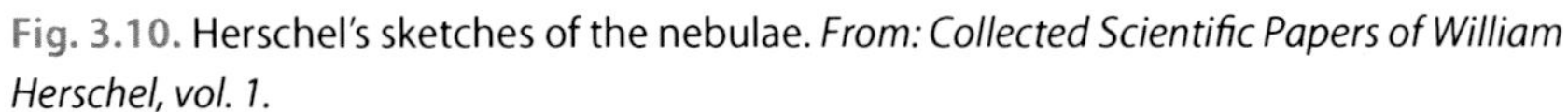

Fig. 3.10. Herschel's sketches of the nebulae. *From: Collected Scientific Papers of William Herschel, vol. 1.*

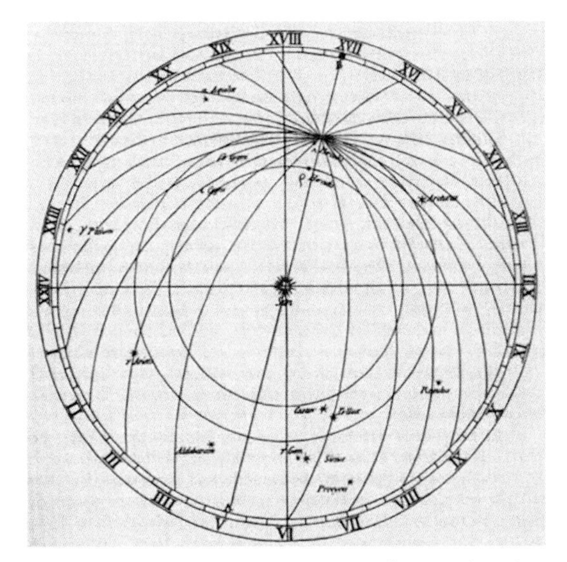

Fig. 3.11. From the proper motions of a mere handful of stars, William Herschel managed to deduce that the Sun and the Solar System were headed in the direction of the star λ Herculis, a result that has essentially held up. *From: Collected Scientific Papers of Sir William Herschel, volume 1.*

the Milky Way just as a count of the (mature) oaks along different lines of sight in a forest would allow one to map the forest's length and breadth.

Indeed, Herschel's method was fundamentally sound. But his diagram is still wrong since his reasoning contained a second, and ultimately a fatal, flaw. It turns out that space between the stars is not empty: there is a veil or haze of interstellar dust causing distant stars to appear fainter than they really are. At very great distances—or in directions where the dust is particularly thick—we cannot see stars at all, at least not in the visible spectrum. (Ironically, Herschel himself, in 1802, was the first person to demonstrate the existence of the infrared, which is able to penetrate that dust.) Unbeknownst to Herschel, his view of the stars along the plane of the Galaxy was being cut off by clouds of interstellar dust at distances of just a few thousand light years. His Grindstone is based on a survey of only about one percent of the stars of the Galaxy, most of which belong to a nearby galactic structure known as the Orion Spur. Because of interstellar dust, he did not even remotely penetrate to the galactic center. The Sun's seemingly exalted position near the center of his flattened disk was an illusion just like that had on a stormy night when, in swinging a flashlight in various directions, one sees the same sheets of blinding rain in

Fig. 3.12. Herschel's grindstone model of the Galaxy. *From: Collected Scientific Papers of William Herschel, vol. 1.*

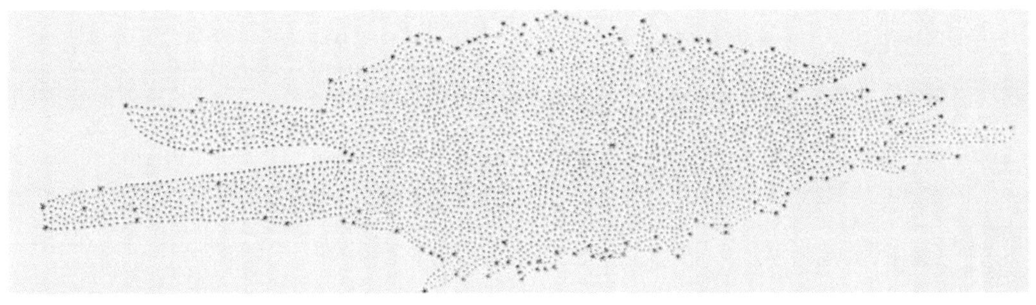

every direction one looks—though one is not justified in thinking that one is therefore standing in the middle of the storm!

With his knack for raising important questions and proposing useful lines of inquiry, Herschel had not only made progress in understanding the vast spatial extension of the Galaxy—its "inner construction"—he was already pondering how that star-system might change over time:

> That the milky way is a most extensive stratum of stars of various sizes admits no longer of the least doubt; and that our sun is actually one of the heavenly bodies belonging to it is as evident. I have now viewed and gaged this shining zone in almost every direction.... In order to develop the ideas of the universe, that have been suggested by my late observations, it will be best to take the subject from a point of view at a considerable distance both of space and time.[39]

Taking the long view across time as he had across space, he imagined cataclysmic events that might produce some of the strange and tortuous forms he visualized in the telescope. Writing about some of the nebulae he had seen, he entertained a conjecture anticipating the modern apprehension of the gigantic collisions of galaxies in the early universe:

> If it were not perhaps too hazardous to pursue a former surmise of a renewal in what I figuratively called the Laboratories of the universe, the stars forming these extraordinary nebulae, by some decay or waste of nature, being no longer fit for their former purposes, and having their projectile forces, if any such they had, retarded in each others' atmosphere, may rush at last together, and either in succession, or by one general tremendous shock, unite into a new body.[40]

This passage, and especially the startling phrase "Laboratories of the universe," contains a poetic leap equal to anything in Milton or Coleridge, and is prophetic of things to come. In a striking metaphor, Herschel charts a course away from familiar moorings in positional astronomy toward future directions that would eventually develop into the discipline of astrophysics.

Some of the cloudy patches – nebulosities – Herschel discovered seemed to consist of small compact clusters of small stars. Others might be vast stellar "strata", in effect whole "milky ways," in their own right. The majority of the nebulae he discovered—and there were thousands of them—seemed to stand well clear of the band of the Milky Way. The Milky Way was, he said, a "Zone of Avoidance" for the nebulae, and Herschel began to suspect that some of them might be milky ways in their own right.[41] Fanny

Burney, the novelist and Second Keeper of the Robes to Queen Charlotte at Windsor, would exclaim, "Herschel has discovered 1500 universes!"

* * *

1. Richard Holmes, *The Age of Wonder: how the romantic generation discovered the beauty and terror of science* (New York: Pantheon, 2009), 74.
2. William Herschel, "Letter to Dr. Hutton," in Dreyer, ed., *Scientific Papers of Sir William Herschel,* vol. 1, xxi.
3. Edward Holden, *Sir William Herschel: his life and works* (New York: Charles Scribner's Sons, 1881), 22.
4. Patrick Moore, *80 Not Out: the autobiography* (London: Contender Books, 2003), 121.
5. W. Davis Jerome, quoted in *Rutgers Focus Magazine,* October 1, 1999.
6. Michael Hoskin, ed., *Caroline Herschel's Autobiographies* (Cambridge: Science History Publications, 2003), 51.
7. Dreyer, *Scientific Papers of Sir William Herschel,* vol. 1, xxiv.
8. Hoskin, *Autobiographies,* 51.
9. Ibid., 53.
10. Herschel Archive, Royal Astronomical Society.
11. For a full account of Herschel's nebular work, see: Michael Hoskin, "William Herschel and the Nebulae," Part 1: 1774-1784," *Journal for the History of Astronomy, xlii* (2011), 178-192 and "William Herschel and the Nebulae, Part 2: 1785-1818," Journal for the History of Astronomy, xlii (2011), 322-338.
12. Hoskin, *Autobiographies,* 51.
13. Michael J. Crowe, *The Extraterrestrial Life Debate 1750-1900: the idea of a plurality of worlds from Kant to Lowell* (Cambridge: Cambridge University Press, 1986), 61.
14. Herschel Archive, Royal Astronomical Society.
15. Hoskin, *Autobiographies,* 127.
16. Michael Hoskin, "Vocations in Conflict: William Herschel in Bath, 1766-1782," *History of Science, 41* (pt. 3), 133 (Sept. 2003): 323.
17. Ibid., 323-324.
18. Hoskin, *Autobiographies,* 57.
19. Dreyer, *Scientific Papers of Sir William Herschel,* vol. 1, xxix. Herschel first introduced the term "planetary nebulae" in 1785 because some of them showed small, hazy greenish disks just like the new planet.
20. William Herschel, "Account of a Comet," read April 26, 1781, *Phil. Trans.,* 1781, 492-501, in *Dreyer, Scientific Papers of Sir William Herschel,* vol. 1, 30-38.
21. Constance A. Lubbock, ed., *The Herschel Chronicle: the life-story of William Herschel and his sister Caroline Herschel* (Cambridge: Cambridge University Press), 40.
22. Dreyer, *Scientific Papers of Sir William Herschel,* vol. 1, xxx.
23. William Herschel to Joseph Banks, Nov. 19, 1781, in *Occasional Notes of the Royal Astronomical Society,* vol. 3 (1949):115.
24. Richard Baum, *The Haunted Observatory: curiosities from the astronomer's cabinet* (Amherst, New York: Prometheus, 2007), 19.
25. Piers Brendon, *The Decline and Fall of the British Empire 1781-1997* (New York: Alfred A. Knopf, 2008), 4.
26. William Herschel to Joseph Banks, *Phil. Trans.,* 1783,1-3, in Dreyer, *Scientific Papers of Sir William Herschel,* vol. 1, 100-101.
27. Holmes, *The Age of Wonder,* 110.
28. Hoskin, *Autobiographies,* 63.
29. Ibid.
30. Michael Hoskin, *The Herschel Partnership: as viewed by Caroline* (Cambridge: Science History Publications, 2003), 60.
31. Hoskin, *Autobiographies,* 71.
32. Ibid., 76.
33. Ibid., 77.
34. Holmes, *The Age of Wonder,* 119.
35. Herschel, "Account of the Changes that have happened, during the last Twenty-five Years, in the relative Situation of Double Stars; with an Investigation of the Cause to which they are owing," read June 9, 1803, *Phil. Trans.,* 1803, 339-382, in Dreyer, *Scientific Papers of Sir William Herschel,* vol. 2, 250-276:264.
36. Herschel, "Account of some Observations tending to investigate the Construction of the Heavens," read June 17, 1784, *Phil. Trans.,* 1784, 437-451, in Dreyer, *Scientific Papers of Sir William Herschel,* vol. 1, 157-166:160.
37. Herschel, "On the Construction of the Heavens," read February 3, 1785, *Phil. Trans.,* 1785, 213-266, in Dreyer, *Scientific Papers of Sir William Herschel,* vol. 1, 223-259.
38. William Herschel to Nevil Maskelyne, April 28, 1782, quoted in Lubbock, *The Herschel Chronicle,* 110-111.
39. Herschel, "On the Construction of the Heavens," 223.
40. Ibid., 259.
41. Michael Hoskin. *William Herschel and the Construction of the Heavens* (New York: W. W. Norton, 1964), 99.

4. Chimneys and Tubules of the Galaxy

While a child, Tennyson said to his brother, who suffered from shyness: "Fred, think of Herschel's great star-patches, and you will soon get over that."
—*Hallam Lord Tennyson, quoted in Michael Crowe, Modern Theories of the Universe*

In addition to everything else he was doing, William Herschel continued to crank out telescopes. He constructed five 10-foot telescopes on the express command of George III, and began to receive orders for seven-foot reflectors similar to the one he had used to discover the "Georgian," eventually producing two hundred of this model alone, on a commercial basis, many to royalty and foreign dignitaries and wealthy amateurs, but at least a few to observatories (Radcliffe, Armagh, Greenwich, Dorpat) or to serious individual astronomers (such as Johann Schroeter at Lilienthal, in Hanover, and Giuseppe Piazzi in Sicily). The income he obtained this way enabled him to make "expensive experiments for polishing mirrors by machinery." Eager to leave the low-lying, damp, and all-but-forgotten place on the north bank of the Thames—today not a rack remains behind, not even a marker to show the spot where he produced the first map of the Galaxy—he and his siblings moved briefly to Clay Hall, Old Windsor. After clashing with a litigious and difficult landlady, he moved in April 1786 to the house where he would remain for the rest of his life—"Observatory House," on Windsor Road, Slough. Parenthetically, Observatory House itself no longer exists; greatly deteriorated, it was pulled down in 1960, and when one of us (W.S.) visited, a few years ago, it had been replaced by a rather nondescript modern office block—rather jarringly called "Observatory House." Indeed, though Herschel's name is remembered everywhere about Slough, there is hardly anything left of Georgian Slough, the city having served as a dump for war surplus materials in the interwar period, and suffering the usual blight from rapid industrialization—until it could serve as the subject of one of the poet John Betjeman's laments:

Come, friendly bombs, and fall on Slough!
It isn't fit for humans now,
There isn't grass to graze a cow.[1]

For some time Herschel's imagination had been fired by the prospect of an even larger telescope. It was ever a feature of Herschel's mind not to recognize its own limitations, and he now dreamed of a giant reflector with a focal-length of 40 feet and a mirror of 48 inches. With financial support from King George, he set to building it. The size and weight of the mirror were such that ten men would be employed in polishing it, but "the whole of it had been designed by the astronomer who was to use it, and every part of it had been made on the spot under his own immediate and never-ceasing supervision, while the amount of personal, mental and manual labor he performed must have been immense."[2] His workmen, impressed with his dexterity in turning metal on the lathe, asked him to what he owed his skill; "to fiddling," he replied. Finally, by February 1787, the giant telescope was far enough along to allow a first test: he had the mirror placed in the tube and went "into the tube, and lying down near the mouth of it," and held an eye-glass in his hand with which he found the place of the focus. He then directed the apparatus toward the Great Nebula of Orion. The mirror, "though far from perfect... showed the four small stars in the nebula and many more. The nebula was extremely bright." Ironically, though the great telescope was built mainly to study nebulae, this would be practically the only nebular observation Herschel ever made with it!

Indeed, the 40-foot proved in the end to be a magnificent failure. It was so cumbersome that it required the assistance of two workmen to set it on a star, and because of the need to ascend a high platform in the darkness to observe with it, it was dangerous to use, rather like shaving with a guillotine. Herschel soon gave up the hope of making extensive use of the telescope from which such miracles were expected, and fell back, for the rest of his career, on the versatile "large" 20-foot. The 40-foot, however, was a wonder of the world to all who visited Slough to see it, and though only a partial realization of its maker's dreams would serve, as we shall see, as the inspiration of the builders of even larger telescopes in the next century. Herschel had thrown his cap over the fence, and others, in due course, would try to follow.

William Herschel: Valedictory

Meanwhile, the circumstances of Herschel's life were changing. On moving to Slough, he had become a regular visitor at the house of a Mr. and

Mrs. Pitt who lived nearby at Upton.[3] When Pitt, a well-to-do solicitor then in failing health, died in 1786, Herschel continued to regularly walk the path from Observatory House to the widow's house. The widow was wealthy, attractive, and amiable. Hitherto Herschel's biography had been almost devoid of romantic interests (he had had a brief interest in a Miss Harper, one of his pupils, when he had first moved to Bath, but it was unreciprocated; if it had been, the whole history of astronomy might have been different!). His life had been too busy, too strenuous, to allow much exercise of the affections, but in due course a romance developed between the widow and the 50 year-old bachelor astronomer. Herschel proposed marriage, and the proposal was accepted. Initially the plan was for the couple to take up residence at Mrs. Pitt's house at Upton. She was shrewd enough to realize, however, that "the Doctor would be principally at [Observatory House], and that Miss Herschel would be mistress of the concern, and considering the matter in all its bearings, she determined upon giving [the engagement] up." Herschel was disappointed, but naturally unwilling to give up his pursuit of astronomy; he also refused to give up the sister who had been trained to be a "most efficient" assistant and was beginning to earn a reputation as an astronomer in her own right as a discoverer of comets. (She had discovered her first "Lady's Comet" at Slough on August 1, 1786, with a small reflector of two foot focal length which she called her "comet sweeper," and would follow it up with seven more, of which two were co-discoveries; the last was made in 1797). Eventually, a compromise was reached: "There were to be two establishments, one at Upton and one at Slough; two maidservants in each, and one footman to go backwards and forwards, with accommodation at both places, and Miss Herschel to have apartments over the workshops [at Observatory House]." The marriage took place on May 8, 1788 at the small church of St. Laurence, Upton. Sir Joseph Banks, President of the Royal Society, served as Best Man.

Fig. 4.1. Herschel's "nebulous star," NGC 1514, imaged with the 32-inch Ritchey-Chrétien telescope at Kitt Peak National Observatory. *Courtesy: NOAO/AURA/NSF.*

Inevitably, William's marriage altered his relationship with Caroline, who had willingly (if sometimes grudgingly) sacrificed her personal happiness in the interests of her idolized brother. She regarded the marriage as a betrayal, and insisted on moving out of "Observatory House," though she settled nearby and continued, not without regular irruptions of bitterness confided to the pages of her diary, to assist her brother in his work.

After his marriage, Herschel's own routine gradually relaxed. He was less driven than before, less eager to persevere through long damp cool nights searching for nebulae. Moreover, his celebrity made him much in demand. Even before the move from Datchet, Caroline had started a double-columned visitors' book of the same kind as that in which she recorded observations. Among those who visited in spring 1784 was Susannah Thrale, the third daughter of Mrs. Thrale, the long-time friend of the celebrated Dr. Johnson. "He can show you in the night sky," Johnson advised, "what no man before has ever seen, by some wonderful improvements he has made in the telescope. What he has to show is indeed a long way off, and perhaps concerns us little, but all truth is valuable and all knowledge pleasing in its first effects, and may subsequently be useful."[4]

Though Herschel was less driven to spend his nights under the stars, he continued his sweeps as before, and in November 1790 made an observation whose implications were particularly portentous. Until then, he had believed that all the nebulae consisted of star clusters, some of which were however too far away to be resolved into stars. He thought this even of the Great Nebula in Orion, which might be a stellar stratum that might, as he wrote, "outvie our milky-way in grandeur." What finally changed his mind was an observation of a much humbler object, a nebula of the smooth round type he had, since 1785, taken to calling "planetary" nebulae, because of their superficial resemblance to the small pale greenish disk of Uranus.

As Herschel had cataloged it, the object was the 19th cluster of his 6th class (now designated NGC 1514 in Taurus). It is much less impressive than such kindred and well-known planetary nebulae as M27 (the "Dumbbell") in Vulpecula or M57 (the "Ring") in Lyra. He examined it carefully with the 20-foot, and entered the following note in his log book:

> November 13, 1790. A most singular phenomenon! A star of about the 8th magnitude, with a faint luminous atmosphere, of a circular form, and of about 3′[of arc] in diameter. The star is perfectly in the center, and the atmosphere is so diluted, faint, and equal throughout, that there can be no surmise of stars; nor can there be a doubt of the evident connection between the atmosphere and the star.[5]

This observation shattered his faith that all the nebulae must necessarily be resolvable into tiny stars. The natural historian of the heavens had, in fact, come across nothing less than a new celestial species. The paper which he published in the *Philosophical Transactions* in 1791, "On the Nebulous Stars, properly so called,"[6] is well worth quoting in part:

> While I pursued [my] researches, I was in the situation of a natural philosopher who follows the various species of animals and insects from the height of their perfection down to the lowest ebb of life; when, arriving at the vegetable kingdom, he can scarcely point out to us the precise boundary where the animal ceases and the plant begins; and may even go so far as to suspect them not to be essentially different. But recollecting himself, he compares, for instance, one of the human species to a tree, and all doubt upon the subject vanishes before him. In the same manner we pass through gentle steps from a coarse cluster of stars, such as the Pleiades, the Praesepe, the milky way, the cluster in the Crab, the nebula in Hercules, that near the preceding hip of Bootes [i.e., M3] etc., etc.... till we find ourselves brought to an object such as the nebula in Orion, where we are still inclined to remain in the once adopted idea, of stars exceedingly remote, and inconceivably crowded, as being the occasion of that remarkable appearance. It seems, therefore, to require a more dissimilar object to set us right again. A glance like that of the naturalist, who casts his eye from the perfect animal to the perfect vegetable, is wanting to remove the veil from the mind of the astronomer. The object I have mentioned above, is the phenomenon that was wanting for this purpose. View, for instance, the 19th cluster of my 6th class [i.e., NGC 1514], and afterwards cast your eye upon this cloudy star, and the result will be no less decisive than that of the naturalist we have alluded to. Our judgment, I may venture to say, will be, that *the nebulosity about the star is not of a starry nature.*[7]

The "cloudy star" led to a paradigm shift in Herschel's thought, and henceforth he allowed the existence of truly nebulous matter—a "shining fluid" or "fire-mist"—into the heart of his cosmogony.

During these years, visitors continued to stream to Observatory House. One of the most famous was the composer, Joseph Haydn, who on his visit to England in 1791 came to Windsor to meet George III and paid a visit to Slough to look through Herschel's telescope. He later claimed that the experience helped him to compose his celebrated oratorio *Creation*, which premiered at the Schwarzenberg Palace in Vienna in 1798 before an audience that possibly included the young Beethoven. (In a sense, Herschel's music—or at least an echo of it—lingers in our ears even today.) The opening orchestral number, "The representation of Chaos," includes a splendid passage which opens in the dark key of C minor and in which the chorus softly sings "And God said, 'Let there be light,'' followed by silence, then an even softer passage, "And there was...," followed by a booming *Light* in the bright key of C major. This was the way Haydn's musical mind responded to the stars and nebulae that had floated before him in the eyepiece of Herschel's telescope. Beethoven biographer Edmund Morris, contrasting Haydn's approach to chaos with what Beethoven would have done with the same theme, declares:

> Miraculously, Haydn had managed to do [it] without afflicting any powdered head with a sense of harmonic or formal disarray. On close analysis, the old man had merely postponed cadences again and again, or only half dissolved them, in order to convey a vague state of disequilibrium. But his modulations fell into logical patterns, and his overall design was sonata form. Eighteenth-century to the core, he could not imagine anything "without form, and void," unless he fixed it inside the frame of Reason.[8]

Well, that was Herschel too.

Meanwhile, Herschel was traveling again—something he had done relatively little since the frenzied days of his early days as a struggling musician. In 1792, he went to Scotland and met James Hutton, who a few years before had enunciated his principle of uniformitarianism—the same geological processes at work on and in the Earth at the present time have been operative throughout many geological ages. Hutton's work was opening up vistas of time as vast as those of space that Herschel was opening with his telescopes. Herschel came to a similar view of the nature of the forces operative in the heavens. Changes were taking place there—in particular, he looked for them in the complex forms of the Orion Nebula, and suspected that the "shining fluid" and "fire mists" in space might supply the raw materials needed for the regeneration of stars. But he was not an evolutionist: he believed that over time, despite cycles of change, the universe remained essentially the same.

In 1802, Herschel visited Paris, where he met Messier, and in the company of the celebrated mathematician Pierre Simon de Laplace, he was received by Napoleon, then First Consul. (It was on this occasion that Laplace, when Napoleon asked him why he had not mentioned God in his great work *Exposition du système du monde,* supposedly gave his famous reply: "Sire, I have no need of that hypothesis!" The story, though a good one, is somewhat apocryphal; Herschel, whose views on religion were conventional and Anglican, must have been shocked—if the incident actually happened.)

Back in England, Herschel was soon sweeping again. But he was now feeling the effects of age, and soon gave up. He recorded his last—Sweep 1112—on September 30, 1803, 20 years after his first.

Despite his herculean efforts, his work remained unfinished, and he knew it. In the study rather than at the telescope, he sifted through the masses of data he had collected, piecing out the taxonomy of the nebulae by arranging them according to their shape, color, and brightness in rather the way a botanist did in classifying plants. Though he had first believed the role of "nebulous matter" was merely to lead to the occasional regeneration of already extant stars, after meeting Laplace he began to take into account the latter's "nebular hypothesis"—the theory that the Sun and planets had condensed out of a rotating cloud of nebulous matter. In an 1811 publication, "Astronomical Observations relating to the Construction of the Heavens," the 72 year-old sage of Slough proposed that most of the nebulae were clouds of "shining fluid" on the way to condensing into new stars.[9]

And what exactly was the shining fluid? This was well before we had any idea of what the most common material or elements in the universe were, or for that matter that nebulous gas or dust existed in space. The science of the day embraced both *ponderable* matter—the ordinary matter that followed the Newtonian laws of motion and gravitation—and *imponderable* fluids. As British chemist Humphrey Davy had jotted in a notebook at about this time: "We cannot account well for the phenomena of electricity without supposing the existence of a peculiar fluid. Is this fluid the same as heat, i.e. are electricity and galvanism currents of a fluid which when expanding in right lines constitutes light."[10] Herschel, having thoroughly absorbed such thinking, imagined that such an imponderable fluid—"widely diffused and chaotic"—permeated the heavens, and by a gradual process of compression condensed into star clusters. In a paper of 1814 he extended his speculative vision even farther, and contemplated the eventual "breaking up of the Milky Way"[11] itself.

Enter the Reluctant Heir

It was clear that William—who became Sir William after 1816, when the Prince Regent George Augustus Frederick made him Knight of the Royal Guelphic Order—was fast running out of time. Where was a successor to carry on the work thus far so nobly advanced?

There was a successor—and he was growing up in the laboratory and mirror-making workshops of Observatory House itself. The Herschels had had their first and only offspring when William was 53. Described by diarist Frances Burney as "a delightful child," John Frederick Herschel was a precocious youth. Apart from a brief period at Eton, the famous public school near Windsor whose harsh punitive methods led his parents to withdraw him, he was educated entirely at Observatory House before going to University. He had prodigious gifts in languages (in addition to the classical languages, Greek and Latin, he learned German in that bilingual household, as well as French and a smattering of other modern languages). He was very keen on mathematics, and enjoyed tinkering in the workshop with his father's workmen. His spinster aunt, one of his only regular playmates when she could spare the time, recalled his fascination with tools. Needless to say, he also took his turns singing and playing musical instruments during the family's musical diversions.

John Herschel grew up, as historian Allan Chapman notes, "lonely, seriously intellectual and, with the exception of visits from his ... cousins Mary and Sophia Baldwin, largely spent in the company of people who were old enough to be his grandparents.... A child less innately gifted, or naturally affectionate, could have been seriously damaged by this upbringing.... Even so, one always feels that John Herschel was old before his time. He never really lost his shyness, while the nervous, bronchitic and rheumatic diseases that plagued him from his twenties onwards suggest that he lacked something of that indomitable toughness possessed by his septuagenerian and octagenerian parents and almost centenarian aunt."[12]

Frightfully erudite, backed by a famous name, he entered St. John's College, Cambridge, at 17 (in 1809; the year Charles Darwin, Abraham Lincoln, and Alfred Tennyson were all born). He rocketed to the top of his class, distinguishing himself in Greek and Latin and becoming Senior Wrangler in the Mathematics Tripos and First Smith's Prizeman in 1813—high honors indeed, since Cambridge University, though academically sluggish during the last sixty years of the 18th Century, took special pride in her prowess in mathematics, which she claimed from her famous son, Isaac Newton. Newton had written the *Principia* laying out the laws of motion and the theory of gravitation while Lucasian professor at Trinity, the college next door to St. John's. Until 1824 anyone seeking academic honors at Cambridge was required to read mathematics, no matter what their other interests.

After finishing his exams, John Herschel moved directly into a Fellowship. In order to remain a Fellow and eventually advance into the long-range prospect of becoming a don, he would have to take Holy Orders. His father strongly urged him in this course. "Such a path," said Sir William, "must surely lead to happiness, or else it would never be so wide and so beaten." Though John was more than pious enough for the Church—that was never his objection—he regarded the teaching duties required of a Fellow unbearably pedestrian. "I am grown fat, dull, and stupid," he told his close Cambridge friend Charles Babbage. "Pupillizing has done this—and I have not made one of my cubs understand what I would drive them at."

Instead he began to pivot toward the law. Though generally tractable and even docile to parental guidance, he was evidently going through a mildly rebellious stage. William remonstrated with him that nothing could be more unsuitable to him than the law, but John proved resolute, and in January 1814 began reading for the bar at Lincoln's Inn, London. Inevitably, he found it to be quite as tedious and unpleasant as his father had warned him it would be. After eighteen months he gave it up; his excuse was that the stuffy chambers aggravated his asthma.

He then returned to Cambridge and immersed himself in what he would always regard as his first and best love, optics. But he was rushing toward a crisis in his affairs, indeed in the affairs of the whole family; it could be postponed but not put off indefinitely, and was reached in the summer of 1816, when John accompanied his parents on a holiday to Dawlish, near Exeter, to visit his father's old friend from the Bath Philosophical Society, Sir William Watson. By then the elder Herschel's health was failing badly; the tremulousness of his hand made writing difficult—all his later letters are in Caroline's hand—and his great work on the stellar system remained incomplete. Clearly he needed assistance, and more than the aging Caroline could provide. By the time the family returned to Slough, John had given in. He agreed to follow the path of duty and to take up at long last the "family business of fathoming the 'length, depth, breadth, and profundity of the Universe.'"[13] Returning to Cambridge only long enough to collect his things, he wrote to Babbage in singularly low spirits:

> Farewell to the University. . . . Now I am about to leave it, my heart dies within me. I am going, under my father's directions, to take up the series of his observations where he has left them.[14]

Years later, after he had completed the astronomical work he had promised his father to carry out, he would tell his wife that all the while he had only been "*dallying* with the stars. *Light* was my first love! In an evil hour I quitted her for those brute & heavy bodies which tumbling thro' ether, startle her from her deep recesses and drive her [light] trembling and sensitive into our view."[15]

So it was that in the summer of 1816, John Herschel, at 24, reluctantly accepted the mantle of heir to his father's work on the stellar system. The wealth he inherited from his mother allowed him to make this choice, as John Herschel was not paid to be an astronomer by the now almost completely mentally disabled king George III, as his father had been.

He began, with Sir William's assistance, to figure an 18-inch mirror for a 20-foot reflector. As yet, however, he was much more often in London than in Slough. Of the latter, he wrote, "I have regularly heard my mother declare that there is not to be found so expensive a place in the Kingdom," he wrote. "But the worst of the story is the society, which owing to the vicinity of the Court is in general peculiarly unpleasant."[16] Indeed, throughout the regime of George III, the atmosphere of the court remained like that of a minor German principality, "stiff, narrow, fusty."[17] The situation did not improve after the first outbreak of King George's madness in 1788, which eventually became unremitting and incurable. After ten years of roaming the drafty corridors of Windsor Castle with a long white beard and purple dressing gown, lamenting the loss of his American colonies and beset with delusions that included seeing his native Hanover through one of Herschel's telescopes, the unhappy king was relieved of his earthly miseries by death in 1820. His successor, George IV, who had served as Prince Regent since 1811, proved to be a rather scandalous figure, and today is best remembered for his dissolute lifestyle.

Compared to Slough—and the Court of Windsor—London was a place of great stimulation and intellectual ferment, and John Herschel seems to have had his fingers in every pie. He was a member of all the leading clubs and societies, acquainted with all the other members who came into contact in formal meetings and over convivial dinners and all living "within walking distance, or at least within a short cab ride"[18] of such venues as the Athenaeum, the Royal Society, the Royal Institution, the Royal Geological Society, the Astronomical Society (later the Royal Astronomical Society). As Secretary of the Royal Society and Foreign Secretary of the Astronomical Society, and through wide-ranging travels, usually with Babbage, he met and corresponded extensively with many of the leading mathematicians and scientists on the Continent. One of his biographers, David Evans, has written, "his mere lists of addresses are of absorbing interest, for they include practically all men of mark in the world."[19] Among those men of mark was James (later Sir James) South, a surgeon who, becoming independently wealthy through marriage to an heiress (one of the most direct routes to independence during the late Regency and early Victorian era), abandoned surgery for astronomy and set up a splendidly appointed private observatory at Blackman Street, Southwark, near the hospital of Guy's and St. Thomas's where he had practiced, and joined forces with John Herschel in observing and calculating orbits of double stars.

Ironically, it was after his father's death in August 1822 that John began spending more time at Slough, and resumed the studies of the nebulae, those mystifying objects that had so perplexed and fascinated Sir William. One of the first orders of business was to carry out a careful study of the Great Nebula in Orion. His father had always regarded it as the "most wonderful object in the heavens" and had concluded that it was perhaps the foremost example of a nebulous "shining fluid," a true "fire-mist" that

Fig. 4.2. (left) Young John Herschel; (right) Margaret Brodie Steward (Lady Herschel). Both by Alfred Edward Chalm (*Flora Herscheliana*, 1829). *From: Wikipedia Commons.*

seemed (as he had long strongly suspected) to be changing form over time as it gravitationally condensed into stars and planetary systems. As soon as he turned his new 20-foot telescope toward it, the younger Herschel was impressed by the uneven texture of the bright region of the nebula around the Trapezium stars (the Huygenian region, so-called since it had been the only part visible to Christiaan Huygens). He compared it to "a curdling liquid, or a surface strewed over with flocks of wool, or to the breaking up of a mackerel sky."[20] Given the suspicion of its undergoing changes, he realized that the accurate depiction of this and other nebulae posed one of the greatest challenges facing astronomers of the day. Photography, to which John Herschel himself would make great contributions, was then literally in its infancy; the "View from the Window at Le Gras," the earliest surviving photograph by France's Nicéphore Niépce made by coating pewter sheets with a solution of bitumen dissolved in lavender oil, dates only to 1826, and it would be decades before the services of the young technology could be entrusted with serving as the artisan of the nebulae. John, by the way, was by far a better artist than his father had been. The images of nebulae that William had published in his 1811 paper, "Astronomical Observations relating to the Construction of the Heavens," were made simply by ruling crossed lines for dark background then stippling with mezzotint to suggest lighter highlights—a technique that had the fault of giving the nebulae "a much better definition than they really possess."[21] From sketches and notes made on several favorable nights of observing with the 20-foot, John Herschel set about making the most accurate drawing of the Great Nebula he could. Intended to serve as a benchmark of later studies, he presented it to the Royal Astronomical Society in 1826—the same year Niépce obtained his primitive but portentous image.

Two years later, we find John still busy with sweeps, and still complaining. "The sacrifice of the night is a very serious evil to me," he wrote, "regarding as I do the completion of this work not as a matter of choice or taste, but a sacred duty."[22] Arduous it was, but not without rewards, as he often enjoyed views of the nebulae superior to any his father had ever had. For instance, writing of M51, which Messier had found between the handle of the Big Dipper and Cor Coroli (the lead-star of Canes Venatici, the Hunting Dogs), Herschel thought it "like our Milky Way," and could not help recalling his father's speculations about island universes as he added, "perhaps ... [it] is our Brother System."

"It is a truth universally acknowledged," Jane Austen reflected in *Pride and Prejudice,* "that a single man in possession of a good fortune must be in want of a wife." Now pushing 40, and a doyen of the British scientific establishment, John was also a very eligible bachelor, a Mr. Darcy possessed of handsome features, a substantial fortune mostly inherited from his mother, and the expectation of more when she died. He had always, it seems, been highly susceptible to female charms. His early letters refer to an unfortunate unreciprocated love followed by a period of intense emotional suffering. In the 1820s the disappointment of a broken engagement was relieved only when Babbage whisked him off to strenuous travels on the Continent including scrambling the slopes of Monte Rosa, near Zermatt, not a bad achievement for a chronic asthmatic. Now he was introduced to Margaret Brodie Stewart, the daughter of a noted Gaelic scholar. She was only 19. Despite the difference in their ages, John and Margaret would enjoy, from first to last, a connection of "unclouded happiness."

Following his marriage, John became even more frightfully productive than he had been as a bachelor. He wrote lucid and influential essays on the undulatory nature of light and the theory of sound, and published a book on the inductive method of science—*A Preliminary Discourse on the Study of Natural Philosophy*—which electrified a 21-year-old former medical student and undistinguished student of divinity at Christ's College, Cambridge, Charles Darwin. Partly on the strength of this performance, in 1831 the King—now William IV— made him Knight of the Royal Hanoverian Guelphic Order, an honor for which his father had had to wait till the last six years of his long life. Henceforth he was Sir John. That year he ran as a candidate for the President of the Royal Society, which had become stagnant and reactionary under the long tenure of the late Sir Joseph Banks;

he ran as a reformer, but was unsuccessful. The next year, when his mother died, he received the remainder of her substantial fortune, and never had to worry about money again.

Financial security and the independence it brought introduced a fresh boldness to Sir John's affairs. While at Slough, immersed in extending his father's observations of double stars, clusters, and nebulae, he realized that the next logical next step to formulating the laws of the heavens by the inductive methods he had championed in the *Preliminary Discourse* would be to extend the survey to the Southern skies. His father had of course studied the construction of the heavens—but only from the northern hemisphere perspective. With a bit of the philosophy of "What shall they know of England who only England know?" he began to think seriously of an extended sabbatical on the other side of the globe. Clearly he was eager for a fresh start, but as a practical matter, he was also one of the only men in England who could finance such a project (the British government's strong embrace of *laissez-faire*—satirized by Dickens in the HOW NOT TO DO IT OFFICE in *Little Dorrit*—meant opposition to state funding for scientific projects on the scale that existed in France and Germany at the time). But there were also personal reasons for leaving England. He was being ground down by frequent trips by stage between Slough, a place to which he no longer had the ties his father as a royal pensioner had, and London, where the many distractions of his life there were, as Lady Herschel recalled afterward, "tearing him to pieces."[23]

Thus, on November 13, 1833, Sir John and his family embarked on the *Monstuart Elphinstone,* a vessel of less than a thousand tons displacement, to one of farthest points of the British empire: South Africa. From Portsmouth to Cape Town they were nine weeks under sail. It is impossible, by the way, not to admire Lady Herschel's fortitude and sheer physical endurance. David Evans writes: "She withstood two long ocean voyages in spite of being a ready victim of seasickness.... In South Africa she was pregnant more than half the time. The home party comprised five adults and three children on arrival, and she had to run an establishment which employed something like ten servants: Clara and Candassa, who did washing; Somai, the (Malay) coachman; David, the groom; Minto, the houseboy; Dawes and Jack, the laborers; Andreas; Jeptha and his wife, who cooked; Hannah; Leah; Catherine Quinn; Nancy, the needlewoman; the cowboy—and as Sir John would say, &c. &c. &c."[24]

Herschel's own diaries make fascinating reading. His entry for December 7, 1833 reads: "Last night saw for the first time the Greater Magellanic Cloud. It is brighter and larger than I expected to see it & a very odd looking object." On December 26, after dining on deck and watching an eclipse of the Moon: "Nothing ever exceeded the magnificence of this night. I watched Jupiter down to setting & after plunging through a small cloud line, saw him appear like a lamp on the very edge of the sea and go down behind the offing like a fireball." On January 15, 1834, coming up on Cape Town, he caught sight of some of the most magnificent scenery on Earth. "Rose & hurried on Deck," he wrote, "whence the whole range of the Mountains of the Cape from Table Bay to the Cape of Good Hope was distinctly seen, as a thin, blue, but clearly defined vapour. The Lions Head was seen as an Island the base being below the horizon."

The Cape had already played an important role in the history of astronomy. It had been the observing station of the French Abbé Nicolas-Louis de Lacaille, who in the 1750s had come here and measured the positions of 10,000 southern stars and discovered two score nebulae. It was now the site of the Royal Observatory, whose first director, Fearon Fallows, had been a classmate of Herschel's at Cambridge. Fallows seems literally to have worked himself to death. Contracting scarlet fever, he insisted on continuing measuring star positions with an instrument called a mural circle—toward the end, he was so weak he had to be carried to the telescope in a blanket! He was succeeded by Thomas Henderson, who accepted only reluctantly after being turned down for the chair of practical astronomy at Edinburgh University and for the post of superintendent of the *Nautical Almanac* office. He found Cape Town a dreadfully unpleasant place—"I

Fig. 4.3. Feldhausen; lithograph by John Ford. *William Sheehan Collection.*

will tell you about my residence in the Dismal Swamp among slaves and savages," he wrote to a colleague back home[25]—and left after only a year, when he returned to Scotland. He stayed long enough to obtain nineteen precise measures with the mural circle of a star, alpha Centauri, which had been found to have a very large proper motion—thus to be presumably one of the Sun's near neighbors in space. Once back in Scotland, he carefully analyzed these measures and was able to obtain a *parallax*—defined as half the small angle of deflection of the star, in seconds of arc, as measured from opposite sides of the Earth's orbit. For alpha Centauri, the parallax was ¾" of arc—equal to the apparent diameter of a U.S. penny seen at a distance of 5 kilometers, which shows why it took so long for astronomers' to make this measurement. From this parallax, Henderson was able to work out the distance to the star—the long-sought grail was at last in the astronomers' grasp. Henderson's star, Alpha Centauri, is indeed a member of the multiple-star system which is the nearest to the Sun at a distance of 42 trillion kilometers. (Since light travels 300,000 kilometers/second, this works out to 4.3 light years; by way of comparison, this is almost exactly 1,000,000 times the distance of the nearest planet, Venus). Henderson published his result just after Friedrich Wilhelm Bessel, another astronomer obsessed with precision who was director of the Royal Observatory at Königsberg (now Kaliningrad, Russia), had done so for another star with large proper motion, 61 Cygni, which had a parallax of 0.3" of arc, making its distance 11 light years. A third parallax, for the bright star Vega, was published in 1840 by Friedrich Wilhelm Struve of the Dorpat Observatory in Estonia; it was less accurate than those published by Bessel and Henderson, for the simple reason that Vega's parallax is much smaller—only 0.13" of arc—making its distance 25 light years.

Because stellar distances are tied to measures of parallax, astronomers often use as a unit of distance the *parsec*—the distance from which the Earth-to-Sun distance would subtend an angle of 1" of arc. One parsec equals 3.26 light years. In what follows, we will sometimes refer to astronomical distances in parsecs.

Henderson had already left the Cape before Herschel arrived. On a temporary basis, Herschel took up quarters at a Boarding House in town, unloaded and carefully unpacked his instruments, and began scouring the countryside for a house. He settled on a place called Feldhausen by the Dutch and "the Grove" by the English, 10 kilometers from Cape Town.

He wrote to aunt Caroline, who had returned to her native Hanover following her brother's death:

> You can imagine nothing more magnificent than its situation which is nearly under the towering precipices of the Table Mountain, and deeply

sheltered in a forest of Oak and Fir which so effectually secure it from the South East Gales, that while it has been blowing a perfect hurricane in Cape Town I have been able to sweep with the 20-feet without inconvenience, & even to light one lamp from another in the open air.

The 20-feet is erected in an orchard, which forms a small part only of the grounds belonging to the place, which are distributed into great squares by long shady avenues of Rich Oak or tall and solemn Pines, and either overgrown with trees or laid out in Gardens.—In the same orchard also stands (or will stand in the course of next week) the [5-feet] Equatorial, for which the house [with a rotating roof] is now building, & will be finished by that time. The sky has been for at least 3 nights out of 5 that we have been here, nearly or quite cloudless & rich in stars, nebulae, and clusters beyond anything you can imagine & of which I trust to give a pretty good account. We are now inhabiting temporarily, a house within half a quarter of a mile of "the Grove."—Not a very good one, but having the advantage of being near enough to allow of my superintending all the repairs &c there, and to go over at night with my man Stone and work with the 20-feet at my sweeps, and return when tired, to sleep here across the rich region of flowers & shrubs which *here* is called a Heath—but in England would be a flower garden or Nursery.[26]

Whereas the Royal Observatory occupied what really was at the time a "dismal swamp," and was frequently buffeted by Southeasterlies, Feldhausen stood on higher ground, and was partly protected so that the blustery gales were reduced to what Sir John called "no more than a refreshing & moderate breeze." The idyllic setting was replete with birds Sir John could shoot at and "glorious flowers and rich aromatic scents" for Lady Herschel to paint; the only drawback seems to have been a pack of savage dogs, which Sir John sometimes encountered in his rides over the countryside. By the end of February 1834, the 20-feet telescope was set up (at a point still marked by an obelisk on the grounds of what is now the Grove Primary School). Commencing in June, Sir John's sweeps of the ravishingly beautiful southern skies continued for the next four years. Just before he had sailed for Cape Town, his aunt Caroline had reminded him to reexamine a peculiar "hole" in the heavens that her brother had encountered while sweeping in the body of the Scorpion. "I wish," she had written, "you would see if there was not something remarkable in the lower part of the Scorpion to be found, for I remember your father returned several nights and years to the same spot, but could not satisfy himself about the uncommon appearance of that part of the heavens. It was something more than a total absence of stars (I believe)."[27] From Cape Town the Scorpion rises high into the sky, and Sir John reported his findings:

> The Stars continue to be propitious, and the nights which follow a shower, or a "black South Easter" are the most beautiful observing nights it is possible to imagine. -I have swept well over Scorpio and have many entries in my sweeping books of the kind you describe—viz: blank spaces in the heavens *without the smallest star. For Example—*
> RA. 16h 15m—NPD 113° 56—a field without the smallest star
> RA. 16h 19m—NPD 116° 3′—*Antares* (alpha Scorpio)
> 16h 23m—114° 25 to 114° 5—field entirely void of stars
> 16h 26—114° 14—not a star 16m.—Nothing!
> 16 27—114° 0—D[itt]o—as far as 114° 10—
> and so on—then come on the Globular Clusters—then more blank fields—then suddenly the Milky Way comes on as here described....[28]

Here in the Scorpion and in other places along the vast extent of the Milky Way he was coming across what he deemed to be true "holes" —

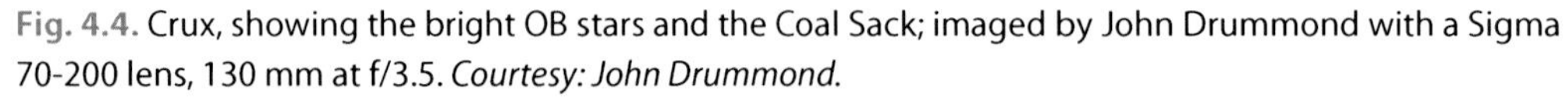

Fig. 4.4. Crux, showing the bright OB stars and the Coal Sack; imaged by John Drummond with a Sigma 70-200 lens, 130 mm at f/3.5. *Courtesy: John Drummond.*

Fig. 4.5. NGC 4755, discovered by the French astronomer Nicolas Louis de Lacaille in 1751-52; it appears as a hazy star about a degree southeast of the 1st magnitude star Beta Crucis, and was given its name the "Jewel Box" by Sir John Herschel (he described it as "an extremely brilliant and beautiful object ... the very different colour of its constituent stars ... gives it the effect of a superb piece of fancy jewellery." It has an obvious A-shape of bright stars, and is one of the youngest open clusters known (estimated age 14 million years); its brightest stars are blue-white supergiants, with one of the stars a red supergiant that stands out at once. *Courtesy: John Drummond.*

Fig. 4.6. The Large Magellanic Cloud. *Courtesy: John Drummond.*

vacant regions, cavernous recesses—in the stratum of stars. The Milky Way appeared to him not as a "stratum of regular thickness and homogeneous formation," but "broken into masses and aggregates of stars," whose structure reminded him of the "cumuli of a mackerel sky." The dark enclosed spaces appeared, he noted in an arresting phrase, as "vast chimney-form or tubular vacancies."[29] There were, in other words, true caves, recesses, in the Milky Way. Herschel, the astronomer, investigating those chimneys and tubules, had become a celestial spelunker.

His researches into the construction of the heavens in the southern sky extended—and in some cases modified, or at least moderated—his father's views. Sir William's interpretation of the circle of the Milky Way as an optical effect due to our immersion in a divided stratum of stars was borne out; but Sir John, in attempting to sketch in the details, appreciated even more fully than his father had done just how curious, complicated and irregular that system was. He noted "the extraordinary display of fine resolved and resolvable globular clusters" toward Sagittarius and along the fringes of the Milky Way, and began to suspect that the stratum of stars might be arranged in the form of a great flat ring. He was also coming across planetary nebulae by the score. In contrast to the half dozen or so known in the northern skies, these glowing "shells" turned up by the score in South Africa.

The question which preoccupied him the most was whether there was a single "galaxy," the Milky Way with all its stars, nebulae and star clusters, or whether there might be similar systems—"island universes"—beyond it. The "true" nebulae, as opposed to the resolved and resolvable star clusters, really did appear to consist of a mysterious "luminous fluid," "masses of glowing vapors," and they puzzled him just as they had his father. He confirmed that they were not irregularly scattered over the visible heavens but collected in a sort of canopy, whose vertex was at the "Galactic Pole" of the vast stratum of the stars of the Milky Way. In the northern hemisphere, they crowded into the constellations Leo, Leo Minor, Ursa Major, Coma Berenices, Canes Venatici, and above all Virgo; in the southern hemisphere, they were more uniformly distributed, with one important exception. In the Magellanic Clouds, the strange objects looking like detached portions of the Milky Way, he found no less than 1,145 nebulous objects between them. The Large Magellanic Cloud, in particular, was "a congeries of stars, clusters of irregular form, globular clusters and nebulae."

One of the enclosed nebulae—now known as the Tarantula Nebula—especially captured his imagination, and he made a careful sketch. As anyone who has spent time observing from the southern hemisphere can attest, scanning the Large Magellanic Cloud, even in binoculars, is almost, as French galactic astronomer Gerard de Vaucouleurs used to say, "like

Fig. 4.7a. The Carina Nebula, also known as the Eta Carinae Nebula (NGC 3372). A huge HII region, surrounded by several spectacular naked-eye star clusters, it contains Eta Carinae and HD 93129A, two of the most massive stars in the Galaxy; and lies at a distance of from 6500 to 10,000 light years. It is located in the Carina-Sagittarius Arm. Photographed from Chile with a Takahashi apochromatic refractor, f/6. *Courtesy: Klaus Brasch.*

Fig. 4.7b. Detail of the above: The star Eta Carinae, which shone as the most brilliant star in the sky apart from Sirius in 1841, just after John Herschel returned from the Cape, is a very unstable massive star (with a mass of from 100 to 150 times that of the Sun, and a luminosity of four million times that of the Sun). The star flirts dangerously close to the so-called Eddington limit, where the outward pressure of the star's radiation is almost strong enough to counteract gravity—thus it is a good candidate to go supernova in the near future. Note the strange bloblike feature surrounding Eta Carinae, the so-called Homunculus Nebula, which appears to have been ejected in the 1841 outburst, and the Keyhole, as Sir John Herschel called the dark dust cloud superimposed on the center of the nebula. Imaged from Gisborne, New Zealand with a 35cm f/10 Meade ACF SCT. *Courtesy: John Drummond.*

viewing plates from the famous Barnard photographic atlas." (For Barnard, see chapter 7.)

Lady Herschel, in a charming letter to Caroline, conveys a sense of Herschel's ongoing routine:

> You wish to hear about Herschel, & in truth I like to write about him—at this moment he is observing, & the bell of his telescope tells me he is very busy—he came in a minute ago to say that it is the most brilliant night he ever saw, & he has had many such of late—His sweeps seem hitherto more productive of Nebulae than Double Stars, & the Mirrors are kept in the highest state of polish, by about ¼ of an hour's rubbing on the very finest powder, so that the figure is not destroyed, & the process can be repeated *very often*—The winter has been most favourable to him & he feels already, that his voyage hitherto has been for good—The Literary & Scientific Institution of the Cape have made him their President (very wisely), & he is setting them all to work to observe the Tides, & collect Meteorological Obsns &c, & I think he ought to consider the members, as *learned Operatives* under his superintendence—Nothing can be better than his health during the whole winter—indeed he looks ten years younger, & I doubt if ever he enjoyed existence so much as now....[30]

In 1837, Sir John tackled yet again the Great Nebula in Orion, which, at a declination of -5°, is well-placed for observation from the Southern Hemisphere. After determining the positions of 150 stars with his 5-inch refractor, so as to furnish a benchmark for later studies, he carefully added in the nebulous features visible with the 18-inch mirror. Referring to this drawing, he later wrote: "In form the brightest portion [of the nebula] offers a resemblance to the head and yawning jaws of some monstrous animal, with a sort of proboscis running out from the snout." Naturally Sir John was eager to compare this drawing with the one he had made at Observatory House 9 years earlier. Though he could vouch for no changes in the interval, the matter could not yet be regarded as settled—there would be, as a later astronomer put it, "strange discrepancies in the drawings of the best hands."[31]

In the gargantuan constellation then known as Argo Navis, the Ship (subsequently broken up into several smaller constellations of which Carina—the Keel— includes the bright star Canopus) he followed the fascinating changes in brightness of the star Eta Argûs (now Eta Carinae). It had first been recorded as a 4th magnitude star by Edmond Halley at the island of St. Helena in 1677. Since 1820 it had been brightening, and in 1837 was entering a particularly dramatic phase, and continued its ascent in brightness after Sir John left South Africa and returned to England, reaching its maximum brightness in 1843: at magnitude -0.8, it surpassed

Fig. 4.8. The Tarantula Nebula (30 Doradus, or NGC 2070), an HII region in the Large Magellanic Cloud, 160,000 light years away. It is so luminous that if it were as close to us as the Orion Nebula, it would cast shadows, and is the largest and most active starburst region known in the Local Group of galaxies. Just on its outskirts the Supernova 1987A was observed, which has the distinction of being the closest supernova observed since the invention of the telescope. It was a Type II supernova, resulting from the collapse of a massive star. *Courtesy: John Drummond.*

Fig. 4.9. NGC 5128 (Centaurus A). The "peculiarity" of the galaxy was first noted by John Herschel. It is one of the closest radio galaxies to the Earth, with radio waves being emitted from its active galactic nucleus, and contains at its center a 55-million solar mass black hole. It is also a "starburst" galaxy, involved in an intense burst of star-formation owing to a collision with a smaller spiral galaxy, which Centaurus A is in the process of "devouring," as has been confirmed in infrared studies with the Spitzer Space Telescope. Imaged from Gisborne, New Zealand with a 35cm SCT, at f/10. *Courtesy: John Drummond.*

Fig. 4.10. The 40-foot telescope at Slough, just before it was dismantled by John Herschel in 1839. This image has been computer-generated positive from the original negative-image cibochrome by Herschel. It was Herschel, by the way, who introduced the terms "positive" and "negative" into photography. *Courtesy: Wikipedia Commons.*

every star in the sky except Sirius. Sir John had noted that the star appeared to be enveloped in a shroud of nebulosity that could be made out even with the naked eye. He could not find words to describe the beauty and sublimity of this nebula, but he did produce a memorable sketch of the dark dust cloud at the center—the "Keyhole."

Parenthetically, since Sir John's time, the nebula has continued to undergo further changes. It now features a small bipolar nebula consisting of an expanding cocoon of gas ejected from the star in the 1830s—the "homunculus." The star itself has appeared very orange in recent years. As a recurring nova, it may begin another rise to brightness at any time, and is a likely candidate to go supernova in the near future.

Sir John would always regard the time he spent in South Africa as the happiest period of his life. On returning to England in May 1838 (just in time for young Queen Victoria's coronation, at which he was created a baronet), he could take satisfaction in having "done his duty. " His survey of the southern skies completed, he never bothered to re-erect the 20-foot telescope in England. Instead he turned with a will to other researches, including photography. Spurred by developments in France where the daguerreotype process had just been invented, he introduced the use of sodium thiosulphate as a fixing agent and, independent of William Fox Talbot, the use of sensitized paper. He also coined such now-familiar terms as positive and negative. As early as 1839, he made several little faint paper photographs of the great 40-foot telescope, which had long since fallen into disrepair in the front yard of Observatory House. These faint paper photographs were, as Tony Simcock has said, "the first

Fig. 4.11. Sir John Herschel, in old age; a wet-process plate by Julia Margaret Cameron. *From: Wikipedia Commons.*

photos ever taken of a scientific instrument and the first architectural record photos ever taken of a structure about to be demolished."[32]

By then, despite the fact that during his absence in South Africa the Great Western Railway had reached Slough as it thrust farther west, making it possible for Herschel to travel from Slough to the Athenaeum (his club on Pall Mall) via Paddington Station in only an hour, he and his growing family—Sir John and Lady Margaret eventually had twelve children together—found themselves increasingly uncomfortable in Observatory House. Moreover, Herschel craved more rural privacy and solitude than he could have at Slough. After knocking down the dilapidated timbers of the 40-foot telescope and sheltering its venerable tube in a shed—in a way, he was symbolically settling a score with the burdens that had so long lain upon his shoulders as the legacy of his famous father—he and his family relocated to Collingwood House, a mansion near Hawkhurst, Kent, 20 kilometers from the nearest train station (Staplehurst), where he would remain for the rest of his life.

At Collingwood, Herschel (alone and unaided) embarked upon the daunting and laborious task of preparing his Cape observations for publication and compiling his great *General Catalogue of Nebulae* (published only in 1864). The GC, in turn, served as the basis of J.L.E. Dreyer's revision of 1888, called the *New General Catalogue* (or to give it its complete name, the *New General Catalogue of Nebulae and Clusters, Being the Catalogue of Sir John Herschel, Revised, Corrected and Enlarged*). Its familiar NGC numbers are still standard today. Herschel also wrote and regularly revised the popular *Outlines of Astronomy* (the first edition appeared in 1849; it would pass through twelve editions in all). Though it had relatively little to say about the nebulae, what little it did say was to be highly influential among those who remained actively involved in this research.

Sir John Herschel lived on happily in the midst of his large family until his death in 1871, amusing himself with many projects, including a rather eccentric translation into English hexameters of Homer's *Iliad,* and posing for one of the pioneering wet-process portraits of Julia Margaret Cameron. But he was never tempted to build bigger mirrors or to make further sieges on the nebulae with new technologies. "One senses," writes Allan Chapman, "that he felt that his own observing work had gone as far as he wished to take it. Now it was up to his younger contemporaries—Lord Rosse, William Lassell, and James Nasmyth, who were also much more compulsive observers than he was—to carry it further."[33]

The worm had finally turned. At last, after two generations in the hands of the Herschels, the study of the nebulae was finally to pass to other hands.

* * *

1. John Betjeman, "Slough," from the collection *Continual Dew.* The poem was published in 1937, before the city was subject to actual air-raids during World War II.
2. Dreyer, *Scientific Papers of Sir William Herschel,* vol. 1, xlviii.
3. The house belonged to Mary's mother, Mrs. Baldwin.
4. Samuel Johnson to Mrs. Thrale, March 24, 1784, in *The Letters of Samuel Johnson,* ed. Bruce Redford (Princeton: Princeton University Press, 1992), vol. 3, 144.
5. Herschel Archive, Royal Astronomical Society.
6. William Herschel, "On Nebulous Stars, properly so called," read Feb. 10, 1781, *Phil. Trans.,* 1791, 71-88, in Dreyer, *Scientific Papers of Sir William Herschel,* vol.1, 415-425.
7. Ibid., 416-417.
8. Edmund Morris, *Beethoven: the universal composer,* New York: HarperCollins, 2005, 74.
9. William Herschel, "Astronomical Observations relating to the Construction of the Heavens, arranged for the Purpose of a critical Examination, the Result of which appears to throw some new Light upon the Organization of the celestial Bodies," read June 20, 1811, *Phil. Trans.,* 1811, 269-336, in Dreyer, *Scientific Papers of Sir William Herschel,* vol. 2, 459-497.
10. Quoted in L. Pearce Williams, *Michael Faraday,* (New York: Basic Books, 1965), 66.
11. William Herschel, "Astronomical Observations Relating to the Sidereal part of the Heavens and its connection with the nebulous part," read February 24, 1814, *Phil. Trans.,* 1814, 248-284, in Dreyer, *Scientific Papers of Sir William Herschel,* vol. 2, 520-541.
12. Allan Chapman, *The Victorian Amateur Astronomer: Independent Astronomical Research in Britain 1820-1920* (Chichester: Praxis, 1998), 55.
13. Ibid., 56.
14. John Herschel to Charles Babbage, October 10, 1816; Royal Society of London. In general, we have relied on Michael J. Crowe, David R. Dyck, and James R. Kevin, eds., *A Calendar of the Correspondence of Sir John Herschel* (Cambridge: Cambridge University Press, 1998), as a useful guide to the whereabouts of John Herschel's vast correspondence.
15. John Herschel to Margaret Herschel, August 10, 1841 or 1843 (date uncertain); quoted in Chapman, *Victorian Amateur Astronomer, 56.*
16. Quoted in Anthony Hyman, *Charles Babbage: pioneer of the computer* (Princeton: Princeton University Press, 1985), 34.
17. Winston S. Churchill, *The Great Democracies* (London: Cassell, 1958), 12.
18. Martin J.S. Rudwick, *The Great Devonian Controversy: the shaping of scientific knowledge among gentlemanly specialists* (Chicago: University of Chicago Press, 1985), 34.
19. David S, Evans, Terence J. Deeming, Betty Hall Evans, Stephen Goldfarb, eds., *Herschel at the Cape: diaries and correspondence of Sir John Herschel, 1834-1838* (Austin and London: University of Texas Press, 1969), xxiii.
20. Quoted in Walter Scott Houston, *Deep-Sky Wonders.* Selections and commentary by Stephen James O'Meara Cambridge, Mass.: Sky Publishing, 1999), 4.
21. Charles Piazzi Smyth, On Astronomical Drawing, *Memoirs of the Royal Astronomical Society,* 15 (1846), 71-82.
22. John Herschel to Francis Bailey, March 12, 1828; Royal Society of London.
23. Lady Margaret Herschel to Caroline Lucretia Herschel, September 29, 1834; University of Texas, Austin.
24. David S. Evans et al., *Herschel at the Cape,* xxliii.
25. Quoted in Alan W. Hirshfield, *Parallax: the race to measure the cosmos* (New York: W. H. Freeman, 2001), 193.
26. Sir John Herschel to Caroline Lucretia Herschel, March 28, 1834; University of Texas, Austin.
27. In a postscript to the letter of Caroline Lucretia Herschel to Lady Margaret Herschel, August 1, 1833; quoted in Lady Herschel, *Memoir and Correspondence of Caroline Herschel* (London: John Murray, 2nd ed. 1879), 258.
28. Sir John Herschel to Caroline Lucretia Herschel, February 22, 1835; University of Texas, Austin.
29. Sir John F.W. Herschel, *Results of Astronomical Observations made during the years 1834, 5, 6, 7, 8 at the Cape of Good Hope, being a completion of a telescopic survey of the whole surface of the visible heavens commenced in 1825* (London: Smith, Elder & Co., 1847), art. 335.
30. Lady Margaret Herschel to Caroline Lucretia Herschel, September 29, 1834; University of Texas, Austin.
31. Rev. T. W. Webb, *Celestial Objects for Common Telescopes,* edited and revised by Margaret W. Mayall (New York: Dover, 1962 reprint of 6th ed., 1917), vol. 2, 193.
32. Tony Simcock to William Sheehan, personal correspondence, December 17, 2009.
33. Chapman, *The Victorian Amateur Astronomer,* 72.

5. Of Leviathans, Spirals, and Fire-Mists

Leviathan, which God of all his works
Created hugest that swim th'Ocean stream:
Him haply slumb'ring on the Norway foam
The Pilot of some small night-foundered Skiff,
Deeming some Island, oft, as Seamen tell,
With fixèd Anchor in his scaly rind
Moors by his side under the Lee, while Night
Invests the Sea, and wishèd Morn delays.

—*John Milton, Paradise Lost, Bk. 1, 201-208*

Between them the Herschels, father and son, bestrode the astronomical world like colossi. For half a century, they commanded the most powerful telescopes in existence. The greatest of all was the 40-foot, with its 48-inch mirror, at Slough. But here William overreached. A glorious failure, difficult, even dangerous, to use, it was never deployed for the study of the nebulae for which it was designed. But it inspired an even more colossal instrument, the "Leviathan," at Parsonstown (now Birr), Ireland. Built by the wealthiest "Grand Amateur" of his day, William Parsons, the 3rd Earl of Rosse, its mirror, 72 inches in diameter, would not be surpassed until the 20th century. The achievement was all the more remarkable, given that at the time Rosse built the telescope, Ireland was an economic basket-case of its day.

This telescope would first reveal that class of nebulae which are among the most imposing, majestic, and characteristic in the universe—the spiral nebulae—amply fulfilling its maker's expectation that with it "data will be collected to afford us some insight into the construction of the material universe."[1]

The 3rd Earl's family, the Parsons, were descendents of Anglo-Irish Protestant landowners who had prospered during the great Irish land confiscations of the sixteenth and seventeenth centuries. The founder of the family fortune was Richard Boyle, who in 1588, the fateful year of the

Spanish Armada, crossed the Irish Sea from England, having, it was said, no more than the shirt on his back. But patronage worked in his favor, and obtaining a position as deputy to the escheator general for Crown lands, he used it, said his enemies, "to defraud Irish landowners, especially in Munster, of their existing titles and to pass title to himself at absurdly deflated prices. He then expelled the Irish tenants and replaced them with more pliable and profitable English settlers."[2] Relentlessly and ruthlessly forging ahead in "the path of Christian virtue and the favor of Providence," he amassed vast fortunes in lands and rents, until his income in rents exceeded those of any other of the Crown's subjects. He was raised to a baronage in 1616, created Viscount Dungarvan and first earl of Cork in 1620. That same year Boyle's nephew, Laurence Parsons, enters the story: having already acquired from Boyle Myrtle Grove (the gabled residence in Youghal where Raleigh had once lived), he now established at Birr the Parsons family's connection with the place; it continues right up to the present day.

A castle had stood at Birr since the twelfth century. Laurence Parsons had a new one built by English workmen, in which he and his successors lived as grandees sequestered behind fifteen-foot high walls (high even by Irish standards) within a showpiece demesne consisting of the castle and well-manicured grounds.

In the eighteenth century, one improving landlord at Birr, another Laurence Parsons, the 5th Baronet, began digging an artificial lake and planting beech trees. His son, the 6th Baronet (and after 1807 2nd Earl of Rosse), turned the old house back to front in order to face the park, heightened and crenellated it in the then-fashionable Gothic style, and added the great "Gothic saloon" whose windows look down on the waterfalls of the river Camcor. He entered politics, and served in Parliament for the next several years. He was, of course, highly conservative, and yet in one respect unconventional—he and Lady Rosse were unusual in the approach they took to the upbringing of their five children, two daughters and three sons, including the eldest son William, the future astronomer, born at York in 1800. Instead of sending them to the public schools such as Eton, they educated them at home with the assistance of tutors, thus affording them a thorough grounding in science and engineering. Laurence himself was of a scientific turn of mind; a suspension bridge he built over the Camcor was the first of its kind in Ireland, described by a visitor as "a curious wire bridge ... suspended in the air just under the castle."

When William (the future 3rd earl) turned 18, he and his younger brother John, who suffered from ill health and died young, were sent to Trinity College, Dublin, though they continued to do most of their work at home as allowed by the regulations of the time. In 1821 they transferred to Magdalen College, Oxford, and William (known as Lord Oxmantown until his father's death) received a First-Class mathematics degree the following year.

He followed his father's footsteps into politics, and served for more than a decade in the House of Lords, voting on the great issues of the day (he was in favor of both Catholic Emancipation and the Reform Bill). But by 1834 he was ready to say farewell to all that; he decided not to stand for election of parliament the next year year, citing as his reasons "defects of the present franchise" and a refusal "to solicit support from the enemies of the Union and of the Established Church."[3] The real reason was becoming mad about astronomy.

As early as 1830, he had completed a telescope with an innovative 15-inch compound speculum mirror (one in which the reflecting surface consisted of strips of bronze soldered to a brass base), and set up a wooden stand for mounting it outdoors. At the same time he jotted an ambitious prospectus for himself:

> The examination of the heavens commenced by the late Sir William Herschel and prosecuted by him with such success, still continues. New facts are recorded: and there can be little doubt that discoveries will multiply in proportion as the telescope may be improved.[4]

His goal was to build ever larger telescopes, until he had surpassed even those of his hero, William Herschel. Perhaps there was something phallic in the aspiration; there was certainly a good deal of the spirit of conquest, for the building of larger telescopes was necessary to achieve the expansion of human knowledge (and perhaps control) into the farther and farther reaches of the universe. Herschel was a kind of Alexander who had pushed the borders of intellectual empire to the Hydaspean limits of his time; others, like William Parsons, wanted to explore territory beyond even the Hydaspes (wherever that might be, cosmically speaking) that Herschel had reached.

Science in the 18th and especially the 19th centuries—the age of the "gentlemanly specialist"—was a rich man's game. John Herschel had won his place at the table by virtue of his outstanding talent and connections but no less so as the heir to a considerable fortune. Despite his not inconsiderable wealth as an Irish landowner, William Parsons's grandiose ambitions could never have been realized on the scale he dreamed were it not for a successful match in 1836 to Mary Wilmer Field, a wealthy heiress who not only brought a significant amount of cash but several valuable properties in the Bradford area of Yorkshire into the family.

Three years later, in 1839—at almost the very moment Sir John was dismantling the 40-foot reflector at Observatory House, Slough, and preparing to move to Collingwood—Parsons set up a new 36-inch reflector, with a speculum-metal mirror weighing 1 ½ tons cast in his own foundry and ground and polished by a steam-powered machine (making it in a real sense the first large telescope fabricated using the methods of the Industrial Revolution). Financed with a large chunk of Mary Wilmer Field's money, it was mounted in Herschelian manner on the extensive lawn in front of the Castle, and it did not disappoint. Turning it to the Moon, Parsons's advisor and friend, Dr. Thomas Romney Robinson of Armagh, a fiercely militant Ulsterman who had already built up Armagh Observatory from a "ramshackle, inactive establishment into a major observatory," and was irascible enough to live to 100, wrote: "It is scarcely possible to preserve the necessary sobriety of language, in speaking of the Moon's appearance." Still, Parsons remained unsatisfied. The Moon was hardly more than a diversion; it was the nebulae he was after, his goal being to solve the problems Herschel had not and to find out whether the nebulae were, as Allan Chapman writes, "luminous 'chevalures' of glowing, perhaps gaseous, material, or ... conglomerations of individual stars.... And why ... the star clusters and nebulae tend to occur most frequently in different regions of the sky."[5] For this he needed even more light-grasp than the 36-inch reflector furnished; the idea of an enormous telescope with a mirror twice the diameter, the Leviathan, was born.

He began work on the great telescope in earnest after 1841, when on his father's death he inherited his title to become 3rd Earl of Rosse and a ranking member of the Irish aristocracy. (Henceforth we will refer to him by his title.) Like Herschel's, Rosse's mirrors were of speculum metal, that finicky alloy of copper and tin. When polished, speculum metal is as reflective as silver (reflecting two-thirds of the light and absorbing another third), but the material is difficult to work with, not easy to cast, prone to crack while cooling, and heavy and ponderous (a 6-foot diameter mirror weighed at least 3 tons.)[6] He cast each of his mirrors—because of their tendency to tarnish, several were needed for a working telescope—in three 24-foot crucibles, heated with peat. Robinson compared the foundry of Rosse to the poet Milton's "splendid description of the infernal palace" in *Paradise Lost.* Though Rosse provided overall direction, much of the actual work was carried out by unsung heroes, including William Coughlan, a local blacksmith, and a skilled workforce drawn from "common Irish laborers."

After several failures, Rosse oversaw the casting of the first successful 6-foot mirror in 1842. A backup was cast the following year, which meant the two mirrors could be exchanged relay-style as each in turn became tarnished and corroded by exposure to the damps and mists of Irish nights (and it often required no more than a couple months for the mirrors to deteriorate). It took another year to figure the mirror, and yet another to polish it—as with the 36-inch, this was managed with the aid of a steam-engine powered machine. By February 1845, the great mirror was placed at the end of a 56-foot tube, supported at its lower end by an enormous universal joint, and slung in chains between two massive stonework walls evoking the remains of a medieval castle. To some, it looked like a "folly," an architectural bauble such as were then fashionable among the idle rich. Others, informed of its purpose as a great engine of research, compared it to "that artillery

Fig. 5.1. (above) William Parsons, Third Earl of Rosse, and Lady Rosse; portrait by Piazzi Smyth. *Courtesy of the Birr Scientific and Heritage Foundation.*

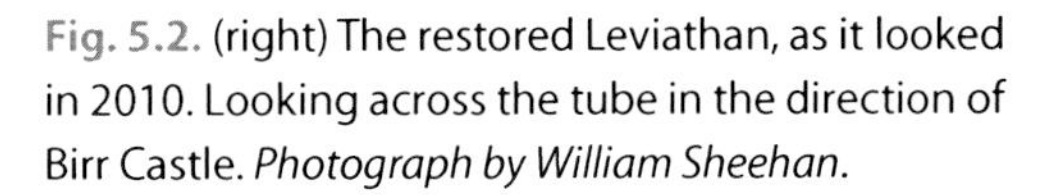

Fig. 5.2. (right) The restored Leviathan, as it looked in 2010. Looking across the tube in the direction of Birr Castle. *Photograph by William Sheehan.*

described by Milton as pointed by the rebellious angels against the host of Heaven," and looked with keen anticipation to its "quiet victory over space."[7]

This sublime and terrifying machine was needed to tackle the fearful immensities of stars and nebulae that, as John Herschel had observed with perhaps unintended understatement, "may be thought to savour of the gigantesque."[8] Like other monsters of the early Victorian period—the iguanadons, ichthysosaurs, and the jawed fishes—the nebulae, with their strange and sometimes grotesque physiognomies, evoked what the poet Tennyson would later call the "Terrible Muses" of geology and astronomy. They were sublime, in the sense of Edmund Burke's influential essay *Philosophical Enquiry into the Origin of our Ideas of the Sublime and the Beautiful* (1757): huge, obscure, capable of stunning the mind with an emotion closely akin to elemental fear. The Herschels and Lord Rosse worked at a time when God was not as neighborly as of yore. Their telescopes revealed something vast and indefinite in the nebulae that paralyzed and astonished the human mind, and showed something of the precariousness (which Burke too had noted) of quaint 18th century ideas of order.

We emphasize here that the 19th-century quest to determine the nature of the nebulae took place without any understanding of the distinction between the true distant galaxies or island universes and the local collections of stars and gas clouds situated within our own Galaxy. It is difficult—but essential—if we are to form an appreciation of the way the pioneers of nebular studies proceeded not to view their efforts from the perspective of current knowledge. What they worked from were merely the more or less indistinct forms seen in the sky—hieroglyphs, but with no Rosetta stone to help puzzle out their composition, sizes or distances (and even the first actual measurements of the distances to the nearer stars had only occurred in the 1820s). We must, in a manner of speaking, in Robert Smith's arresting phrase, "put the strangeness back in."

When Rosse began to employ his gigantic machine on the problem of the nebulae, there were essentially two views as to their nature. The first was that all of them would, when seen with sufficient telescopic power, resolve into stars. This had been the view of William Herschel up to his crucial "nebulous star" observation of 1790. Then he had deserted that position and believed right up to his death in 1822 that, in addition to resolvable nebulae (star clusters), there were others that were irresolvable, consisting not of stars but of a mysterious "shining fluid," the substance of which might rejuvenate dying suns and even perhaps—as argued by the French mathematician Pierre Simon Laplace, working up the idea partly from Herschel's own nebular studies—give rise, through the flattening of a slowly rotating nebula into a thin disk followed by collapse and fragmentation under the influence of gravity—to the formation of new suns and solar systems.

It was these questions—which were not without terrestrial implications, since the nebular hypothesis was "French," and in England and Ireland at least regarded as associated with atheism, the French Revolution, and other such ideas dangerous to the established order—that Rosse and Robinson yearned to tackle with the great telescope. Finally, in February 1845, Rosse proclaimed that the Leviathan was capable of being used "without personal danger." It was officially opened in a memorable ceremony in which George Peacock, one of John Herschel's best friends at Cambridge and now Dean of Ely, wearing his top hat and deploying an umbrella, walked the length of the tube to demonstrate its enormous size. The observing platform at the mouth of the great tube was raised by a winch, and reached by a system of staircases and galleries. An ingenious machinery of cables and chains was worked by two men to point the tube toward objects within forty minutes on either side of their culmination along the meridian—the only objects accessible given the telescope's more or less rigid north-south orientation between its massive stone walls. Rosse and Robinson were joined by Sir James South, John Herschel's former companion in double star observations, who came up from London eager for the telescope's inaugural views. They were a revelation. One can imagine what they must have experienced from the notes that the Astronomer Royal at Greenwich, Sir George Airy, made during his own subsequent visit to Birr Castle. He described how "the orb of Jupiter produced an effect compared to that of the introduction of a coach-lamp into the telescope," while the approach of Sirius was "like the dawn of morning," its entry into the field accompanied "with all the splendor of the rising Sun."[9]

Of course, the Leviathan had not been built to show Jupiter or Sirius. It was built for the nebulae, and its first priority was the Great Nebula of Orion. Unfortunately, the weather during February was "of the worst astronomical character," as Robinson groused, and whenever the Great Nebula was in range of the great telescope, clouds interfered. Soon it was sliding hopelessly out of range, vanishing early into the evening twilight. There was no alternative but to try for something else. The mirror was therefore removed from the telescope, and repolished on March 3. The weather then turned clear (but very cold) during the week of March 4-13. South exclaimed that the night of March 5 was "one of the finest I ever saw in Ireland. Many nebulae were observed by Lord Rosse, Dr. Robinson and myself."[10] The three men carried out a first survey of 40 some of the brighter nebulae catalogued by Sir John Herschel. They worked at great speed; their excitement must have been almost unbearable, and the preliminary results were everything they could have hoped. "We have," Robinson privately exulted, "... worked to some purpose, for of the 43 nebulae which we have examined *All have been Resolved* [into stars]."[11]

Among the nebulae that seemed to have been resolved, one of the most interesting was M51 in Canes Venatici, the Hunting Dogs. It was one of the entries in Messier's catalog that was actually discovered by him, though it was Méchain who first recognized it as a double nebula, "each having a brilliant center [with] the two atmospheres touching."[12] On examining it in 1833, John Herschel had shown a bifurcated ring swirling around a central condensation, and in this form it had also appeared to Rosse himself with the 3-foot. In the first view of it through the Leviathan, Robinson seemed to peg another notable success: "Here ... the central nebula is a globe of large stars; as indeed had previously been discovered with the three feet telescope: but it is also seen with [a magnifying power of] 560 that the exterior stars, instead of being uniformly distributed ... are condensed into a ring, although many are also spread over its interior."

In fact, in reporting stars, Rosse and Robinson were not merely falling victim to the illusion of expectation; what they saw was real enough—M51, and similar objects such as M33 and M101, do show starlike objects in their outer parts, which appear in better instruments as nebulous knots of glowing gaseous matter (now known as HII regions). These are much brighter than individual stars. Of course, we shall say more to say about all this later, but for now let's stay with Rosse and Robinson.

In these observers' first remarks on M51, no comment was made on what was to be its most notable feature—its spiral structure. When, and under what circumstances, was the discovery made? Apparently, in March 1845, the observers' attention was so totally concentrated on resolving it into

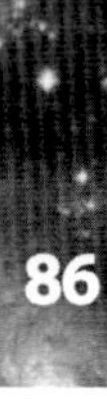

Fig. 5.3a. Lord Rosse's paradigm-shifting sketch of M51, a galaxy located 37 million light years away, made with the Leviathan in April 1845. *Photographed from the original notebook by William Sheehan and reproduced with the permission of the Birr Scientific and Heritage Foundation.*

stars that the novelty of the spiral structure eluded them. Only in April 1845 did Rosse, now observing alone, discern that M51's structure was "more complicated than had hitherto been appreciated," its form consisting of "a curvilinear arrangement not consisting of regular re-entering curves."In other words, it was a double spiral, with one arm curling out as far as the secondary nucleus. And so within only two months of Peacock's top-hatted traverse of the great tube, Rosse had made a discovery that would introduce a sea-change in the way astronomers engaged the nebulae at the time and, as Sir Bernard Lovell would remark a century later, "open an avenue of exploration which would lead into the inconceivable depths of space and time."[13]

An excellent artist, who left many examples of his skill (some from boyhood) in the archives at Birr Castle, Rosse was more than capable of recording with perfect assurance the whorls of M51 and its companion. Indeed, his sketch from spring 1845 compares favorably with modern photographic and CCD images.

He passed the sketch around at the meeting of the British Association for the Advancement of Science, held in Cambridge in June 1845. It was received with "much acclaim." Henceforth, Rosse proclaimed decisively, "our attention was ... directed to the form of nebulae, the question of resolvability being a secondary object."[14]

The Great Nebula in Orion, meanwhile, was not well placed enough to try at with the Leviathan until Christmas 1845, when John Pringle Nichol,

Fig. 5.3b. A modern amateur image of M51, for comparison. Imaged by Mike Hatcher with a C-11 SCT. At the time this image was taken, on June 10, 2011, a Type II supernova was visible about a third of the way down the spiral arm extending toward the satellite galaxy NGC 5195. A classic example of interacting galaxies. *Courtesy: Mike Hatcher.*

Glasgow University astronomer and radical economist (as such, a fervent supporter of the nebular hypothesis) joined Rosse on the platform of the great telescope as it was successful pointed to it for the first time. Nichol expressed the hope that the observation would, "in so far as human insight could ever confirm it," either show the correctness of Herschel's hypothesis that it was a shining fluid or else dispel that notion once and for all by resolving into discrete stars.

The best laid plans of mice and men oft go awry; and in this case, the Irish weather wreaked havoc with their plans. The mists and agitated air of the Irish winter did not permit a decisive result. Nichol reported, heavily, "Not yet the veriest trace of a star."

Rosse, undaunted, persisted, and on March 19, 1846 believed he had finally succeeded in achieving a definitive result.

Ere we go further, however, we need to interpose something about the appearance of the Great Nebula in a good telescope. Sir John Herschel gave a minutely exact description in his *Outlines of Astronomy:*

> The general aspect of the less luminous and cirrus portion is simply nebulous and irresolvable, but the brighter portion immediately adjacent to the [multi-star system] the trapezium … is shown with the 18-inch reflector broken up into masses … whose mottled and curdling light evidently indicates by a sort of granular texture its existing of stars.[15]

But it was just in this granular texture around the Trapezium that Rosse inferred the resolution into stars. As he wrote to Nichol: "I may safely say there can be little, if any, doubt as to the resolvability of the nebula; all about the trapezium is a mass of stars, the rest of the nebula also abounding with stars, and exhibiting the characteristics of resolvability strongly marked."[16] Based on Rosse's apparent success, but without the qualifications that the resolution was achieved only in parts of the nebula, the report was passed on to foreign journals, and brought apparent confirmations, including, most notably, one from the just unveiled 15-inch Merz refractor at Harvard. Robinson rejoiced that the powerful telescope seemed set to fulfill the utmost dreams of its maker by shedding "a light hitherto not merely unattainable, but unhoped, on the constitution of the sidereal universe"?[17]

Interlude

A candid vignette of Rosse's activities during the industrious inaugural year of his telescope is provided by Thomas Langlois Lefroy, an Irish judge who early in his career had had a brief and not very serious romantic flirtation with the novelist Jane Austen (he has even been invoked as the possible inspiration for Mr. Darcy in *Pride and Prejudice*). During a visit to Parsonstown, Lefroy was cordially entertained by Lord and Lady Rosse, and shown the magnificent telescope. He afterward recalled:

> The wonders of his telescope are not to be told. He says – with as much ease as another man would say, "Come and I'll show you a beautiful prospect" – "Come and I'll show you a universe, one of a countless multitude of universes, each larger than the whole universe hitherto known to astronomers." The planet Jupiter, which through an ordinary glass is no larger than a good star, is seen twice as large as the moon appears to the naked eye. It was all true what Doherty [a fellow justice] said, that he walked upright in the tube with an umbrella over his head before it was set. But the genius displayed in all the contrivances for wielding this mighty monster even surpasses the design and execution of it. The telescope weighs sixteen tons, and yet Lord Rosse raised it single-handed off its resting place, and two men with ease raised it to any height.[18]

Hiatus: the Great Famine

Just then, Ireland plunged into the catastrophe of the potato famine, which saw a third of the indispensable crop (and foundation of the ill-famed "cottier system") fail in 1845, and the entire crop in 1846. Though it affected most of Ireland, and the devastation was especially severe in the West, it was the product of factors whose consequences (though anticipated by some of the more far-seeing among them, including Rosse, who had read his Malthus) could not be prevented by any of them. Rosse did what he could, on a local basis. Indeed, he was completely absorbed in the whole affair—as of course it was his duty to be. From early 1846, when the most acute phase of the crisis began, the Leviathan was sidelined. Its last pre-Famine discovery, in the spring of 1846, was of the spiral structure of M99 in Coma Berenices.

For the next two years the monstrous tube loomed shadowy and unused beneath the cold, uncaring stars, idled within the walled and manicured grounds of the Parsonstown demesne as the huddled masses outside the walls grew pale and spectre-thin and died.

Judging the worst to have passed by the winter of 1848-49, Rosse at last returned to the activities suspended during the depths of the crisis—including the promising work with the Leviathan. (Even so, aftershocks of the humanitarian crisis continued. As late as May 1849 fifty-seven "wretched paupers" at the cholera hospital in Parsonstown "breathed their last on the same day.")[19] There was still the unfinished business of the Orion Nebula—the resolution question seemingly settled, next was the question that had so stirred William Herschel over several decades: did it change?—and there were many other nebulae to be examined. By 1850, fourteen had been identified as spirals by Rosse and his assistants, including such well-known examples as M33 in Triangulum; M65, the edgewise spiral in Leo; M77, the "blue" spiral in Cetus; and M95, the barred spiral in Leo. (A few others, such as the globular clusters M12 and M30, were also reported as spirals; in these cases, later studies, with better instruments and photography, would show this suspected structure to be illusory.)

The enormous light-grasp of the Leviathan showed haunting details in many objects which had hitherto seemed little more than nondescript blurs of light. Thus Messier's premier object, M1, showed "streams running out like claws" (as Robinson described it); possibly this was the basis of its name of "Crab nebula." Rosse also first detected the two faint stars that form the "eyes" in M97, the planetary nebula in Ursa Major, and gave it its popular name, the "Owl."

Rosse puzzled over the explanation of these strange forms. He wrote in 1850 that their discovery

> may be calculated to excite our curiosity, and to awaken an intense desire to learn something of the laws which give order to these wonderful systems [but] as yet, I think, we have no fair ground even for plausible conjecture; and as observations have accumulated the subject has become, to my mind at least, more mysterious and more unapproachable.[20]

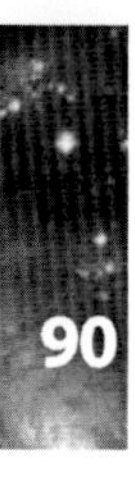

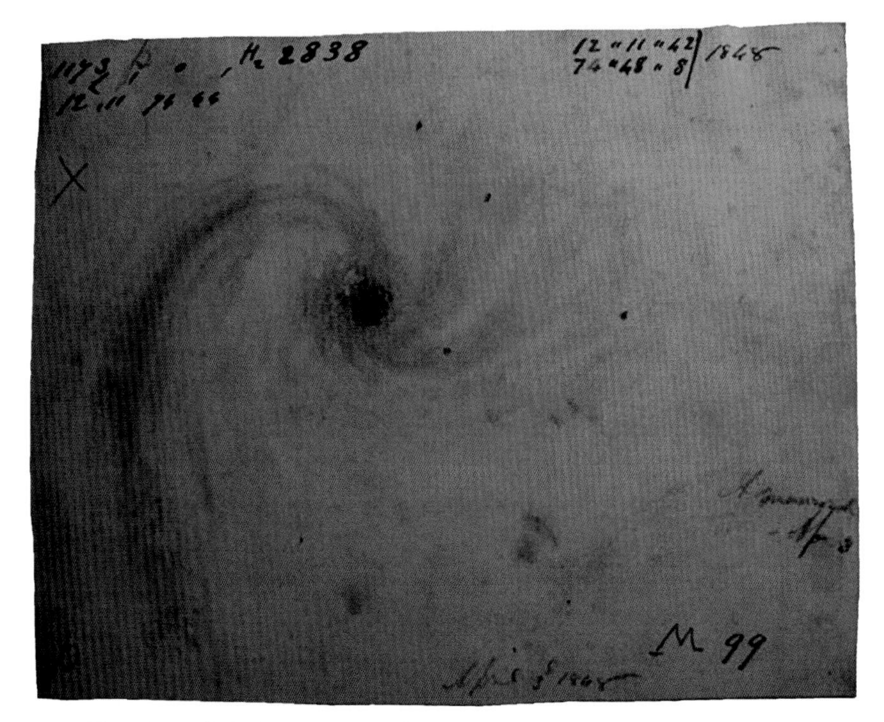

Fig. 5.4. The other nebula that Lord Rosse found to be spiral before the Potato Famine broke out: M99, at a distance of 50 million light years, in Coma Berenices. *Photographed from the original notebook by William Sheehan and reproduced courtesy of the Birr Scientific and Heritage Foundation.*

He was coming to terms with the complexity of the nebulae. By 1850, he had realized that M51—the term "Whirlpool," by the way, was first applied to the object by Ormsby MacKnight Mitchel, director of the Cincinnati Observatory and later Civil War general, in 1848, though it did not become prevalent until the early 20th century)[21]—as well as the other spirals could not be systems in "static equilibrium." Such forms could only result from internal movements of rotation. He was leaving a crack open for the nebular hypothesis to creep back in to respectability (as indeed it would).

Rosse himself was by now too busy to devote much time to observing. It has been said that the "Rosse telescope would have been more effectively used had its maker been as interested in observing as he was in instrument-making. His special interest in the instrument seems to have ceased when the last nail was driven into it."[22] There is a grain of truth to this: Rosse was an engineer of genius, and what chiefly engaged him was the challenge of building such an instrument, then using it to solve the problem of the nebulae. But he was also an extremely overworked and heavily burdened man, both in and out of science; not only responsible for his demesne and his tenants, but for many years president of the Royal Society of London. Moreover, by the 1860s his health was beginning to fail (he died in 1867; his son, Laurence, the 4th Earl of Rosse, was also an astronomer, whose specialty was the Moon). Despite his other preoccupations, he was determined that the great engine of research did not fall into disuse, and employed a series of gifted assistants who included George Johnstone Stoney (1848-52), the Rev. William Hautenville Rambaut (1848-

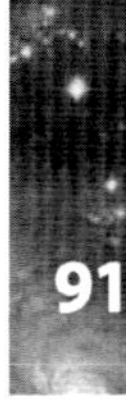

50), Johnstone Stoney's brother Bindon (1850-52), R.J. Mitchell (1852-5), Samuel Hunter (1860-64), and Robert Stawell Ball (1865-67).

Their *chef d'oeuvre* was, naturally, still the Great Nebula of Orion. Since at least the time it was observed by Huygens and called by William Derham a "window into the empyrean," no other object of the heavens had seemed more likely than this, with its complicated mixture of small stars and regions of milky smoothness, to be the Rosetta Stone of the nebulae, the one that would provide the means to decipher all the rest. Rosse himself had set up the result seemingly beyond question that the inner regions around the trapezium were resolvable into stars; but this still left the greenish outer regions, that looked for all the world like a cloud of fire-mist. Was the region around the trapezium, which appeared clear of nebulosity, one thing, the outer regions another, a true shining fluid? And what of the other great question, which had bearing on both possibilities: the question of change.

And here we see that, if in fact the Great Nebula consisted of a glittering dust of small stars, resolvable only in the largest instruments (and then only in part), from the great distance with which it was viewed, the nebula's structure ought to maintain a more or less permanent form. On the other hand, if it were an insubstantial mist or vapor, it might shift like cloud-forms on a windy day. The only way to settle the matter seemed to be through construction of an accurate map—a "physiognomy"—of the singular object for the purpose of answering, once and for all, the question of its changeability. But the problem was that, in the pre-photographic era, when the observer had only his eyes, his brain, and a pencil and sketchpad, the mapping of such a complex object was extremely laborious, and ultimately likely to be inconclusive.

Next to the studies of John Herschel, who had completed a drawing at the Cape which he compared with one made at Observatory House, Slough, nine years earlier, and Rosse himself, the next most widely mooted investigation regarding the nature of the Great Nebula was made in March 1846—just months before a new planet (Neptune) was discovered in Berlin as it lurked among the stars near the Ophiuchus-Capricorn border. It was the work of an American, William Cranch Bond at Harvard College Observatory, using the just unveiled 15-inch Merz refractor at Harvard College Observatory. Though conceding the Leviathan's superior light grasp, the Harvard achromat was unrivalled for exquisite defining power (not to say in need of an observation that might justify its great expense), and Bond, eager (overeager) to put Harvard College Observatory on the map of leading astronomical institutions, wrote to the College president Edward Everett on September 22, 1847:

> You will rejoice with me that the great nebula in Orion has yielded to the power of our incomparable telescope.[23]

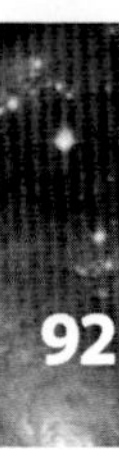

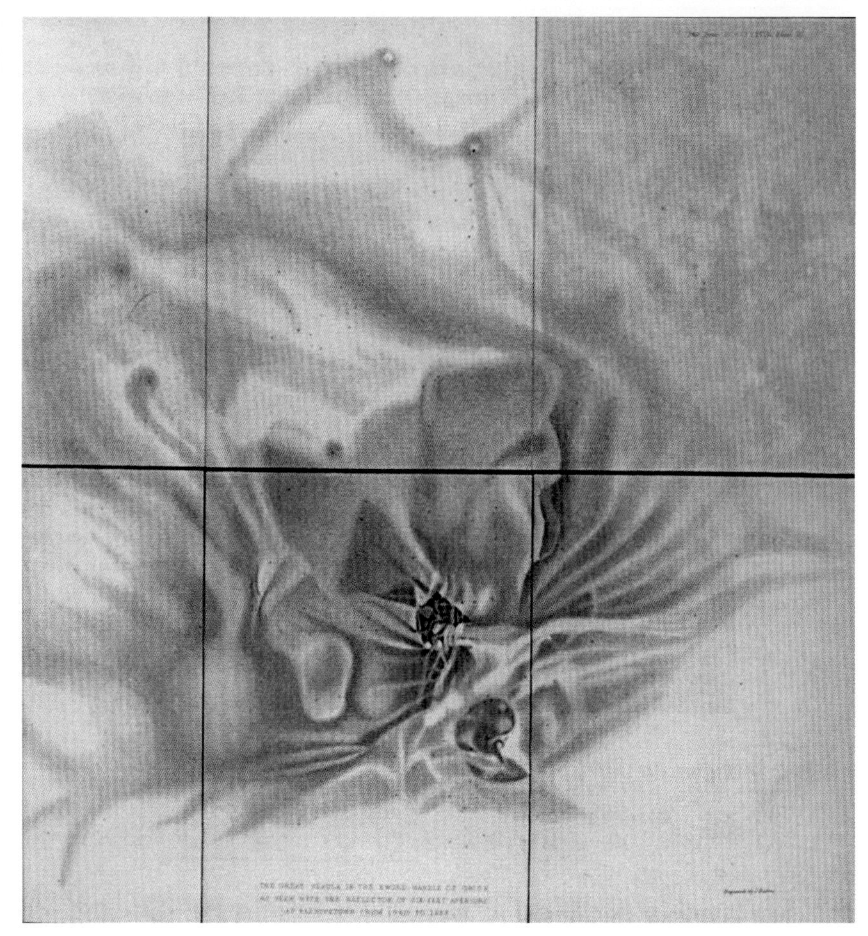

Fig. 5.5. A detailed map of the Orion Nebula, made by Samuel Hunter from observations by himself and several other observers with the Leviathan at Parsonstown between 1860 and 1864. *From: Lord Oxmantown (later the Fourth Earl of Rosse, "Account of the Observations of the Great Nebula in Orion," Philosophical Transactions of the Royal Society (1868).*

The Parsonstown group began their own massive study in the winter of 1848-49. They began by carefully laying down a grid of measured stars, based on Sir John Herschel's Cape Observations and on the catalogues of leading stellar positional astronomers Wilhelm and Otto Struve of Pulkovo Observatory in Russia. Upon this grid, the observers registered their fluctuating impressions of the nebula, which were found to be extremely variable depending on the steadiness and transparency of the air and the condition of the mirror. Thus George Johnstone Stoney, observing on February 17, 1849, rejoiced in the revelation of a "multitude of stars," only to discover, as soon as he sat down to attempt a drawing, "the state of the air" had changed, leaving him with only nine sure and five suspected stars. His brother Bindon, engaged in taking observations during the whole winter of 1851-52 of the bright inner Huygenian region, the part of the nebula where resolution was most strongly suspected, noted the persistent impression of faint nebulosity but explained it away: "an observer might suppose milkiness to exist," he wrote, "on unfavourable nights for definition or with an imperfect speculum, where in fact no nebulosity would appear with a freshly polished speculum free from all tarnish and on a first class night." Unfortunately, the speculum-mirror was seldom freshly polished

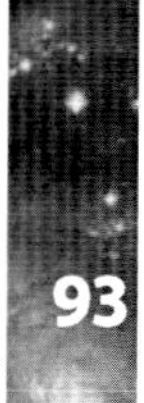

Fig. 5.6. Brendan, Seventh Earl of Rosse, and William Sheehan, in front of the restored Leviathan, at Birr Castle, 2010. *Photograph by Deborah Sheehan.*

and free from tarnish, and as always, truly fine nights for observing were exceedingly rare in Ireland.

Samuel Hunter, the most gifted artist among Rosse's assistants, devoted four years between 1860 and 1864 to making a drawing that would be suitable for a steel engraving. He explained why it took so long: on any given night he could gaze at Orion only for fifty minutes, when the tube stood in the right position between its two stone walls. Moreover, conditions in Ireland were such that over the course of four years he enjoyed only five "really good" and twelve "fair nights" for observing it.

As the Parsonstown observers continued their long labors to record the "physiognomy" of the Orion Nebula, an even more heroic attempt—more heroic because it proceeded as the work of a lone observer—had been advancing slowly over the course of several years at the Harvard College Observatory. The observer was George Philips Bond, an astronomer-son of William Cranch Bond whose report of the resolution of the nebula with the 15-inch Merz had attracted so much attention a decade earlier.[24] Motivated initially by filial devotion to redeem his

Fig. 5.7. A cartoon of G.P. Bond and W.C. Bond by volunteer observer Sidney Coolidge. This sketch is the only likeness that exists of G.P. Bond, the obsessive observer of the Orion Nebula. *Photograph by William Sheehan with the permission of the Harvard College Observatory.*

Fig. 5.8. Comet Donati, seen above Notre-Dame de Paris, October 4, 1858. This, one of the most beautiful comets ever observed, was studied intensely by G.P. Bond, and distracted him from his great work on the Orion Nebula. *From: E. Weiss, Bilderatlas der Sternenwelt, 1888.*

father's work, which had been harshly criticized by Otto Struve, the younger Bond labored with the monomania of obsession between 1859 and 1865, distracted only by the visitation, *in seriatim,* of several unusually brilliant and beautiful comets, beginning with Donati's of 1858. (It is this comet, by the way, which Walt Whitman seems to be referring to in his poem "Year of Meteors," on the events of the fateful year in which John Brown was hanged and America drew steadily toward Civil War. At the time, comets were widely believed, as J.P. Nichol had maintained, to be themselves "nothing but nebulosities, small portions of a substance precisely similar in physical constitution to that which our hypothesis assumes. Even their nuclei dissolve into a fog under the inspection of the telescope.")[25] Bond published a lavish monograph on Donati's Comet,[26] and made careful observations of the Great Comet of 1861 as well as the pair of great comets of 1862.[27] But nothing on earth or in the heavens could distract him for long from his lengthy and arduous siege on the Great Nebula of Orion.

The harsh New England winters took their toll on him and added their rigors to the work. Thus Asaph Hall, assistant at the observatory and later renowned for his discovery of the two satellites of Mars, recalled how cold my feet were when he was making his winter observations of Orion. I sat in the small alcove of the great dome behind a black curtain, and noted on the chronometer the transits of stars when Professor Bond called them out, and wrote down also the readings for declination.... Sometimes I was called to examine a very faint star, or some configuration of the Nebula. Prof. Bond had one of the keenest eyes I have ever met with. His work on this great nebula forms an epoch in its history.[28]

A perfectionist, Bond "made scores of drawings, in white on black, and the reverse, in colors, etc.," in preparation for it, as his cousin and later first director of the Lick Observatory, Edward S. Holden, would later write. Each of these, Holden continues, "was revised and re-revised many times. The revision of the ... plate lasted many months, and I have myself examined from fifteen to twenty 'final' revisions of the plate. Color, form, and relative brilliancy were all successively and exhaustively criticized."[29] (Note: some of these drawings, each lovely in its way, are being published here for the first time.) In the end, Bond was to prove a martyr to the Great Nebula. He had suffered from tuberculosis since 1858. After his father's death in

Fig. 5.9. (left) Owen Gingerich and Rich Schmidt seated at the eyepiece of the Merz refractor, with its mahogany-veneered tube, at Harvard College Observatory. *Photograph by Rich Schmidt. Courtesy: Rich Schmidt.*

Fig. 5.10. (right) The Great Nebula of Orion. Engraving by J.W. Watts, based on visual observations made by G.P. Bond between 1858 and 1863 with the 15-inch Merz refractor of the Harvard College Observatory. *From: "Observations of the Great Nebula of Orion," ed. Truman Henry Safford, Annals of the Harvard College Observatory, vol. 5, 1867, frontispiece.*

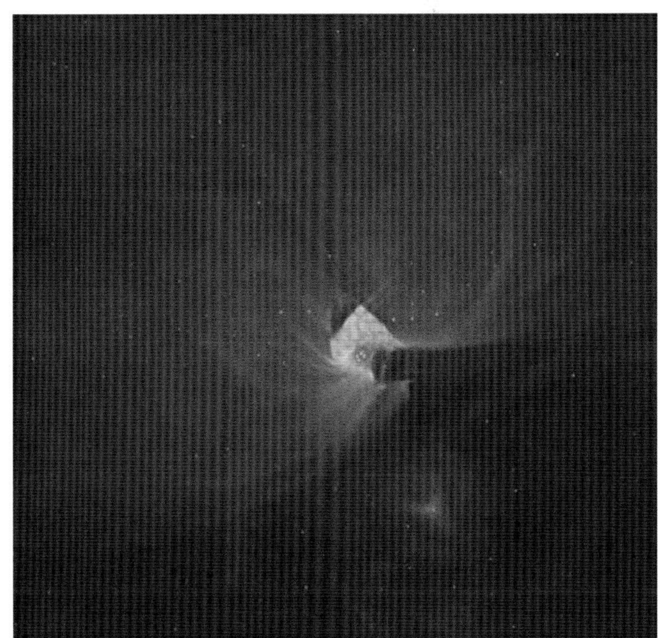

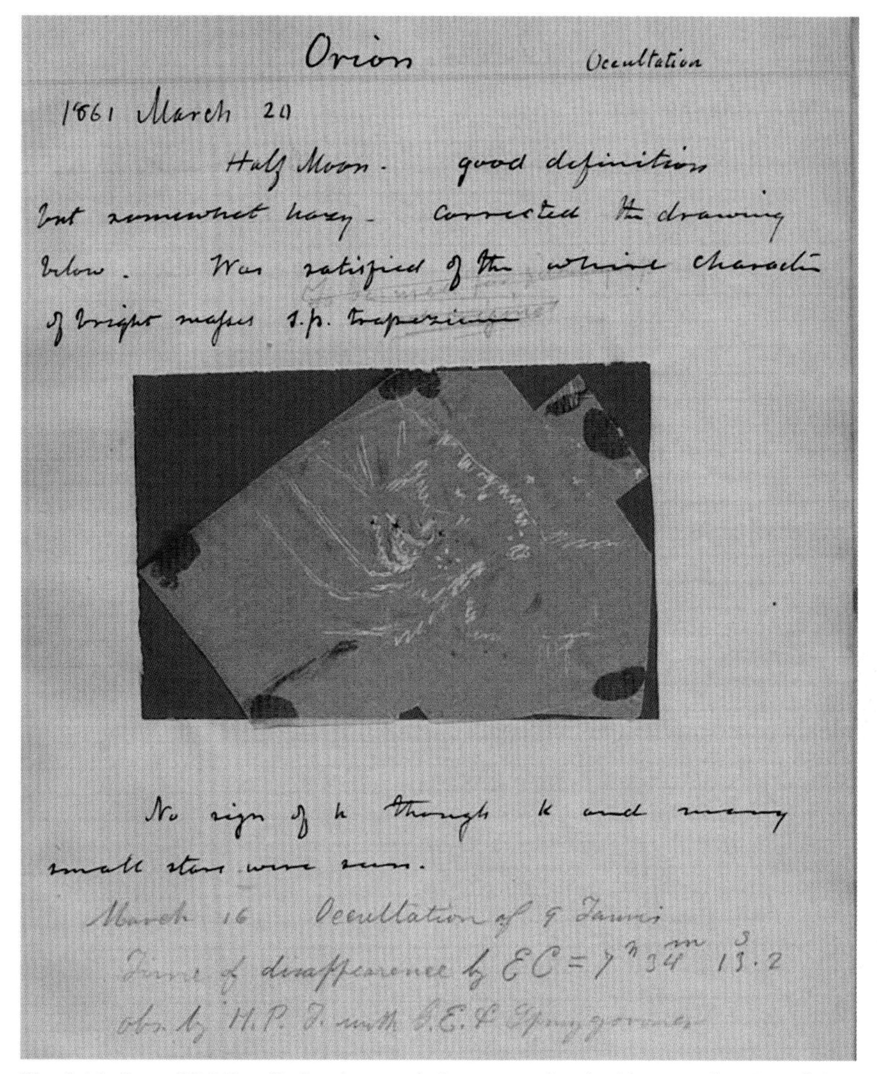

Orion Occultation

1861 March 20

Half Moon. good definition
but somewhat hazy. Corrected the drawing
below. Was satisfied of the [illegible] charact-
of bright masses s.p. trapezium

No sign of h though k and many
small stars were seen.

March 16 Occultation of 9 Tauri
Time of disappearance by EC = 7h 34m 13s.2
obs. by H.P. T. with S.E. & [illegible]

Fig. 5.11. One of G.P. Bond's sketches made in preparation for his great drawing of the Orion Nebula. *Photographed by William Sheehan with the permission of the Harvard College Observatory.*

1859, he was appointed director of the observatory in his stead, but immediately was faced with difficulties. With the onset of the Civil War, funds were diverted to the war effort, and the observatory's financial situation suffered; federal contracts and income ended completely by 1862.[30] Moreover, he complained that all but one of his assistants had enlisted or been drafted into the army. Nevertheless, in spite of failing health, he persevered. He gave up work at the telescope in August 1864. As the winter of 1864-65 came on, and Orion once more commanded foremost place in the skies, he studied only an effigy of the nebula lying before him on his desk. He was busy reducing his earlier observations, which included measured positions of 800 or 900 stars. After a brief rest in Maine urged on him by friends growing alarmed by his pale complexion and gaunt appearance, he rallied for a final effort, and spent the last weeks of his short life "forced to give up everything but the one obsessive thought ... [of completing] his long work on the nebula in Orion."[31] On January 7, 1865, he admitted to Hall:

> My disease makes progress, and leaves me little hope of putting the materials of my work on Orion—to which I had devoted so much labor—into condition such that another could prepare them for the press. In truth, I am becoming resigned to the idea that most of it is destined to oblivion.
>
> I had planned to accomplish something considerable, and this is the end. 'It is not in man that walketh to direct his steps.'[32]

He died on February 17, 1865, at the age of only 39. His great monograph on the Orion Nebula was published posthumously.

In 1864, Sir John Herschel published his Catalogue of Nebulae (the General Catalogue, or GC, which was later absorbed into the still-standard New General Catalogue, or NGC). That same year Edward Sabine, then President of the Royal Society, opined that nebular astronomy had reached a crisis. Observations of the nebulae with powerful telescopes, he wrote, had "revealed many strange arrangements ... some of which are scarcely reconcilable with ordinary dynamics."[33]

Sabine's point was that, despite the efforts of the Herschels, Rosse, and the Bonds, the nebulae remained as shrouded in mystery as ever. Were they truly fire-mists? Were they resolvable, with sufficient telescopic power, into stars? Did they lie within the star-system of the Galaxy? Were they island universes far beyond the Milky Way?

What were those questions the Almighty hurled at Job out of the whirlwind?

* * *

1. W. Garrett Scaife, From *Galaxies to Turbines: Science, Technology and the Parsons Family* (Bristol and Philadelphia: Institute of Physics, 2000), 12.
2. Steven Shapin, *A Social History of Truth: civility and science in seventeenth-century England* (Chicago and London: University of Chicago Press, 1994), 132.
3. Scaife, From *Galaxies to Turbines,* 31.
4. Ibid., 31.
5. Chapman, *Victorian Amateur Astronomer,* 96.
6. The first mirror cast ("on the 13th of April 1842 at 9 in the evening") weighed 3 tons, the second 4 tons, according to Thomas Woods, *The Monster Telescopes Erected by the Earl of Rosse... etc,* 1st edition, 1844.
7. Quoted in Simon Schaffer, "On Astronomical Drawing," in Caroline A. Jones and Peter Galison, eds., *Picturing Science: Producing Art,* Caroline A. Jones and Peter Galison (New York and London: Routledge, 1998), 456.
8. Ibid., "On Astronomical Drawing," 461.
9. G. B. Airy, "Notice of the Substance of a Lecture on the Large Telescopes of the Earl of Rosse and Mr. Lassell," *Monthly Notices of the Royal Astronomical Society,* 9 (1849):120.
10. J. South, *The Times,* April 16, 1845.
11. T.R. Robinson, manuscript entry in Astronomical

Scrapbook: 3rd and 4th Earl of Rosse, quoted in Michael Hoskin, *Stellar Astronomy* (Chalfont St. Giles: Science History Publications, 1982), 145.

12. S.J. O'Meara, "Méchain's unsung discovery," *Sky & Telescope*, September 2006, 62-63.
13. Sir Bernard Lovell, quoted in Patrick Moore, *The Astronomy of Birr Castle* (Birr: The Telescope Trust, 1971), 49.
14. William Parsons, "On the construction of specula of six-feet diameter; and a selection of observations of nebulae made with them," *Phil. Trans.*, 1861, 681-745.
15. Sir John Herschel, *Outlines of Astronomy* (Philadelphia: Blanchard and Lea, 1861), 512.
16. Schaffer, "On Astronomical Drawing," 462.
17. Hoskin, *Stellar Astronomy*, 145.
18. *Memoirs of Chief Justice Lefroy*, Dublin: Hodges, Foster & Co., 1871.
19. Brendon, *Decline and Fall*, 122.
20. Rosse, "Observations on the Nebulae," *Phil. Trans.*, 1850, 499-514.
21. For a complete discussion of the use of the term "whirlpool" for this object, see: William Tobin, "Full-text search capability: a new tool for researching the development of scientific language. The 'Whirlpool Nebula' as a case study." *Notes and Records of the Royal Society, 62*, 2 (20 June 2008),187-196.
22. Henry C. King, *The History of the Telescope* (New York: Dover, 1979 reprint of 1955 edition), 212.
23. William Cranch Bond, "[Letter to President Everett]," *American Journal of Science and Arts*, Series 2, no. 4 (1847), 427.
24. W.C. Bond had three sons and two daughters. The sons were W.C., Jr., who died while a student at Harvard or just afterward, G.P., whose fame as an astronomer equaled his father's, and Richard Fifield Bond, who observed with the 15-inch at Harvard and helped run the family chronometer business; the daughters were Elizabeth Lidstone and Selina Cranch.
25. John Pringle Nichol, *Views of the Architecture of the Heavens in a Series of Letters to a Lady* (Edinburgh: William Tait, 1837), 129.
26. George Phillips Bond, "Account of the Great Comet of 1858," *Annals of the Astronomical Observatory of Harvard College, 3* (1862).
27. The last two were discovered by Harvard College Observatory assistant Horace P. Tuttle. A great deal of useful information about Tuttle and the early history of Harvard College Observatory is found in Richard E. Schmidt, "The Tuttles of Harvard College Observatory: 1850-1862," The *Antiquarian Astronomer, 6* (January 2012), 74-104.
28. Asaph Hall, "My Connection with the Harvard College Observatory and the Bonds—1857-62," in Edward Singleton Holden, *Memorials of William Cranch Bond and of His Son George Phillips Bond* (San Francisco: Holden and Day, 1897), 78-79.
29. Edward Singleton Holden, "Monograph of the Central Parts of the Nebula of Orion," *Washington Astronomical Observations for 1878*, Appendix I (Washington, D.C.: Government Printing Office, 1882), 227.
30. During that year 1862, Alvan Clark, the telescope-maker, was testing the 18.5-inch objective that eventually went to Dearborn Observatory, in a temporary mount at his workshop in Cambridge. It was used to discover the companion of Sirius, and later was turned to the Orion Nebula. One of the observers, H.P. Tuttle, described the occasion: "Mr. Clark at once adjusted the telescope on the celebrated nebula in the sword handle of Orion. With a far less powerful telescope than this, the most wonderful of all nebulous objects presents a spectacle surpassing all description; but the view which this telescope revealed was gorgeous and striking far beyond any view ever before obtained of it. Stars of every conceivable telescopic magnitude were scattered over the brilliant mottled nebulous mass, and blaze upon the eyes with great distinctness and splendor. The remarkable sextuple star θ^1 Orionis, disclosed to view its two minutes members, which stood out clear and prominent...How many of the innumerable stars in the field had been before noted by astronomers, it was impossible to tell from so general an observation. The grand nebula was most admirably defined, and the luminous ridges of its granular structure seemed almost ready to bloom into swarms of individual minute stars, and especially so in that part where its form assumes a kind of rude resemblance to the head and yawning jaws of a monster."
31. Bessie Zaban Jones and Lyle Gifford Boyd, *The Harvard College Observatory: the first four directorships, 1839-1919* (Cambridge, Mass.: The Belknap Press of Harvard University Press, 1971), 133.
32. Ibid., 133. The line quoted is from Jeremiah 10:23: "O Lord, I know that the way of man is not in himself: it is not in man that walketh to direct his steps."
33. Edward Sabine, "President's Address," *Proceedings of the Royal Society of London, 13* (1863-64):500.

6. The Various Twine of Light

Meantime, refracted from yon eastern cloud,
Bestriding earth, the grand ethereal bow
Shoots up immense; and every hue unfolds,
In fair proportion running from the red
To where the violet fades into the sky.
Here, awful Newton, the dissolving clouds
Form, fronting the Sun, thy showery prism,
And to the sage-restricted eye unfold
The various twine of light, by thee disclosed
From the white mingling maze.

—*James Thomson, Spring (1728), from The Seasons*

The Leviathan's example inevitably stirred attempts to emulate its power without suffering the disadvantages of its situation. William Lassell, a wealthy Liverpool brewer who had visited Rosse's workshops in Parsonstown in 1844 and had partly witnessed the six-foot reflector's erection, set to building a grinding machine and polishing machine similar to (but improving upon) those Rosse had used, and constructed a 24-inch reflector, mounted equatorially rather than on an alt-azimuth, at his residence near Liverpool. He demonstrated its power with the discovery of Triton, Neptune's largest satellite, soon after the discovery of the planet itself in September 1846. Increasingly dissatisfied with the poor observing conditions in Liverpool—this was, after all, the period of the Industrial Revolution, and Liverpool was in the heart of the north of England with their "dark Satanic mills" and choking air pollution–Lassell in 1852 transported the 24-inch to the clearer skies of Valetta, Malta, chiefly to observe the nebulae. He followed up with construction of a 48-inch reflector, also mounted equatorially, which he took to Malta in 1861. It was never installed in a building, but remained in the open air when not in use. Until his return to England in 1865, Lassell announced the discovery of six hundred new nebulae, and published wonderful drawings of many of the more notable ones, including the Great Nebula of Orion. However, he found no signs of the latter's resolvability (to the contrary, it appeared to him that the nebula actually "retreated" than became more "concentrated" around the stars). Lassell wrote to fellow amateur and retired naval admiral W. H. Smyth:

> All the stars I see [in the Orion Nebula] are individual, isolated, and rather unusually brilliant points, without apparently any connection with it. Examined under good circumstances, with a power of 1018, the brightest parts of the nebula look like masses of wool ... one layer seemingly laying partly over another, so as to give the idea of great thickness or depth in the stratum.[1]

By the time Lassell wrote these words—and as the debate over the nature of the Orion Nebula ground on—a startling announcement came from John Russell Hind, Superintendent of the Nautical Almanac Office and in charge of the private observatory of George Bishop, who had made a fortune in the wine business, at South Villa, London. Hind wrote a letter to *the Times* in 1862, asserting that a small nebula (now known as NGC 1555), discovered a decade earlier, had disappeared. The German astronomer Heinrich d'Arrest in Leipzig, who had observed it several times since Hind discovered it, found no trace of it in October 1861. By the end of the year Otto Struve at Pulkova could barely make it out, but by March it appeared to have brightened again, and "Hind's wonderful nebula" was cited as an example of a change in a nebula that had been established on grounds of unimpeachable authority.[2] Such a change could be expected of a shining fluid but hardly of a remote system of stars! Not only did the astronomers of the time not know the distances to these nebulae, they had to address the issue of what was the source of the light from these systems, with either heated gas or unresolved star light the two leading ideas.

The curious nebula discovered by Hind, disappearing from the sky, might have provided the same impetus to the study of the nebulae as the apparent disappearance of the small crater Linné did to studies of the Moon, setting off decades of inconclusive controversy.[3] It might have, except that just then the spectroscope intervened as an active tool of astronomical research. With its aid, the existence of "true nebulosity," the "shining fluid" of William Herschel, in space was definitively proved, the longstanding debate over the nature of the Orion Nebula was finally settled, and the way was opened up to the analysis of the chemical composition of the stars. This is how it came about.

Spectra had been studied since Isaac Newton's famous studies with the prism in the 17th century. A particularly important development occurred in 1814, when a Bavarian optician, Joseph von Fraunhofer, placed a slit in front of the collimating lens used to concentrate light onto a prism—this gave a cleaner image of the spectrum, and allowed him to perceive and map 600 dark lines (still known as Fraunhofer lines) in the otherwise continuous spectrum of the Sun.

Fig. 6.1. Fraunhofer. *From: Wikipedia Commons/ small portraits collection, History of Science Collection, University of Oklahoma Library.*

Fraunhofer himself was not particularly curious about the origin of these lines—he was a master optician, and his purposes were more practical. He noted that the dark lines, corresponding in fact to the image of the slit, appeared always in the same position when dispersed by his prism. Thus he could use them to determine precisely the refractive indices of glass used to produce his superb achromatic lenses. The lines most useful for his lens calibration work were given letter labels—for instance, a pair of closely spaced lines in the yellow part of the spectrum (at wavelengths 589.6 and 589.0 nanometers) were assigned the letter *D*, another pair in the red were designated H and K (at 396.8 and 393.4 nanometers), and so on. We mention these particular lines because they would prove very important to the work of astronomers later, when they began to examine the spectra of distant galaxies in detail.

Tragically, Fraunhofer died of lung disease in 1826, at the age of 39, either of tuberculosis or deterioration of his lungs caused by incessant exposure to furnace heat and lead oxide, such as was suffered by many glassmakers at the time.

The explanation of Fraunhofer's lines proved elusive until 1859 (the year of *Darwin's Origin of Species*), when the German physicist Gustav Kirchhoff presented an important paper to the Berlin Academy. At the time Germany was bursting with vigor and material success, well on the way to becoming a leading European military power that would defeat France with shocking ease in the Franco-Prussian War of 1870. Its universities and technical schools were greatly admired, and German methods were the most thorough anywhere (it was in Germany, for instance, that the Ph.D degree was introduced, which was soon adopted elsewhere). For some years Kirchhoff and a colleague, the chemist Robert Bunsen, had been studying the spectra of heated salts. Recalling that Fraunhofer himself had observed that heating table salt in a flame produces two bright lines in the same positions as the two dark lines D in the solar spectrum, Kirchhoff hit on the idea of passing sunlight through a flame containing table salt. "I ... let the solar rays, ... before they fell on the slit [of the spectroscope] pass through a strong table salt flame," he wrote. "If the sunlight was sufficiently weakened, two bright lines appeared in place of the two dark D lines; if the intensity of sunlight exceeded a certain limit, the two dark D lines appeared in much greater distinctness, than without the presence of the table salt flame."[4]

Physicists were then actively investigating the nature of light, electricity, and magnetism. This proved to be the breakthrough that would eventually lead to the chemistry of the stars (and the foundation of the discipline now known as astrophysics). Only a few years before, in 1835, the French philosopher August Comte had written that though it was possible to determine the shapes, distances, sizes and movements of the stars, "we would never know how to study by any means their chemical compo-

Fig. 6.2. Fraunhofer lines on the Sun. *Credit: Wikipedia Commons.*

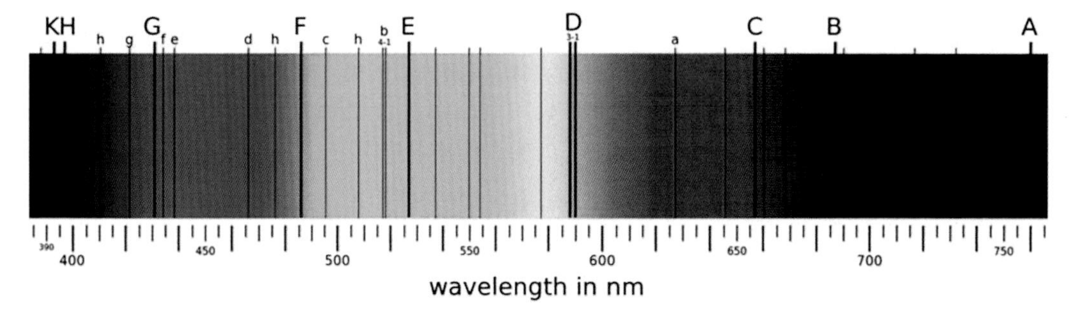

Fig. 6.3. Emission spectrum: hydrogen, including the H-alpha line in the red. *Credit: Wikipedia Commons.*

Fig. 6.4. iron. *Credit: Wikipedia Commons.*

sition, or their mineralogical structure, and, even more so, the nature of any organized beings that might live on their surface."[5] It was one of the most infamous wrong predictions in history, ranking with the astronomer Simon Newcomb's claim in 1903 that humans would never achieve heavier-than-air flight. Mercifully, Comte, who had died in 1857, did not live to learn of his misfire.

Kirchhoff and Bunsen established spectroscopy as an empirical science by enunciating three laws of spectroscopy:

1. incandescent solids or liquids (or stars, which are of gas under high pressure) emit a continuous spectrum;
2. the spectra of a heated gas (like the atmosphere around a star) consists of bright lines, with the lines appearing at wavelengths that are characteristic of the gas;
3. when light from an incandescent gas or liquid traverses a heated gas, the gas absorbs light at the same wavelength as it emits when heated to the same temperature.

On the basis of these three laws, Kirchhoff could explain the origin of the Fraunhofer lines in the solar spectrum. The continuous spectrum of the Sun is absorbed in discrete wavelengths by elements in the gas that made up the Sun's atmosphere. This is the reason the Fraunhofer lines are dark—they represent gaps where the continuous spectrum has been absorbed. It is also the reason they are located in the same positions as the bright lines of spectra of elements heated in the laboratory. A given set of lines is, moreover, a unique signature of the presence of a given element.

After Kirchhoff had recognized that the pair of dark Fraunhofer lines, D, in the solar spectrum corresponded to the pair of bright yellow lines given by sodium, he resolved to compare the spectra of other elements with the solar dark lines. In so doing, he passed from the spectrum of sodium, to others—that of hydrogen was much simpler, while that of iron was of much greater complexity, yielding more than 2000 lines, irregularly distributed through all the colors from deep-red to lavender. To create a comparison spectrum for iron, for instance, he needed to resort to the spark of an electric arc, which could produce a much higher temperature, or the spark of an induction coil; in the latter case, a vacuum tube with a trace of the gas of the substance under scrutiny (here, volatilized iron) was submitted to the discharge of an induction coil charged by a series of Leyden jars. Directing his spectroscope toward the Sun and exposing the lower half of its slit to its light, and then by means of a reflecting prism placed in front of the upper half bringing the image of the spark into juxtaposition with it in the field of view, he could compare the positions of the dark Fraunhofer lines directly with the bright lines due to vaporized iron. In this way, Kirchhoff determined that not only sodium but also iron and a number of other terrestrial elements, calcium (which produces the H and K lines), barium, strontium, magnesium, nickel, copper, cobalt and zinc, were present in the Sun's atmosphere.

Meanwhile, there was another development, owing to the Austrian physicist Christian Doppler, which allowed spectroscopy to be applied not only to the important problem of the composition of the Sun and stars but also to the analysis of their motions toward or away from the Earth. Born in 1803, Doppler was the son of a stonemason in Salzburg, who couldn't follow his father's business because of poor health. After completing his education in Salzburg and Vienna, he found it difficult to obtain a permanent academic position in Austria. He briefly considered emigrating to the United States before obtaining a post in Prague, where his teaching and examination duties were onerous and he was regarded as a harsh (and hence unpopular) examiner. It was in Prague that he did the work for which he will always be famous. In a May 1842 lecture to the Royal Bohemian Scientific Society, "On the colored light of double stars and some other stars in the heavens," he announced a formula showing how the frequency (or wavelength) of sound waves changed depending on the radial motion of either the source or the observer. This is the effect that explains the change of pitch as a train is approaching and then leaving a platform: as it approaches, the pitch of the wailing siren is higher than when it moves away.

Believing that light was a longitudinal wave, like sound, rather than a transverse wave, he tried to use his theory to explain the colors of double stars. His analysis was, however, incorrect, as Doppler thought that the motion of stars along the line of sight towards or away from us would result in their colors being shifted towards the red or blue end of the spectrum (thus white stars would become colored), and didn't seem to allow for invisible radiation from stars being shifted into the visible part of the spectrum and thereby compensating for the other colors being shifted out. After Doppler's death, of tuberculosis in 1853, the French physicist Armand Fizeau—who is best remembered for being the first person to succeed in measuring the speed of light—concentrated not on changing hues of stars, due to the Doppler effect, but upon the shift of identifiable spectral lines toward the blue or red, which was much more useful and measurable. (For future reference, we note that the change in wavelength is equal to $z = v/c$, where v is the velocity of the emitting source relative to the observer and c is the speed of light. This formula applies when the emitting source has velocities of less than about a tenth the speed of light; for higher velocities, such as those encountered in remote galaxies due to the expansion of the universe, the velocities can be much higher, and since Einstein's special theory of relativity does not allow anything to travel faster than the speed of light, a relativistic correction is needed.)

It should be emphasized that the Doppler effect does not give us any information about motions perpendicular to the line of sight; it is used only to determine motions along the line of sight. In the late 19th and early 20th centuries, the determination of the radial velocities of stars was to become a major research program at several observatories, and later the principle was applied to the nebulae, leading to the recognition of the expanding universe in the early 20th century.

Spectroscopist of the Nebulae

By 1861—the same year that Hind's nebula disappeared—the ideas from Germany were coming to the attention of British scientists. Warren De la Rue, whose father had founded a highly lucrative stationary and printing business but who like a number of scions of wealth preferred the pastimes of science to the toil of business and became an avid amateur astronomer, commented on a talk at the Royal Institution of London by Henry Roscoe, a professor of chemistry in Manchester who had once worked with Bunsen:

> The physicist [Kirchhoff] and the chemist [Bunsen] have brought before us a means of analysis that, as Dr. [Michael] Faraday recently said, if we were go to the Sun, and to bring away some portions of it and analyze them in our laboratories, we could not examine them more accurately than we can by this new mode of spectrum analysis.[6]

De la Rue grasped the potential spectrum analysis had for astronomical uses. So did W. Allen Miller, professor of chemistry at King's College, London, who discussed "the new method of spectrum analysis" at a meeting of the British Association for the Advancement of Science in Manchester. He reprised the lecture at a meeting of the Pharmaceutical Society in London in January 1862 at which William Huggins, another young man with money and time on his hands to devote to technical avocational interests, was in attendance.

Still in his 30s, Huggins had retired from running the family silk business in order to devote himself to astronomy. Like Lassell, George Bishop, De la Rue and others, he would become an independent "Grand Amateur," to use a phrase coined by historian Allan Chapman, able to pursue his own research.

It is important to point out that at this time there existed no "scientific profession" as such, and very few men were able to earn a living doing science. Chapman reminds us that professional astronomers, like George Airy, the Astronomer Royal at Greenwich, "received a stipend … [but] were either under-resourced or else kept too busy with routine or teaching

Fig. 6.5. William Huggins. Caricature by Leslie Ward (Spy); Vanity Fair, April 9, 1903. *From: Wikipedia Commons.*

or administrative duties to think seriously of research. In fact, the ex-Royal Navy officer Grand Amateur, Admiral William Henry Smyth … likened amateurs such as himself—who did astronomy for love—to the initiative-taking officers in the armed forces; whilst the professionals—who needed a job and a salary—resembled the 'other ranks,' who obediently followed where the officers led!" A profoundly different situation from that prevailing today.[7]

At first Huggins tried his hand as an enthusiastic observer of the planets, using an 8-inch refractor built by the Cambridgeport, Massachussetts, telescope-maker Alvan Clark. But he soon became dissatisfied with the routine character of this work. "In a vague way," he wrote, "I sought about in my mind for the possibility of research upon the heavens in a new direction."[8] He seems to have been at least vaguely aware of Kirchhoff's "great discovery" when he attended Miller's lecture, but now he grasped its potentialities for the first time. "This news," he recalled afterward, "was to me like a coming upon a spring of water in a dry and thirsty land."

He was fortunate in that Miller was not only an experienced laboratory spectroscopist, well connected in the highest scientific circles (he was both treasurer and vice-president of the Royal Society), he was also a neighbor—the two men lived across the street from one another at Upper Tulse Hill, a posh and rapidly growing London suburb on the south side of the Thames. As soon as Miller finished lecturing, Huggins wrote, "A sudden impulse seized me to suggest to him that we should return home together. On our way home I told him of what was in my mind, and asked him to join me in the attempt I was about to make, to apply Kirchhoff's methods to the stars."[9]

What happened next is known only in outline, since Huggins's records from this period of intense activity are incomplete. Clearly, however, in the spring and summer of 1862, Miller, who had been trying to photograph metallic spark spectra in his laboratory, began to work in Huggins's observatory. They adapted a laboratory spectroscope (of the kind used by Kirchhoff and Bunsen and still familiar in physics labs today) to the telescope, and subjected spectra of the Moon, Mars, Jupiter, and several

bright stars to the same kind of analysis that physicists and chemists did the substances they heated in the flame of the laboratory. They did not have the field to themselves. Others were already working along similar lines, including Giovan Battista Donati in Florence, Lewis M. Rutherfurd in New York City, and Father Angelo Secchi in Rome. By February 1863, Huggins and Miller published detailed maps of dark lines in the spectra of Betelguese, Aldebaran, and Sirius, and in a paper, "Spectra of the Fixed Stars," published the following year, they pondered the implications of the fact that they were finding elements common to life on Earth in the stars:

It is remarkable that the elements most widely diffused through the host of stars are some of those most closely connected with the constitution of the living organisms of our globe, including hydrogen, sodium, magnesium, and iron. Of oxygen and nitrogen we could scarcely hope to have any decisive indications since these bodies have spectra of different orders. These forms of elementary matter, when influenced by heat, light, and chemical force, all of which we have certain knowledge are radiated from the stars, afford some of the most important conditions which we know to be indispensable to the existence of living organisms such as those with which we are acquainted.[10]

Huggins made his most important discovery soon afterwards. The spectroscope was demonstrating the similarity in the constitution of stars and the Sun from the similarity of their spectra, and now he wanted to see if the same result could be established for the "distinct and remarkable class of bodies known as nebulae."[11] Rather than begin with a large and extended object, like the Great Nebula in Orion or the one in Andromeda, Huggins decided to investigate the planetary nebulae, which he regarded as the most enigmatic of all. As the first such object of spectroscopic scrutiny, on August 29, 1864, he set the slit of his spectroscope on a nebula, the so-called "Cat's Eye" (NGC 6543) in Draco.

In "A Personal Retrospect" penned thirty years later, which Huggins's biographer Barbara Becker calls a "suspenseful and well-crafted narrative,"[12] even though one which, through golden mists of nostalgia, perhaps glosses over the details, Huggins takes us with him to his observatory as a virtual eyewitness. He asks us

> to picture ... the feeling of excited suspense, mingled with a degree of awe, with which, after a few moments of hesitation, I put my eye to the spectroscope. Was I not about to look into a secret place of creation?... I looked into the spectroscope. No spectrum such as I expected! A single bright line only! At first, I suspected some displacement of the [spectroscope's] prism, and that I was looking at a reflection of the illuminated slit from one of its faces. This thought was scarcely more than momentary; the true interpretation flashed upon me. The light of the

nebula was monochromatic, and so, unlike any other light I had as yet subjected to prismatic examination, could not be extended out to form a complete spectrum... A little closer looking showed two other bright lines on the side towards the blue, all the three lines being separated by intervals relatively dark. The riddle of the nebulae was solved. The answer, which had come to us in the light itself, read: Not an aggregation of stars, but a luminous gas. Stars after the order of our own sun, and of the brighter stars, would give a different spectrum; the light of the nebula had clearly been emitted by a luminous gas.[13]

Huggins presented the solution of the problem of the nebulae as an intuitive leap. The reality of how he came to his conclusion was probably more complicated, but in the end he managed to satisfy himself that NGC 6543 consisted of luminous gas or vapor. It was certainly not a distant cluster of stars. As usual old William Herschel had been right after all. (The identity of the vapor was unknown. There were two bright green lines in the spectrum appearing at wavelengths of 495.9 and 500.7 nanometers, but they did not appear to be in positions corresponding to any known elements. When, at the total solar eclipse of 1868, Jules Janssen and J. Norman Lockyer discovered the 587.7 nm line in Sun's spectrum—another line that did not correspond with that of any known element—Lockyer deduced that it was a new element found only in the Sun, and named it Helium. Huggins, by a similar line of reasoning, concluded that his green lines were also due to an unknown element, and in 1898 or so began to refer to it as Nebulium. But Nebulium does not exist. (Parenthetically, Huggins's green lines were explained only in 1927, when Ira Sprague Bowen, an assistant professor at Caltech, who specialized in laboratory spectroscopy, used quantum mechanics to show that they were emission lines from twice-ionized oxygen raised to an excited state by collisions with other atoms, and deexciting only after a very long time. They are examples of what are known as "forbidden lines," because they are never seen in laboratory samples, but they are important in nonterrestrial environments where extreme conditions, such as low density, prevail.)

After his success with the Cat's Eye, Huggins launched a systematic examination of the spectra of sixty of the brighter nebulae and clusters. This grand project occupied him for the next two years. He found that a third resembled the planetary nebulae in showing the signature bright green-line spectrum (and became known as "green" nebulae).

In 1865, he classed among the "green" nebulae the Great Nebula of Orion itself. On whatever part of the nebula he placed the slit of his spectroscope, he saw only the signature bright green emission lines of luminous vapor or gas. The conclusion was obvious:

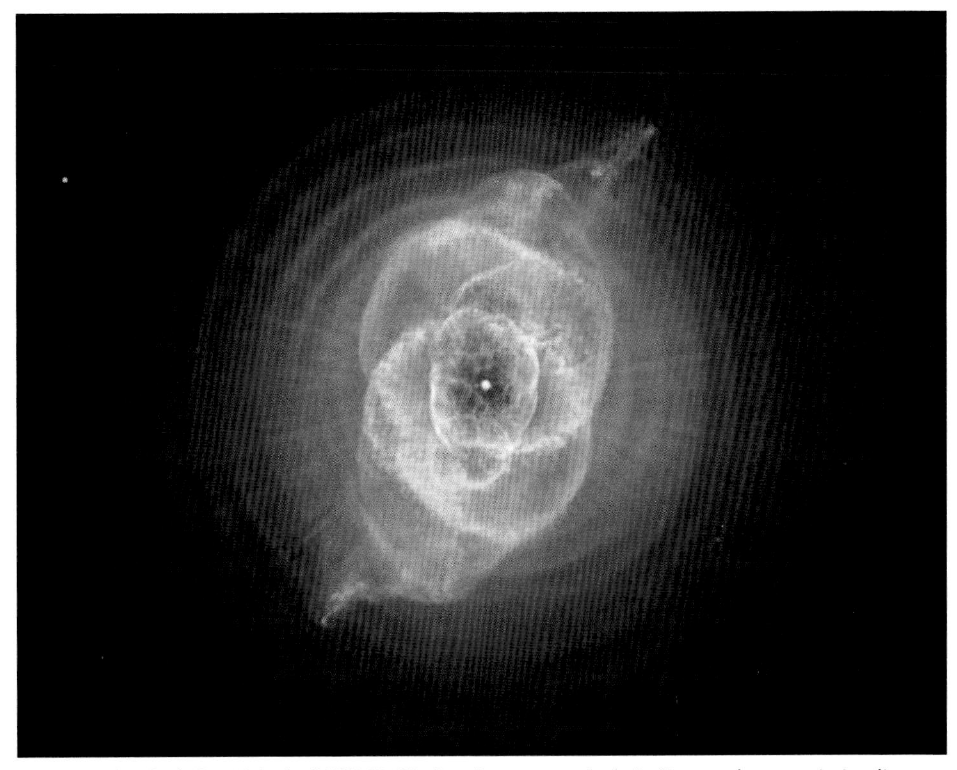

Fig. 6.6. The Cat's Eye Nebula (NGC 6543), the planetary nebula in Draco whose emission lines first revealed its gaseous nature to William Huggins in 1864. The complicated structure surrounding the star has been produced by ejection of the star's outer gaseous layers, in a process through which the Sun will eventually pass late in its evolutionary history. *Courtesy: Hubble Heritage Team (STScI/AURA).*

> the detection in a nebula of minute closely associated points of light, which has hitherto been considered as a certain indication of a stellar constitution, can no longer be accepted as a trustworthy proof that the object consists of true stars. These luminous points, in some nebulae at least, must be regarded as themselves gaseous bodies, denser portions, probably, of the great nebulous mass, since they exhibit a constitution which is identical with the fainter and outlying parts have not been resolved. These nebulae are shown by the prism to be enormous gaseous systems.[14]

The vast beast with its proboscis running out from its snout did not consist of stars but was an infernal region consisting of, to borrow the poet Milton's phrases describing the infernal regions, "surging smoke" and a "cloudy chair."

The other two-thirds of the nebulae examined by Huggins showed, as he found in the case of the most famous example, M31, the Great

Fig. 6.7. M31, the classic example of a "white" nebula, i.e., one showing a continuous (starlike) spectrum, now known to be the nearest spiral galaxy to the Earth. The dark lanes below the brilliant nucleus were known as "Bond's canals," after W.C. Bond who recorded them with the 15-inch Merz refractor at Harvard College Observatory. Imaged with a TMB-130 apo refractor at f./7 by Klaus Brasch. *Courtesy: Klaus Brasch.*

Nebula of Andromeda, "light ... distributed throughout the spectrum, and consequently extremely faint.... Though continuous, the spectrum did not look uniform in brightness, but its extreme feebleness made it uncertain whether the irregularities were due to certain parts being enhanced by bright lines, or the other parts enfeebled by dark lines."[15] In contrast to "green" nebulae like the Cat's Eye or the Orion Nebula, the Andromeda Nebula became the leading object belonging to the group of "white" nebulae.

The interpretation of the meaning of the Andromeda Nebula's spectrum was not entirely straightforward at the time. The hazy oval of the nebula was seen through a rich stratum of 1500 stars, but no telescope had ever resolved its nebulosity into stars—not even the great 15-inch Merz at Harvard, with which William Cranch Bond had made out a nearly stellar nucleus at the center and, along one side of the nebulous oval, two narrow mysterious dark lanes ("Bond's canals"). The ghostly vision of the Andromeda Nebula in the telescope was of one of Rosse's "spirals" seen along an oblique line of sight. Inevitably it conjured up an image of Laplace's nebular hypothesis in action—perhaps a Solar System in formation, a cloud of gas and dust swirling around an already condensed center that was just begin-

ning to shine as a sun. This vision of its nature seemed however no more probable than the idea that it might be a mass of distant suns in its own right, grown misty with excessive distance. Indeed, both points of view had its adherents. How could one decide between them?

In 1882, the Rev. T.W. Webb summed up the situation as it appeared to him at the time when he called the Andromeda Nebula "a mystery in all probability never to be penetrated by man."[16]

Then, in August 1885, a new and unexpected clue came within sight of astronomers which, if only it could be deciphered, might say something of the true nature of the Andromeda nebula. A nova or "temporary" star appeared only 16 seconds of arc (the separation of a close double star) from its star-like nucleus, thereby producing "one of the strangest sights in the heavens," in the words of one witness. Another, E. Walter Maunder of the Royal Observatory at Greenwich, pronounced that "the strange and beautiful object has broken silence at last, though its utterance may be difficult to interpret."[17]

Once before a nova had appeared in a nebula—in May 1860, when the German astronomer Arthur Auwers had discovered a nova in M80, a globular cluster in Scorpio. But M80 was a different case; it was not strictly a nebula at all, having been resolved by William Herschel in 1785 into a globular cluster that was "the richest and most condensed mass of stars which the firmament can offer to the contemplation of astronomers." The new star in the Great Nebula of Andromeda stood out as a singular stellar point against what to all intents and purposes was a cloud.

The new star, "S Andromedae"—the designation "S" signified that, for want of a better name, it was considered a new "variable" star, the letters from R onwards being assigned to variable stars in each constellation; this, then, was the second variable star to be noticed in Andromeda—had definitely not been present on the evening of August 16, 1885, when the nebula was examined by several skillful European observers. First noticed by a French observer on August 17, he thought it too strange to countenance. On August 20 was it finally seen by someone confident enough to accept it as reality, Ernst Hartwig at the Dorpat Observatory in Estonia, who thinking in terms of Laplace's nebular hypothesis imagined it was the central sun forming out of the nebular mist. ("There's the beautiful central Sun shining through the fog!" he exclaimed.) When Hartwig saw it, it was possibly as bright as the 6th magnitude, but the observatory's conservative director refused to allow him to announce the discovery until its existence could be confirmed in a moonless sky, and because of a run of clouds, this did not happen until August 31. By then the star had already begun to fade. It was then 8th magnitude. Within only a few weeks it had dropped to the 11th magnitude, and when seen for the last time, on February 7, 1886, by Asaph Hall with the great 26-inch refractor of the U.S. Naval Observatory in Washington, D.C., it had faded to 16th magnitude—the limit of visibility with that telescope. Then it vanished forever.

The eruption of "S Andromedae" led to a brief but vigorous discussion regarding its implications. The irrepressible French astronomer Camille Flammarion, favoring the "island universe" theory, enthused: "Probably the Andromeda Nebula is a cluster of stars, and the star which has flared up close to the center is one of its suns of which the photosphere has undergone a sudden conflagration.... [I]t is a huge sun, without doubt thousands of times more gigantic than our own! The heavens are immense; man is insignificant."[18] Writing eight years after the event, the judicious Agnes M. Clerke, a great Irish popularizer of astronomy, called attention to the fact that the outburst in the Great Nebula in Andromeda had occurred only a few years after that in M80, and that they could hardly have occurred accidentally just on the line of sight between the Earth and the central portions of those two objects. Thus, she argued, they had to be associated with the objects in the midst of which they had flared up. If so, however, she concluded it was

> eminently rash to conclude that [these nebulae] are really aggregations of sun-like bodies.... For it is practically certain that, however distant the nebulae, the stars were equally remote; hence, if the constituent particles of the former be suns, the incomparably vaster orbs by which their feeble light was well-nigh obliterated must ... have been on such a scale as the imagination recoils from contemplating.[19]

Man had entered the 19th century using "only his own animal power, supplemented by that of wind and water, much as he had entered the Thirteenth, or, for that matter the First." At least in the industrial world, he bade farewell to all that "with his capacities for transportation, communication, production, manufacture and weaponry multiplied a thousandfold by the energy of machines."[20] Even so, it was impossible at the time to imagine any source of energy capable of keeping a star like the Sun burning long enough to make it last for the aeons of time required by Darwin's theory of evolution much less a source capable of powering the stupendous outburst of a new star erupting in an island universe located far beyond our Milky Way. As Kenneth Glyn Jones writes:

> This one item of astronomical evidence remained an obstacle in the way of progress to a firmer grasp of the scale of the universe for nearly forty years, and the recognition of the "island universe" status of the great spiral nebula in which the nova had appeared, was perennially hampered by the incredible implication that a single star could, if only temporarily

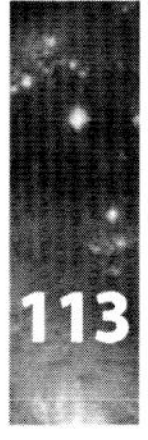

equal in brightness a sizeable fraction of a huge system of millions of stars. [21]

Instead, S Andromeda became a stumbling block, and a seemingly insurmountable one, to the island universe theory. Miss Clerke noted in her inimitable way, "The conception of the nebulae as remote galaxies ... began to withdraw into the region of discarded and half-forgotten speculations."[22]

* * *

1. W. H. Smyth, *A Cycle of Celestial Objects continued at the Hartwell Observatory to 1859* (London: J.W. Parker, 1860), 165.
2. J. R. Hind, "Changes among the stars," The Times, February 4, 1862.
3. For the Linné controversy, see: William P. Sheehan and Thomas A. Dobbins, *Epic Moon: a history of lunar exploration in the age of the telescope.* Richmond, Virginia: Willmann-Bell, 2001.
4. G. Kirchhoff, *Ueber die Fraunhoferschen Linien. Monatsbericht der Akademie Wissenschaften Berlin,* 1859, 662-665. Within six months, it was translated into English by George Stokes, "On the simultaneous emission and absorption of rays of the same definite refrangibility...," *Philosophical Magazine,* Series 4, no. 21 (1860), 195-196.
5. Auguste Comte, *Cours de Philosophie Positive, II, 19th lesson.* Paris: Bachelier, 1835.
6. "Proceedings of the Chemical Society," *Chemical News,* 4 (1861), 130-133.
7. Allan Chapman, "Airy's Greenwich Staff," *The Antiquarian Astronomer,* issue 6 (January 2012), 4-18:7.
8. William Huggins, "The New Astronomy: a personal retrospect," *Nineteenth Century,* 41 (1897), 907-929:911.
9. Ibid.
10. Sir William Huggins and Lady Huggins, eds., *The Scientific Papers of Sir William Huggins* (London: Publications of Sir William Huggins's Observatory, 1909), 60.
11. William Huggins, "On the Spectra of some of the Nebulae," *Phil. Trans.,* 1864, 437-444:437.
12. Barbara Becker, *Unravelling Starlight: William and Margaret Huggins and the Rise of the New Astronomy* (Cambridge: Cambridge University Press, 2011), 73.
13. Huggins, "New Astronomy," 916.
14. William Huggins, "On the spectrum of the Great Nebula in the Sword Handle of Orion," *Proceedings of the Royal Society, 14* (1865), 39-42: 41.
15. Ibid.
16. Quoted in Gerard de Vaucouleurs, "Discovering M31's Spiral Shape," *Sky and Telescope,* 74 (1987):596.
17. Quoted in William Sheehan, *The Immortal Fire Within: the life and work of Edward Emerson Barnard* (Cambridge: Cambridge University Press, 1995), 85.
18. Camille Flammarion, "Apparition d'une Étoile dans la Nébuleuse d'Androméde," *L'Astronomie,* 4 (1885), 361-367.
19. A.M. Clerke, *A Popular History of Astronomy During the Nineteenth Century* (London: Adam & Charles Black, 3rd ed., 1893), 493.
20. Barbara Tuchman, *The Proud Tower: a portrait of the world before the war,* 1890-1914 (New York: Macmillan, 1966), xii.
21. Kenneth Glyn Jones, "S Andromedae, 1885: an analysis of contemporary reports and a reconstruction." *Journal for the History of Astronomy,* 7 (1976), 27-40:27.
22. Clerke, *A Popular History,* 5.

7. Fields of Glory

Sweep on through glittering star fields and long for endless night! More nebulae, more stars. Here a bright and beautiful star overpowering in its brilliancy, and there close to it a tiny point of light seen with the greatest difficulty, a large star and its companion. How plentiful the stars now appear. Each sweep increases their number. The field is sprinkled with them, and now we suddenly sweep into myriads and swarms of glittering, sparkling points of brilliancy — we have entered the Milky Way. We are in the midst of millions and millions of suns — we are in the jewel house of the Maker, and our soul mounts up, up to that wonderful Creator, and we adore the hand that scattered the jewels of heaven so lavishly in this one vast region. No pen can describe the wonderful scene that the swinging tube reveals as it sweeps among that vast array of suns.

—Edward Emerson Barnard, The Nashville Artisan (1883)

Even from the northern hemisphere, the Milky Way, seen from a clear dark site, exhilarates with its magnificence. A northern hemisphere witness to its grandeur, the Japanese haiku writer and pilgrim Basho (1644-94), looked across the sea to Sado Island, where political exiles were confined, and exclaimed:

High over wild seas,
Surrounding Sado Island —
The River of Heaven![1]

The River of Heaven was, of course, the Milky Way. Basho wrote that haiku on the eve of Tanabata Matsuri (the "Evening of the Seventh"), a Japanese summer star-festival celebrating the reunion (for one night) of Orihime and Hikoboshi, represented by the stars Vega and Altair. According to a legend originally imported into Japan from China, these personali-

ties were separated from one another by the river Amanogawa (literally, the "heavenly river," i.e., the Milky Way) but were permitted to meet again for one night a year, on the seventh day of the seventh lunar month of the luni-solar calendar.[2]

For the lover of the Galaxy, it is not separation that is the difficulty; it is obscuration. The overpowering lights of cities keeps him or her from the objects of desire. Though the center of the Galaxy (best seen from the Southern Hemisphere, since it lies at 20 degrees south latitude) is, rather surprisingly, bright enough to cast a shadow on a clear dark night—and is one of only a few celestial objects able to do so; the others include the Moon, Venus, Jupiter, and Mars at its very best—it cannot be seen at all from cities. It glows with a subtle light, only fifty percent brighter than the ambient background of the sky brightness furnished by the Zodiacal Light and airglow. City lights brighten the back-sky glow by factors of 10 to 100 or more, causing the Milky Way to fall below the threshold of visibility. Thus the Milky Way for the city-dweller resembles a Renaissance fresco whose vivid hues are obscured under a centuries-old coat of grime and soot.

Sadly, the glories of the night sky have been expunged by light pollution in the vast cities in which most of the world's population lives. The habitat of amateur astronomers is vanishing. It is probably fair to say that most of the world's population has never experienced a pristine view of the Milky Way. No wonder those who take an interest in such things are now a graying, affluent, and predominantly male demographic (as surveys of the readership of popular astronomy magazines like *Sky & Telescope* and *Astronomy* show over and over) whose enthusiasm was sparked in another era and who can afford to travel to dark sites with portable telescopes to feed their appetite for galactic wonders.

One can well appreciate their determination to do so, for under conditions of transparent and dark skies, the Milky Way opens up as a grand and glorious avenue of bright stars, powdered with open clusters, diffuse nebulae, and—most wonderful of all—a fretted lattice-work of strange dark markings. Under such conditions, the Galaxy—our own vast star-system—presents what is probably the most awe-inspiring and sublime sight a mortal can behold.

The structure of the Milky Way Galaxy, which was already hinted at in Herschel's snapping alligator depiction, as mentioned in early chapters, resembles two fried eggs clapped together. By this analogy, the yolks make up the "bulge." Luminous with the massed effect of billions of small, old, distant suns, it extends across some 30 degrees of the sky. Dark markings superimposed upon the distant stars form an intricate meshwork and include a vast tongue-like extension to the south, extending another 30 degrees or more. The dark markings do not form a uniform band, and even the naked-eye reveals that they are full of complications. As one studies, a few of the better-defined forms emerge. Dark tubules reach from the Antares, the red star that marks the heart of the Scorpion, toward the claws. (Here, the globular cluster M4 is visible as a hazy patch with the naked eye and becomes big and bright in binoculars). There is a kind of heavenly stairway consisting of dark forms winding upward through Sagittarius; the curving Pipe nebula scoops up in its bowl immense vacuities. (And how many light years, we may ask, does that bowl span?)

Relative to these dark monstrosities, which have a sort of impenetrable emptiness, the brighter stars — especially alpha and beta Centauri and the Southern Cross that straddles the sharp outline of the Coal Sack – stand out strikingly by contrast.

The Milky Way's river of light winds its way, a seeming backdrop to all of these, from Canis Major, whose leader Sirius stands almost overhead in the Southern Hemisphere and appears as brilliant as a planet, through the brilliant stars and profuse nebulae of Monoceros and Orion; upward along the stars of the Summer Triangle, Vega, Deneb, and Altair; till it arrives at the Great Rift in Cygnus, a great black slipstream cleaving the river, and farther on past the glittering Scutum star-cloud, through the treasure houses of Scorpio and Sagittarius, through Norma, Centaurus, and Crux, with all their wonders, and peters out at last among the mellow glories of Carina and Vela and Puppis. The brighter stars — jewel-like, glistening — hug the plane of the Galaxy along this mysterious trail.

Turn binoculars on some of these objects and what an extension of the wonderful appears! Scores of diffuse nebulae, open clusters, globular clusters swim through the field. And always there is the fretwork of dark markings, the strange, mysterious, ominous abysses leading to where God only knows.

Are they abysses? Dark clouds? In some places they seem to tower into black nimbuses; in other places they become tarry horsetail-bobs, or thin coal-dust cirrus. In still others they stretch into thick taffy-pulls. Some are distinctly spheroidal blobs; others are elongated and wisp-like. But they are everywhere — one could not view them for long without realizing that space — the darkness of space — is filled with what appears to be a mysterious substance, thickly or thinly spread. The impression of these dark markings is of a puzzled and involved topography. They form an intricate landscape – they feature ruts, crannies, outcrops — from which, once lost, it might never be possible to emerge again. They form a rugged Peloponnesus shaped into jutties, islands, inlets, coves, peninsulas. Or is it a landscape at all, not rather a depth below the ocean — a seascape full of darkling forms, sprawling kelp or tangled seaweed? Are these lurking and drifting shadowy immensities, monsters of sea-bottoms, sulking among cavernous recesses, lurching after their prey in the depths as upon their surface the ephemeral light-forms, stirred by the bow-wake of a ship, churn phosphorescent microorganisms into quickening forms send them skidding about like hosts of dancing fireflies?

Admittedly it is difficult to form a clear idea of the relation of all these different objects. Though they appear to be sparks and inkblots on the curved surface of a great dome — the ceiling, if you will, of a great and mysterious cave – they are seen in projection. We must conjure up for ourselves the third dimension and realize that they are arranged along curvilinear arc-swaths. Viewed along a line of sight extending from the Sun to the Galactic Center (toward Sagittarius, as we now know), foreshortened and silhouetted along more or less oblique lines of sight, so that objects at different distances are superimposed upon one another in confused and contortionate projection.

Viewing the distant parts of the Galaxy, we face the same problem as a tourist wishing to view the forest on a distant mountain ridge from within a pine wood. The star masses and dark blobs appear like the ridges of mountains viewed through a foreground of mist in a complicated — but flat — picture. Everything is foreshortened or obscured or otherwise muddled. We are left with a view rather like that of intussuscepted human bowels as envisaged by a swallowed flea.

The nature of the dark markings of the Milky Way was long one of the seemingly insoluble problems of astronomy. No other objects were more obscure or menacing or sublime. The Herschels, father and son, concluded they were unplumbable abysses. William had called the starless patch in the body of the Scorpion a "true hole in the heavens," John had seen the dark markings as the tubules and chimneys of vast cavernous strata. In his charming novel Two on a Tower, Thomas Hardy takes chapter and verse from John Herschel in making the ambitious young astronomer Swithin St.-Cleeve conjure for his patron Lady Constantine something of the terror of these darknesses visible:

> "You would hardly think at first that horrid monsters lie up there waiting to be discovered by any moderately penetrating mind—monsters to which those of the oceans bear no sort of comparison."
>
> "What monsters may they be?"
>
> "Impersonal monsters, namely, Immensities. Until a person has thought out the stars and their interspaces he has hardly learnt that there are things much more terrible than monsters of shape; namely,

> monsters of magnitude without known shape. Such monsters are the voids and waste places of the sky. Look ... at those pieces of darkness in the milky way," he went on, pointing with his finger to where the galaxy stretched across over their heads with the luminousness of a frosted web: "You see that dark opening in it near the Swan? There is a still more remarkable one south of the Equator called the Coal Sack, as a sort of nickname that has a farcical force from its very inadequacy. In these our sight plunges quite beyond any twinkle we have yet visited. Those are deep wells for the human mind to let itself down into, leave alone the human body! And think of the side caverns and secondary abysses to right and left as you pass on."[3]

As John Herschel had realized, the Milky Way consisted of a complicated structure traversed, in places, by what he thought were "conical or tubular hollows."[4] And here we revert to an image with which we began: the cave of the night. Herschel and others of his generation could be considered explorers, spelunkers, of the chimneys and tubules of the Galaxy, structures remote in space—as men like Marcelino Sanzo de Sautuola, Marcel Ravidat, and Jean-Martin Chauvet explored the Upper Paleolithic caves of northern Spain and southern France, structures remote in time.

And, of course, because light travels at the vast but finite speed of 300,000 kilometers/second (186,000 miles/second), structures remote in space are also remote in time. It is curious to ponder that the monster Black Hole which, as we now know, sits black-widow like at the center of the Galaxy in the direction of but far beyond the foreground stars that outline the constellation Sagittarius, lurks some 27,000 light years away, which means that light from stars in that neighborhood left them almost as far back as when the "world's oldest paintings" were being painted at Chauvet cave, 30 to 32,000 years ago. Time on such a grand scale is, if not abolished, at least no longer really comprehensible. And how is the human mind to grasp such enormous distances? One light year is equal to 10 trillion kilometers (6 trillion miles). The galactic center is so remote that in all the time since the end of the Last Glacial Maximum, light has just barely had time enough to traverse that unimaginable distance.

Astronomical Photography

Attempts to produce adequate impressions of the visual appearance of the Milky Way, such as the evocative but stylized print by the French astronomer-artist Leopold Trouvelot, were even more unsatisfactory than attempts to render the Orion Nebula had been. Ultimately, both objects defied human skill with the pencil and the brush.

They would finally yield only to the advancing arts of photography. Photography was one of the great triumphs of 19th century technology, and—with spectroscopy—utterly transformed astronomy. We briefly advert again to the pioneering experiments of Nicéphore Niépce and his collaborator and successor Louis-Jacques-Mandé Daguerre in the 1820s and 30s, as well as the first astronomical applications of photography using the primitive daguerreotype process, which deposited positive images directly on a metal plate. Daguerre publicized his process in 1839. Almost at once, it was taken up by an American, John William Draper, professor of chemistry and botany at New York University, who in March 1840 obtained the first daguerreotype showing features on the Moon. A decade later, George Phillips Bond imaged the star Vega; while on July 28, 1851, at the Royal Observatory in Königsberg, Prussia (now Kaliningrad, Russia), a local daguerreotypist named Berkowski (his first name has not been recorded), obtained the first daguerreotype of a total eclipse of the Sun.

A great step forward was taken with the introduction of the wet-collodion process, which involved coating a glass plate with gun cotton and potassium iodide dissolved in alcohol or ether, letting this collodion dry, and then, just before use, dipping the prepared plate into a silver nitrate solution. Introduced in England by Scott Archer in 1851, the wet-collodion process offered amazingly good resolution of detail, conveniently produced negative rather than positive images, and allowed exposures of up to 10 seconds. By the mid-1850s, wet-collodion plates had rendered the daguerreotype obsolete, but the method was still difficult to use, as the exposure had to be made while the plate was still wet. Pioneering photographers such as Mathew Brady, Alexander Gardner, and Timothy O'Sullivan, who followed the clashing armies of the North and South during the American Civil War and developed their plates while still in the field, demonstrated enormous technical skill, and through their efforts, the "shockingly realistic medium" of photography showed citizens at home the carnage of far-away battlefields and stripped away the Victorian-era romance of warfare forever ("into the valley of death," anyone?). But what the wet-collodion process did for far-away battlefields it could not do for the far-away stars.

As early as 1857, G.P. Bond obtained a wet-collodion plate of the double star Mizar and Alcor, in the handle of the Big Dipper, and confidently predicted the future application of photography in astronomy on a magnificent scale. Dying tragically young, he did not live to see it. Even wet-collodion plates were only to capture images of the brighter celestial objects, such as the Sun (photographed on every clear day from the King's Observatory at Kew between 1858 and 1872, when the program was transferred to the Royal Observatory at Greenwich), the Moon, and the brighter planets and stars. The first photograph of a spectrum of a star—Vega again—was obtained by Henry Draper, John W.'s son, on a wet-collodion plate in 1872. Trained as a physician, just as his father had been, he married an heiress whose father happened to own a significant part of the island of Manhattan. This allowed him to devote full-time to his real passion, astronomy, from 1873 until the end of his life. He was a "Grand Amateur" in every sense of the word, setting up a private observatory at Hastings-on-Hudson (near Dobb's Ferry, New York) equipped with a homebuilt 28-inch reflector with a silver-on-glass mirror, and an 11-inch Clark refractor especially designed for photography.

During a visit to London in 1879, Draper called on Sir William Huggins at Tulse Hill. By now, Huggins had upgraded his own observatory, replacing the 8-inch Clark refractor he had used to obtain spectra of the nebulae with two instruments provided by the Royal Society of London: an 18-inch reflector and a 15-inch refractor, which he set up in an imposing new dome in his garden. He had also, in 1875, at age 51, married Margaret Lindsay Murray of Dublin, 25 years his junior, a highly intelligent woman who shared his enthusiasm for astronomy. Despite the difference in their ages, the marriage was a success, and Lady Huggins entered fully and equally in the work. In consequence of the new techniques he was pioneering, Huggins was, moreover, transforming the nature of the astronomical observatory. Thus, the establishment that Draper saw was nothing if not futuristic. To give Huggins's own description:

> [The] astronomical observatory began, for the first time, to take on the appearance of a laboratory. Primary batteries, giving forth noxious gases, were arranged outside one of the windows; a large induction coil stood mounted on a stand on wheels so as to follow the positions of the eye-end of the telescope, together with a battery of several Leyden jars; shelves with Bunsen burners, vacuum tubes, and bottles of chemicals … lined its walls.[5]

Fig. 7.1. William Huggins in his laboratory-observatory. *From: Publications of Sir William Huggins's Observatory, vol. II (1909).*

Draper learned from Huggins of the commercial availability of the gelatino-bromide dry plate, which would render the wet-collodion plates obsolete. Here, photoactive chemicals were suspended in a solution of gelatin and spread in a thin layer on a glass plate. The chemicals remained sensitive even after the gelatin had dried, allowing long exposures to be obtained. In addition, the new process was incalculably more convenient than wet plates, since exposed gelatino-bromide dry plates did not need to be processed immediately. This made the new plates decidedly more useful than the wet-process plates in astronomy. After Draper returned to Hastings-on-Hudson, he made an exposure of 50 minutes on the Great Nebula of Orion, on September 30, 1880. A better image, with a 137-minute exposure, was obtained on March 14, 1882. Of the latter, Edward Singleton Holden, in his *Monograph of the Central Parts of the Nebula Orion,* noted that it showed the Nebula, "for nearly every purpose incomparably better"[6] than the careful drawing, the work of so many years, by G. P. Bond. A new era—the photographic study of the nebulae—had commenced.

Draper, alas, did not live long enough to fulfill his own nearly limitless potential. Soon after he obtained this image of the Orion Nebula, he went on a hunting trip to the Rocky Mountains, was caught out in the open on a stormy night, and fell ill with double pneumonia. He died on November 20, 1882, at the age of 45.

Fig. 7.2. Henry Draper's 137-minute exposure on a gelatino-bromide dry plate, taken on March 14, 1882. This pioneering effort already shows more detail than G.P. Bond's masterpiece of visual observations. *Courtesy: C. Robert O'Dell.*

By this time the name of Edward Emerson Barnard, who would revolutionize this field, was beginning to echo around the astronomical world. Barnard would become the foremost practitioner of celestial photography of his era, and will always be remembered for his pioneering (and in some ways still unsurpassed) wide-angle photographs of the Milky Way. He was also, in contrast to most of the figures we've encountered so far, decidedly not born with a silver spoon in his mouth. To the contrary, his childhood was spent in abject poverty, and he had only two months of formal schooling.[7]

Observer of all that Shines and Obscures

One of the most appealing things about Barnard was his ability to overcome so much childhood hardship. Though nineteenth century readers might have thought of him in terms of the Horatio Alger "rags to riches" stories, we are more likely today to see him as an object-study in resiliency. Psychologists are eager to discover factors allowing someone like Barnard to overcome a level of hardship that would defeat most of us. In Barnard's case, those factors included a keen sense of wonder, extraordinary motivation and drive (what one of his youthful acquaintances described as "the immortal fire within himself"), patience and endurance, including the ability to get by on very little sleep, and perhaps most important, the kindly interest of adult mentors who encouraged his interests and helped him acquire technical mastery in an area—photography—that would soon, and largely in his hands, revolutionize astronomy.

Barnard was born in Nashville in December 1857. His father, Reuben, an illiterate laborer, died before he was born, leaving Edward to be brought up by his mother, Elizabeth. (He had an older brother too, who appears to have been mentally deficient.) At the time of Edward's birth, she was 42 years old, struggling to eke out a living modeling flowers in wax.

In the 1850s, Nashville was a place of culture and cultivation. It was then the largest and most important American city south of the Ohio River with the exception of New Orleans, with a population of 17,000, of whom 4,000 were black slaves. Wide turnpikes and busy railroad lines radiated from it; the coiling Cumberland River, draining Cumberland Gap on its way to the Mississippi, was crowded with river traffic carrying goods from the city's foundries, ironworks, and small manufactures to Memphis and New Orleans. With its Female Academy and Nashville University, it was a center of education. Its free public school system was the first in the south.

But, for all its dazzling appearances, Nashville in the 1850s was a troubled city. The decade had begun with the Convention of Southern States, which convened in the Nashville Capitol in 1850 with the express purpose of forming a Southern sectional party to protest the attempt being made in Washington by President Zachary Taylor's administration to exclude Southern men with their slaves from territories recently ceded to the U.S. by Mexico. Though there was far from a unanimous viewpoint represented at the Convention, and it adjourned without reaching any definite conclusions, its members planned to reconvene in November if Congress failed to respond to the Southerners' demands. It never did; Taylor died suddenly, and his successor, Millard Fillmore—often rated by historians as one of the worst presidents in American history—was eager for appeasement, and worked out with Congress the Compromise of 1850 whereby popular sovereignty was proposed as a way of deciding the issue of slavery in the territories.

The Compromise unraveled with the election of Abraham Lincoln in 1860, when ten southern states withdrew from the Union. In April 1861, Fort Sumter fell, and Tennessee, throwing in its lot with the seceded states in June, almost at once became a theater of war. General Albert Sidney Johnston was placed in command of the Confederate defense line running from the mountains around Cumberland Gap to the Mississippi, and Fort Donelson was built to defend the vital supply lines where they crossed the Cumberland. After a few days of assault, the fort surrendered to Ulysses S. Grant on February 16, 1862, with the capture of 15,000 men, thereby opening Nashville itself to the invading Union troops. Confederate soldiers meanwhile cut the cables and burned the railroad bridge on abandoning the city. Within two days, Nashville had fallen into Union hands, and remained under occupation for the duration of the war. The Capitol was surrounded with temporary stockades made of cotton-bales, and an extensive system of fortifications was built up throughout the city. That was Nashville as young Ed Barnard would have known it.

The dreadful privations that befell the South during and just following the war only intensified the misery of the impoverished Barnard family. Many years afterward, Edward stated that his early youth was "so sad and bitter that even now I cannot look back to it without a shudder."[8]

But the first two years of the Civil War also happened to coincide with the appearance of spectacular comets—two in 1861, one in 1862. Many an astronomer has been captured for his or her vocation by the experience, at a susceptible age (almost always in pre-adolescence, sometimes even in early childhood) of an awesome spectacle in the sky—usually an eclipse or a brilliant comet. Barnard was one of them. Long afterward he recalled: "When I was very small I saw a comet; and I have a vague remembrance that the neighbors spoke of this comet as having something to do with the terrible war that was then devastating the south."[9] Possibly it was the Great Comet of 1861, discovered by amateur astronomer John Tebbutt in New South Wales (Australia) but not making an appearance in the northern sky until June 30, when it emerged spectacularly out of the evening twilight with a head brighter than Jupiter and its tail stretching across a swath of 105° of the sky. Though fading fast, it was still visible to the naked-eye on July 21, when the Union and Confederate armies grappled in Virginia at the First Battle of Bull Run. Or it may have been Comet Swift-Tuttle. With a head reaching the 2nd magnitude and sporting a tail 25 to 30 degrees long in late August and early September 1862, it was fading (but still visible to the naked-eye) on September 17, when the Union and Confederate armies were engaged at Antietam, in Maryland, in what was till then, and remains still, the bloodiest day in American history.

Whichever it was, it clearly awakened the child's sense of wonder—surprise at the rare and unexpected; perhaps also of awe, the quintessential emotion of astronomy, which exists, psychologists say, "in the upper reaches of pleasure and on the boundary of fear" and is associated with the contemplation of whatever is vast and powerful and obscure. "Fleeting and rare, ... [it] can change the course of a life in profound and permanent ways."[10] Then and later, the small, half-starved and frightened youngster must have reveled in such emotions.

During the years the war was raging, Barnard used to lie out in the open air on summer nights, flat on his back in an old wagon bed, watching the stars (and the Milky Way, which was still brilliant in Nashville), a pastime which, he afterwards recalled, "helped to soften the sadness of my childhood."[11] Among the stars he saw from that wagon was "a very bright one, which during the summer months shone directly overhead in the early hours of the evening." It would be many years before he learned its name—Vega.

On Barnard's seventh birthday, December 16, 1864, the fighting between the Union and Confederate armies came within a few miles of Nashville. Barnard himself could hear cannons booming in the distance. General George Thomas's Union forces routed the Confederate troops under General John Hood. Hood's troops had already been decimated at the battle of Franklin two weeks earlier, when the Confederate General had ordered a desperate frontal assault on the Union line, "sending 18,000 men forward through the haze of an Indian-summer afternoon in an attack as spectacular, and as hopeless, as Pickett's famous charge at Gettysburg," in the words of Civil War historian Bruce Catton.[12] The Confederacy was wobbling toward inevitable defeat; for all practical purposes the war was over. Lee's surrender to Grant at Appomattox only months away.

Unfortunately, the Barnards' situation did not improve much under Reconstruction—"the last battle of the Civil War." Nashville remained occupied as the major supply base for Union forces in Tennessee (the last troops were not withdrawn until 1877). Nashville, like the rest of the South, was economically depressed. A cholera epidemic swept through the city in 1866, claiming 800 lives. Meanwhile, in the North, an unprecedented period of expansion began, in which opportunities seemed to open on every hand; the country was made "safe—and lucrative—for the capitalist."[13] Government was a paid agent. Scandals and deals became blatant. A few men made vast fortunes. It was the Gilded Age, the Age of the Tycoon. Some of them would endow institutions of learning. A few would even support the progress of astronomy.

Meanwhile, the nightmare of Ed Barnard's childhood was coming to an end. His mother came upon a photograph of a man now working as a photographer in Nashville, whom she had known quite well in Ohio during better days. The man was John H. Van Stavoren, who since the beginning of the Civil War had owned a photograph gallery in Nashville. When

Elizabeth called on her old acquaintance, she found that he was looking for a boy to run errands and to guide an immense solar camera he had mounted on the studio roof. The camera, called "Jupiter" because of its enormous size, was used to make life-sized enlargements on silvered paper from negatives. Because of the relative insensitivity of the silvered paper then in use, intense sunlight was needed, and was produced by concentrating the sunlight with a condensing lens. An artist could also paint from an image formed in this way. Van Stavoren himself often used it this way for portrait painting.

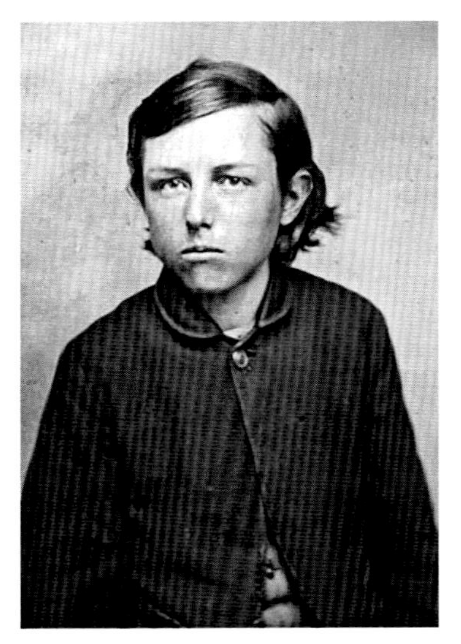

Fig. 7.3. E.E. Barnard, at about the time he started work at John van Stavoren's Photograph Gallery in Nashville. Note the ulcerated area on his right jaw. *Courtesy: Yerkes Observatory.*

Van Stavoren needed an assistant to keep the camera moving precisely with the Sun by turning a pair of hand wheels. One wheel turned the camera west, the other tilted it up or down. Because of the camera's size, a small boy could reach the hand wheels only by standing on top of a stepladder. It was tedious work. A number of boys tried in the position had nearly burned down the house falling asleep in the warm sunshine. Van Stavoren asked Elizabeth: "Will your boy keep awake?" She responded without hesitation, "My son will not go to sleep!" Barnard, who was then nine years old, was duly hired, and he never did go to sleep while on duty. As he wrote long afterwards, "Through summer's heat and winter's cold I stood upon the roof of that house and kept the great instrument directed to the Sun. It was sleepy work and required great patience and endurance for one so young, and at this distant day I realize that this training doubtless developed those qualities—patience, care and endurance so necessary to an astronomer's success."[14]

Fig. 7.4. The Jupiter Camera on the roof of Van Stavoren's Photograph Gallery. *Courtesy: Yerkes Observatory.*

Barnard's association with the Photograph Gallery would be long and productive; he remained there as an employee until he was 25. As a result, we have more portraits of him, in childhood and as a young man, than of any 19th-century astronomer. He was far more often photographed than even Abraham Lincoln.

His interest in astronomy, awakened by the comet and by watching the stars from his wagon bed, received a further stimulus with the eclipse of the Sun on August 7, 1869. It was almost total from Nashville (the path of totality ran from Alaska to North Carolina), and Barnard noted that "the Sun was so nearly hidden that the spectacle presented some of the awe and sublimity of the total phase," and increased his "wonder at the phenomena of nature."[15] He was then living in "Varmint Town," a rough part of South Nashville, putting in long hours at the Photograph Gallery. It was often dark when he walked home (remember, those were days long before electric lights). "Away back yonder," he afterwards recalled in a letter to the assistant postmaster Joseph S. Carels,

> … when I was small and ragged and sick and desolate just after the close of the war, when even those who had rolled in wealth but a few years before were struggling for subsistence, and few there were who could bestow even a kind word—so terrible had been the desolation and its effect on the people—in these times when I used to trudge home some two miles every night from work, timid and frightened, I frequently would meet a gentleman who always had a nod and a smile for me—in bad or cold weather he always wore a cloak. Sometimes he would stop me and ask how I was getting on but he never passed me without a recognition. I did not know who that man was, but his smile lighted up my heart, for years he never failed to greet me. Soon I learned to my awe that he was assistant Postmaster! Had he been President [of the United States] his position would not have appeared higher and more exalted to me—and that he should notice me and should stop to speak to me—I could not understand it, and I cannot understand it to this day—unless it was indeed an inborn desire in him to sympathise with the friendless and wretched for friendless and wretched I was in those days if any one was ever friendless and wretched.[16]

During this sad time Barnard had a sore on his face that refused to heal and afterwards left a prominent scar on his right jaw. Whenever Carels stopped Barnard, he would always ask how that sore was coming along. To Barnard's niece Mary Calvert, this episode always "seemed eloquent of the loneliness of his life in those days."[17]

Apart from Carels, the other friends Barnard remembered from this lonely period were the stars. "I often noticed in my long walks homeward in the early night an ordinary yellowish star which, to my surprise, seemed to be moving eastward among the others stars":

> This attracted my attention because in all the time I had noticed the stars, though they came and went with the seasons, they seemed all to keep to their same relative positions. This one must be quite different from the others though it resembled them.... I watched it night after night and saw that though it moved eastward with reference to the other stars, it also partook of their general drift westward and was finally lost with them in the rays of the sun. In later years when I had become more familiar with astronomy it occurred to me to look up this moving star, and I found then that what had attracted my boyish attention was the wonderful ringed world of Saturn.[18]

So far Barnard's interest in astronomy was of a completely uninformed kind. Though a sense of wonder and awe had been often awakened by celestial phenomena such as comets, eclipses, and wandering planets, like William Herschel he was anything but precocious in his formal study of astronomy. That began only at the rather advanced age of 19! By then he was no longer living in "Varmint Town" but had taken a room in the top story of the Hotel St. Charles, in downtown Nashville, which was more convenient to the Photograph Gallery. One night a young man came to his room:

> We had been children together, but he was a born thief. As a boy he stole, and when he got older the law often laid its hands upon him. On several occasions I had helped him out from my meager earnings for my sympathies were easily worked upon. At one time I had paid his fine when a policeman brought him around where I was at work. On this night in particular I was in no mood to be gracious; for he had come to borrow money from me, which I knew from previous experience would never come back. As security for the return of the money he had brought a large book. This I refused to look at; and finally, to get rid of him, I gave him two dollars (which was the amount asked for). I never saw him or the money again. Shortly after he had gone I noticed he had left the book lying upon my table. I felt very angry, because the money was a large amount to me then, and it was some time before I would open the book.[19]

The book contained the *Works* of the Scots writer, the Reverend Thomas Dick, known as the "Christian philosopher of Dundee," who tried to blend fundamentalist Christianity with astronomy and who had included a set of rudimentary star charts in one of his books. Soon Barnard set out to compare these charts with the patch of sky visible from the open window of his small apartment:

> In less than an hour I had learned the names of a number of my old friends; for there was Vega and the stars in the Cross of Cygnus and Altair and others that I had known from childhood. This was my first intelligent glimpse into astronomy. It is to be hoped that my sins may be forgiven me for never having sought out the rightful owner of that book in all these years.[20]

By now, Barnard was finding new mentors. After Van Stavoren went bankrupt, the Photograph Gallery passed into the hands of one Rodney Poole. Barnard had long since graduated from guiding "Jupiter." Having inherited some of his mother's artistic talent, he was employed for a time as the Gallery's artist until Poole hired Peter Ross Calvert, a Yorkshire immigrant who had studied art in London. Calvert took over as Gallery's artist, and Barnard became chief assistant to James Braid, Poole's chief photographer and later a prominent electrician in Nashville.

One day Braid found a one-inch spyglass lens on the street leading to the Union barracks and helped Barnard assemble it into a small telescope; he wrote of it as "a paper tube and lenses that looked as If they had been chipped out of a tumbler by an Indian in the days of the Mound Builders. Yet it filled my soul with enthusiasm when I detected the larger lunar mountains and craters, and caught a glimpse of one of the moons of Jupiter."[21] Eventually Braid helped him to get a better telescope, with a 2 1/4-inch lens, and Barnard began using it in the spring of 1876, when the brilliant planet Venus was a exhibiting a crescent phase as it dropped toward the Sun in the evening sky. His telescopic sighting of this phase "made a more profound and pleasing impression," he later recalled, than his celebrated discovery of the fifth moon of Jupiter. Henceforth he lived heart and soul for astronomy.

Braid learned from relatives in New York City that John Byrne, a skillful optician, was offering first-rate refracting telescopes for sale. Now Barnard began saving from his slender salary enough money to acquire a 5-inch refractor. Byrne listed it for $550 but was willing to part with it (as a favor to Braid) for only $380—still two-thirds of Barnard's annual salary at the time.

Barnard had hardly inaugurated his work with the new telescope when members of the American Association for the Advancement of Science, the premiere scientific organization in the U.S., gathered at the State Capitol in Nashville for their annual meeting. The most anticipated speaker was O.C. Marsh, paleontologist of Yale, who had just pieced together an extinct land beast, *Titanosaurus montanus,* found in the sandstone hog-backs of Morrison, Colorado (near Denver), but the meeting was presided over by an astronomer, Simon Newcomb, president-elect of the Association and so influential he has been called "America's Astronomer Royal." Barnard's friends were able to arrange a meeting with Newcomb. Though he must have petrified with fright, Barnard managed to ask the great man how someone with a small telescope might make himself useful in astronomy. Newcomb made the usual suggestions—searching for comets and nebulae. Then suddenly he changed tack and asked Barnard if he was acquainted with mathematics. He was not. Newcomb then advised him, "Lay aside at once that telescope and master mathematics, for you will never be what you seek to become without this mastery." Barnard was so depressed at this that he got behind one of the columns of the State Capitol and had a good cry. Though Barnard did his best to follow Newcomb's stern admonition (and hired a mathematics tutor, possibly the very next day), he eventually decided that Newcomb's injunction was too severe, and decided on a compromise—he would devote all clear and moonless nights to working with the telescope, the rest to the study of mathematics. Despite his efforts in mathematics, he never advanced very far in that line. As an observer, however, he was a natural, and possessed perhaps the greatest talent since William Herschel.

He spent his nights observing Jupiter, whose Great Red Spot was then becoming prominent, and searching for comets. The discovery of a new comet was then rewarded with a handsome monetary prize offered by a New York patent-medicine (i.e., snake-oil) manufacturer, H. H. Warner. Barnard, frankly, needed the money; he had his invalid mother to care for, and he was also courting Peter Calvert's sister Rhoda. She was seven years his senior, but they were already very devoted to each other, and hoped to marry as soon as they could afford it. Barnard's enthusiasm and diligence were soon rewarded; he discovered his first comet in 1881, and between then and 1887 discovered a total of 10. By then he was, along with Louis Pons, a one-time doorman of the Marseilles Observatory, Messier, and Méchain among the greatest visual comet discoverers of all time. Money earned through the Warner prizes helped him to acquire a plot of land and to build his own cottage, "Comet House," where he lived with his wife and mother. He was a local hero now, and received a fellowship to the recently opened Vanderbilt University (funded by the uneducated transportation genius "Commodore" Vanderbilt) where he took charge of the college observatory's 6-inch refractor—though he continued, when comet-seeking, to use his own 5-inch refractor.

Comet-seeking was laborious, routine work, not without its share of drudgery. Nowadays, the era of visual comet discovery is all

Fig. 7.5. E.L. Trouvelot's evocative rendering of the Great Comet of 1881 (Tebbutt's Comet), the brightest of the comets of the year that saw E. E. Barnard's first discovery with the 5-inch Byrne refractor at Nashville. *From: Wikipedia Commons.*

but over, and most discoveries are made by automated surveys, such as Project LINEAR. But for Barnard, at least, "sweet were the uses of adversity." He learned a great deal as he swept the sky for comets. "To me," he later recalled, "the views of the Galaxy were the most fascinating part of comet-seeking, and more than paid me for the many nights of unsuccessful work." As Messier had done, he came across new nebulae in the course of searching for comets. Among his notable finds were the dwarf galaxy N.G.C. 6822, now known as Barnard's galaxy, in Sagittarius, N.G.C. 1499 (the "California" nebula) in Perseus, and N.G.C. 281 (the "Pacman" nebula; obviously not Barnard's name for it!) in Cassiopeia.

Barnard's fellowship allowed him experience in practical astronomy without leading to a degree, and in 1887 he received an invitation to join the staff of the Lick Observatory, then under construction on 4200-foot high Mt. Hamilton, near San Jose, California. The Lick Observatory marked the beginning of a new era in astronomy. Hitherto, professional astronomers and their assistants at the great national observatories like those of Paris and Greenwich and Washington had largely been occupied with routine work such as measuring the positions of the Moon and planets relative to the background stars, in the service of time-keeping and navigation. The work of discovery was largely the province of the "Grand Amateurs," like Rosse, Lassell, De la Rue, Huggins, Draper, men who possessed independent means and were able to equip themselves with the best equipment and pursue whatever investigation they wished. The Lick Observatory, on the other hand, was built through the philanthropic endowment of James Lick, who had made fortune in real estate speculation in California during the Gold Rush, and who was, therefore, even richer than the Grand Amateurs.

Fig. 7.6. An invitation to the night-sky. The Moon appears with Venus and Jupiter in the evening sky, as the shutter of the dome of the Lick Observatory's 36-inch refractor opens. This telescope was used by E. E. Barnard to make many spectacular discoveries, including the fifth satellite of Jupiter, which he discovered visually in September 1892. *Copyrighted image by Laurie Hatch, used with permission.*

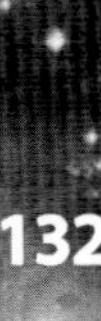

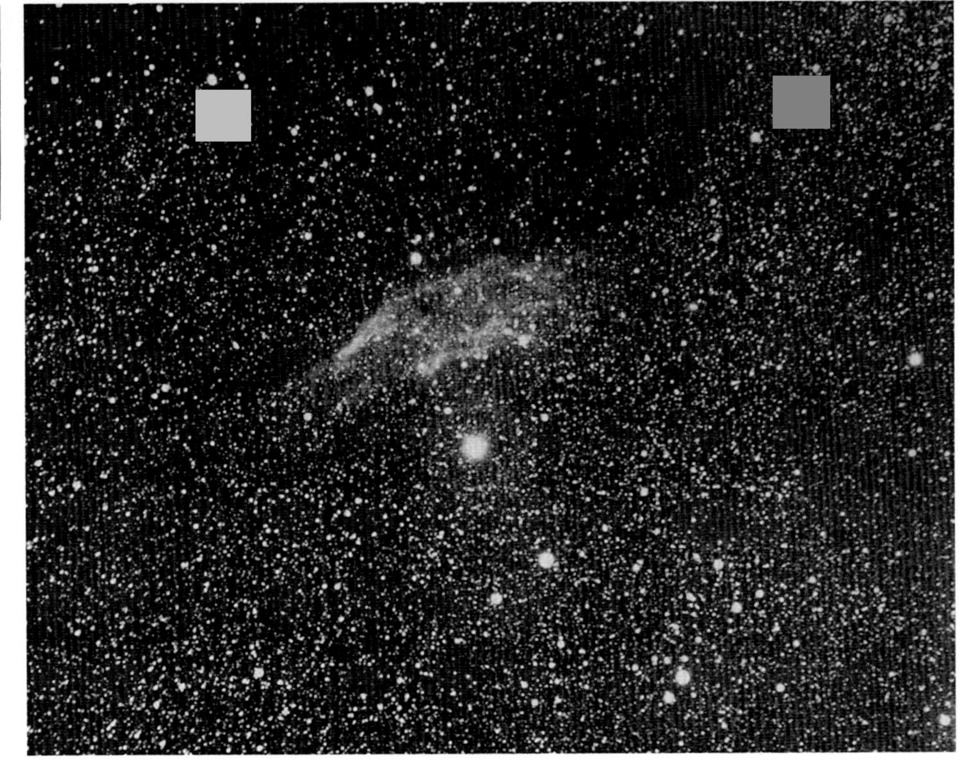

Fig. 7.7. California Nebula (NGC 1499), an HII region in the Perseus Arm, discovered by E. E. Barnard at Vanderbilt University Observatory in Nashville, in 1884. This image was obtained by Barnard with the 6-inch "Willard" portrait lens at Lick Observatory. *Courtesy: Yerkes Observatory.*

Having no heir (he refused to the end to legally recognize a natural son from a disappointed love affair), he began to cast around for a way to dispose of his fortune. After a couple of miscues—for a while he favored a plan to set up colossal figures of himself and his parents bestriding San Francisco Bay, and next considered building a pyramid larger than that of Cheops on the corner of Fourth and Market Streets—he was finally persuaded that it would be more useful to use the money to build the largest and most powerful telescope in the world. In the end a 36-inch refractor was commissioned from the Clark firm, and it was decided—at a time when most observatories were located at sea-level and in cities—to set it up on a mountain-top, where it could "see" rather than be "seen."

Lick died in 1876, but the observatory was not finished until 1888. Meanwhile, a staff of astronomers had to be selected. The director was to be West Point trained Edward Singleton Holden, a former astronomer at the U.S. Naval Observatory. Other staff members included James Keeler, a superb spectroscopist, Sherburne Wesley Burnham, by profession a Court Reporter from Chicago who had made a name for himself as the discoverer of new double stars, John Martin Schaeberle, a professor of astronomy at the University of Michigan, and Barnard, who was chosen expressly for his proficiency in discovering comets.

Barnard, however, was not interested exclusively in comets. He proved to be an all-around observer, interested in every conceivable class of object in the heavens—planets, comets, double stars, star clusters, nebulae.

His greatest preoccupation was with the Milky Way, which was first fostered during his comet-sweeps in Nashville if not even earlier during his

Fig. 7.8. Barnard, setting up to photograph with the Bruce photographic telescope at Mount Wilson in 1905. *Courtesy: Yerkes Observatory.*

childhood vigils from the old wagon-bed. In August 1889, he began drawing on his long experience in photography by strapping a wide-angle portrait lens (a six-inch "Willard" lens, obtained for a pittance to take photographs of the total solar eclipse visible in California on New Year's Day, 1889) onto an equatorially mounted telescope, and meticulously guiding it for three hours on the star clouds of Sagittarius. It was the first of scores of plates he obtained between then and 1895, when he left Mt Hamilton, largely owing to difficulties with Holden whom he regarded as an insufferable autocrat; he was not alone—there was a general exodus of astronomers from the mountain, beginning with Keeler in 1891. Among the terms they used to describe their director were "the Devil," "the Czar," "an unmitigated blackguard," "the Dictator," "Prince Holden," "the great I am," "the Great Mahatma," "that humbug," ""that contemptible brute," "that immoral and incompetent man," and "our former colleague and fake."[22]

But before Barnard left, he had accomplished some of his greatest work. The plates obtained with the humble Willard lens were scientifically rich, and showed a large number of mysterious and hitherto unappreciated dark features. Barnard described the area of sky shown in a plate taken in 1894 to the English barrister and amateur astronomer Arthur Ranyard: "It is essentially a region of vacancies. There is a great chasm here in the Milky Way."[23] Another plate centered on the star Rho Ophiuchi, the very place where William Herschel had exclaimed to his sister Caroline, "Here surely is a hole in the heavens." For a long time Barnard lacked the confidence to challenge the conventional interpretation of these dark markings as holes. But Ranyard offered a fresh perspective: "The dark vacant areas or channels ... seem to me to be undoubtedly dark structures, or absorbing masses in space, which cut out the light from a nebulous or stellar region beyond them."[24] Barnard did not yet agree, and could not bring himself to relinquish the old idea that they were true vacancies in the sky.

Nevertheless, the issue continued to nag at him. After leaving Lick, Barnard took a job at the University of Chicago's Yerkes Observatory. Named for Charles Tyson Yerkes, a Chicago elevated train tycoon who provided some of the funds, the principal instrument was a 40-inch refractor, with which the observatory's founder and first director, George Ellery Hale, wanted to study the spectra of the Sun and stars. Barnard used the great refractor whenever he could, and also began casting around for a new photographic telescope that would be better suited for his Milky Way photography than the one built around the primitive Willard lens. Eventually Catherine Bruce, a spinster heiress, whose father was the inventor of a process for molding metal type, agreed to provide funds for such a telescope. The 10-inch Bruce Photographic Telescope was finished by 1904, and mounted under a tin dome on the grounds at Yerkes on the path between the main building and Barnard's house on the shore of Lake Geneva.

Hale, not yet 30 when the Yerkes Observatory was finished, was a restless soul (see chapter 10). Even a 40-inch refractor could not satisfy him, and he began planning for an even larger telescope in a better climate. Mt. Wilson, a 5,886-foot peak near Pasadena, was known to be an excellent site for an observatory, and Hale, receiving funding from the Carnegie Institution (founded by the steel magnate and philanthropist Andrew Carnegie), resigned from Yerkes and began building the Mt. Wilson Solar Observatory there. The observatory was not very much developed when in 1904 Barnard received permission from Edwin B. Frost, who took over as the director of Yerkes after Hale's departure, to send the Bruce telescope to Mt. Wilson to photograph the Milky Way from its clear dark skies. An added advantage of the Mt. Wilson site was that it was ten degrees farther south than Yerkes.

On January 10, 1905, Barnard, pushing 50 and gaining weight, led a mule saddled with the 10-inch lenses on a five-hour trek up what was little more than a footpath, from Pasadena to the "Monastery," the astronomers' residence on the mountain. A horse-drawn wagon hauled the rest of the telescope to the summit. As soon as he reached the summit Barnard, beside himself with excitement, worked relentlessly, seldom bothering to sleep. In only two weeks he had the telescope up and running, and was already exposing the first plates to the skies.

During the winter of 1905, Mt. Wilson was still a primitive wilderness, and Barnard was often the only human on the mountain (his wife Rhoda had remained with relatives back East). He later recalled:

> I must confess that at times, especially in the winter months, the loneliness of the night became oppressive, and the dead silence, broken only by the ghastly cry of some stray owl winging its way over the canyon, produced an uncanny terror in me, and I could not avoid the dread feeling that I might be prey at any moment to a roving mountain lion.... So

> lonely was I at first that when I entered the Bruce house and shoved the roof back I locked the door and did not open it again until I was forced to go out.[25]

While guiding exposures, Barnard usually let his legs dangle through a trap door in the floor of this building. One night he heard rustling noises. On investigating, he discovered that a rattlesnake had been making its home there. "It must have been a friendly snake and was there for the purpose of warming the observer's feet," he quipped.[26] Friendly or not, the snake was killed; the next night the trap door was open as usual.

There was no running water on the mountain, so the water used to develop his plates had to be carried up the peak by a burro named Pinto. Barnard became especially fond of Pinto when he discovered that a burro's hair is finer than a human's, which made it eminently suitable for making the crosshairs of a guiding telescope.

Barnard maintained an inhuman work pace on Mt. Wilson. He was a tireless observer — an "observaholic" — and long since has been known as the man "never known to sleep"—who night after night, with superhuman patience, guided the Bruce Telescope on the magnificent star-clouds and dark markings of the Milky Way. As the French writer Balzac had done, he kept himself going by consuming enormous quantities of coffee. As spring came on, the crickets and other insects began their nightly serenades, and Barnard felt less alone on the mountain. Also, one by one, other astronomers began taking up residence there. One of them, Walter Sydney Adams, noted, "Barnard's hours of work would have horrified any medical man."[27]

During eight months at Mt. Wilson, Barnard exposed 500 Milky Way plates. The best of them were included in the *Atlas of Selected Regions of the Milky Way,* and have never been surpassed.

Their most remarkable features were what Barnard called the "dark markings." Some — dark lanes — ran through Taurus and Perseus. Others straddled Sagittarius. Still others, east of Rho Ophiuchi, "shattered into fragments." The "wild region" north of Theta Ophiuchi was "fuller of strange and curious things than any other … with which my photographs have made me familiar." Barnard's descriptions are wonderfully evocative and romantic. One dark area, he thought, looked like a lizard; another like a parrot's head; yet another resembled "a beast, with round head, nose, mouth, ears, and great staring eyes."[28]

The appearance of these dark markings strongly supported Ranyard's contention that light-absorbing material existed in interstellar space. Of the photograph of the region north of Theta Ophiuchi, for instance, Barnard wrote: "that most of these dark markings … are real dark bodies and not open space can scarcely be questioned. There seems to be every

Fig. 7.9. (left) Rho Ophiuchi. Plate 13 from *The Photographic Atlas of Selected Regions of the Milky Way.* "The region of Rho Ophiuchi is one of the most extraordinary in the sky. The nebula itself is a beautiful object. With its outlying connections and the dark spot in which it is placed and the vacant lanes running to the east of it, it makes a picture almost unequaled in interest in the entire heavens." *Courtesy: Yerkes Observatory.*

Fig. 7.10. (right) Plate 20 from *The Photographic Atlas of Selected Regions of the Milky Way:* "This picture was especially made to show the remarkable blank space east and south of Theta Ophiuchi... Remarkable as is the region south and east of Theta Ophiuchi, the most curious of the dark markings lie to the north of it... The S-shaped figure, B 72, just above the center of the picture, is much darker than the great dark region southeast of Theta... The curled figure, B 75, is, perhaps, as strange an object as can be found anywhere in the sky... Certainly this is one of the most surprising and curious regions in the sky." *Courtesy: Yerkes Observatory.*

evidence of their reality. What their true nature is does not seem clear. That they are some form of nebulosity is possible if not probable; but that they are real obscuring bodies is evident."[29] A photograph Barnard took in January 1907 – by then the Bruce was back in its tin dome at Yerkes — of a region in Taurus, in the Northern Milky Way, showed, he wrote, "curious dark lanes" whose feeble light covered "an abrupt, irregular, vacant hole in the dense background of small stars. But there is every evidence that this is not a hole in the stratum of stars, for over its entire extent is a feeble veil of nebulosity with small, round dark spots in it." Though edging up to the

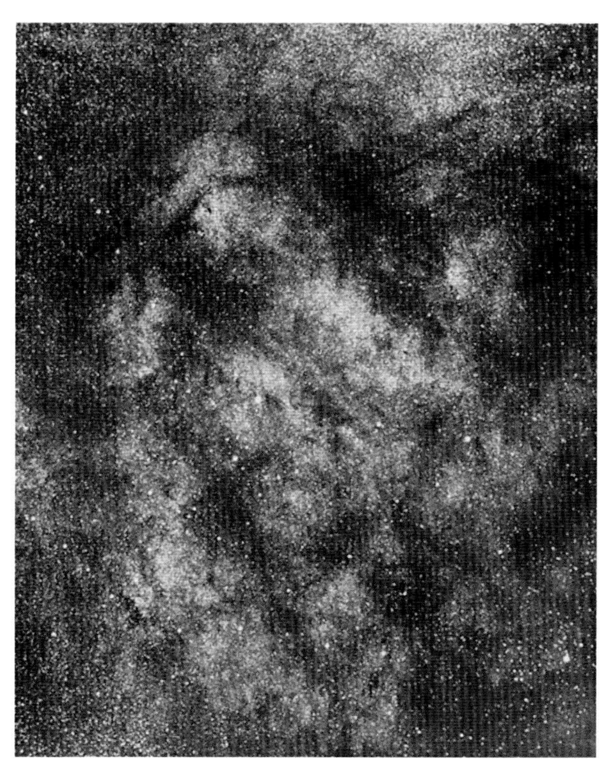

Fig. 7.11. (left) Plate 23 from *The Photographic Atlas of Selected Regions of the Milky Way*. Region of 58 Ophiuchi. *Courtesy: Yerkes Observatory.*

Fig. 7.12. (right) Plate 31 from *The Photographic Atlas of Selected Regions of the Milky Way*. Sagittarius Star Cloud. "This splendid star cloud, with its dark spots and lanes, with its straight and curved lines of stars, is well shown on this plate. The region to the northwest of it apparently consists of a relatively thin stratum of stars." *Courtesy: Yerkes Observatory.*

recognition of some kind of interstellar matter, Barnard was still not able to commit himself—quite. He was still thinking that some of the dark nebulae were burned out examples of luminous nebulae, of similar form, that appeared in the sky.

His moment of epiphany did not finally come until a "beautiful transparent moonless night" in the summer of 1913, when the brilliant star clouds of Sagittarius were in view. He had often wondered at the brilliance of those star clouds when, on heavily clouded moonless nights, small breaks appeared in the clouds allowing the Milky Way to be visualized beyond. On this particular night he was treated to the reverse spectacle, as a few small cumulus clouds, each a round globose about the size of the Moon, stood in projection against the great star clouds. Barnard recalled:

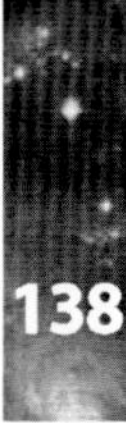

Fig. 7.13. (left) Plate 5 from The Photographic Atlas of Selected Regions of the Milky Way. Of this plate, Barnard wrote: "Very few regions are so remarkable as this one. Indeed this photograph is one of the most important of the collection, and bears the strongest proof of the existence of obscuring matter in space." *Courtesy: Yerkes Observatory.*

Fig. 7.14a. (right) Plate 40 from The Photographic Atlas of Selected Regions of the Milky Way. In Aquila, northeast of the star cloud in Scutum. "The curved object, looks like a great, black lizard crawling south. Its body is curved toward the west, and its head is a sharp, dark projection, B 137." Barnard thought this perhaps a burnt-out nebula of the type that is shown in Fig. 7.14b; but in fact, the resemblance is entirely superficial. *Courtesy: Yerkes Observatory.*

Fig. 7.14b. (below) The Veil Nebula in Cygnus (NGC 6960, the western veil which passes through the foreground star 52 Cygni; it was discovered by William Herschel in 1784), and the Eastern Veil (NGC 6992), and various other filaments and knots in the area, are remnants of a supernova explosion that occurred 5,000 to 8,000 years ago. The morphological resemblance Barnard saw in this to his "lizard" is entirely accidental, as in fact the two objects are of entirely different type. Imaged by Klaus Brasch with a TeleVue 101 apochromat at f/5/5. *Courtesy: Klaus Brasch.*

I was struck with the presence of a group of tiny cumulus clouds scattered over the rich star-clouds of Sagittarius. They were remarkable for their smallness and definite outlines.... Against the bright background they appeared as conspicuous and black as drops of ink. They were in every way like the black spots shown on photographs of the Milky Way.... The phenomenon was impressive and full of suggestion. One could not resist the impression that many of the small spots in the Milky Way are due to a cause similar to that of the small black clouds mentioned above — that is, to more or less opaque masses between us and the Milky Way.[30]

In later years Barnard suffered from diabetes. Because of declining health, he was expressly forbidden by Frost from working on the 40-inch refractor for the whole year of 1914-1915. He no longer had his erstwhile stamina, but he soldiered on as best he could, and among his most important achievements was completion of his famous catalog of dark nebulae. It is to that class of obscuring objects what Messier's had been for the shining ones. The first list of Barnard objects, carrying the designation B, was published in 1919, and contains 182 objects; more would be added later, bringing the total to 349.

Fig. 7.15. Barnard and Frost, on the gallery of the 40-inch refractor looking out toward Lake Geneva. Authors Sheehan and Conselice stood in the very spot in 1993, after observing an eclipse of the satellite Iapetus by the A-ring of Saturn. *Courtesy: Yerkes Observatory.*

Though most of the Bruce plates had been taken on Mt. Wilson in 1905, and their publication was assured by a grant from the Carnegie Institution in 1907, Barnard did not get around to publishing the work he originally planned to call *An Atlas of the Milky Way* for many years. There were several reasons for the delay. First, as Frost pointed out, his "well-known eagerness to observe the heavens whenever the sky was clear left him little time for the remainder of the preparation of the work for publication."[31] But he was also a perfectionist, and could not find means of reproducing them that were up to his high standards. (Despite time-consuming experiments, neither the collotype nor the photogravure met his approval). At last he decided on nothing less than an edition of actual photographic prints. He had to write up descriptions of the photographs and personally prepare, inspect, and pass each individual print —35,700 in all. After Rhoda, his wife, died in 1922, he rapidly faded, and never lived to see the publication of his masterpiece. He died in February 1923. The work was finished after his death by Frost and Barnard's niece, Mary Calvert, who utilized not only the master's drafts but many scattered notes on which he had indicated his intentions. Though originally to appear under the title *An Atlas of the Milky Way,* Barnard abandoned that grandiose title for *An Atlas of Selected Regions of the Milky Way.* It appeared in 1927, just as prohibition-era Chicago was in the midst of gang wars; a truck carrying copies of the *Atlas* was caught in the crossfire — one copy, riddled with a bullet, was long on display at the observatory.

The *Atlas* is today a collector's item, one of the most highly sought-after books in astronomy. No reprint, no matter how carefully prepared, can do it justice.[32] How could it? Only the original volumes carry the Master's seal.

Though Barnard's plates were taken over a century ago, and technology has continued its inevitable march since, they remain in a class by themselves. They are works of art as much as of science. The black-and-white medium Barnard used — even the aberration of the Bruce telescope that softened and rounded up the stars — bring out in inimitable fashion the "dark Monsters of Immensity" with which he first communed on wild and lonely nights on Mt. Wilson.

After his night of epiphany in 1913, Barnard was never again to doubt the existence of vast clouds of interstellar matter in space, though only later did astronomers fully appreciate their significance. By 1930, it was apparent that dust exists not only in the thicker clouds Barnard photographed but also as an exiguous veil permeating the entire Galaxy. Thus the light of all the stars is reddened to some extent by the intervening dust, causing astronomers to believe them fainter, and thus more distant, than they really are.

Some of this dust has been shot through interstellar space by the violent supernova explosions of massive stars, but most of it comes from more ordinary stars like the Sun which, during their expiring phases, blow silicate and carbon dust grains into the galactic medium, where it accumulates into dusty cocoons. Sheltered from ultraviolet radiation coursing through the Galaxy, complicated molecules, such as hydroxyl, water, carbon monoxide, formaldehyde, are able to form as ices around the grains, and within these cocoons Molecular Clouds begin to form. These interstellar beasts are distributed patchily along the galactic plane, especially along the inner edges of the Galaxy's great spiral arms. We see their greatest concentration in the general direction of the galactic center in Sagittarius, which is

Fig. 7.16 (left) Horsehead Nebula (Barnard 33, NGC 2070), located just south of Alnitak, the easternmost star in Orion's belt. First recorded by Williamina Fleming in 1888, it is visible due to ionization by the nearby star Sigma Orionis (just out of the field in this image). *Image by Michael Conley and William Sheehan with an RC-10 telescope and an Orion Parsec imaging system.*

Fig. 7.17 (right) The Horsehead, as imaged in the infrared with the Wide Field Camera 3; in which it "rises like an apparition rising from whitecaps of interstellar foam." *Courtesy: NASA/ESA/Hubble Heritage Team (AURA/STScI).*

Fig. 7.18 (below) The Trifid Nebula (M20) in Sagittarius, so-called by John Herschel because of its three bands of interstellar dust, located some 9000 light years away. It is a region of star formation. The O-type stars illuminate the pillar of gas and dust to the right of this image; these young hot stars release a flood of ultraviolet radiation and erode the dust. *Courtesy: F. Ysef-Zadeh, NASA, ESA and the Hubble Heritage Team (AURA/STScI).*

located 27,000 light years away. At about half the Sun's distance between us and the galactic center circulatesa massive ring of molecular clouds, containing enough gas and dust to produce more than a billion suns; it looms in projection along the moth-eaten, mottled, and rifted outline of the Milky Way, and Barnard imaged many of its features in loving detail. The Orion Nebula is itself a centerpiece of such a giant molecular cloud.

The Milky Way is not unique. We see similar structures in the spiral arms of other galaxies. They are especially striking as the dark obscuring bands that bisect galaxies that, like the Milky Way itself, are viewed edgewise to our line of sight, such as NGC 891 in Andromeda, NGC 4565 in Coma Berenices, and M104 in Virgo. They are awesome specters of otherworldly beauty.

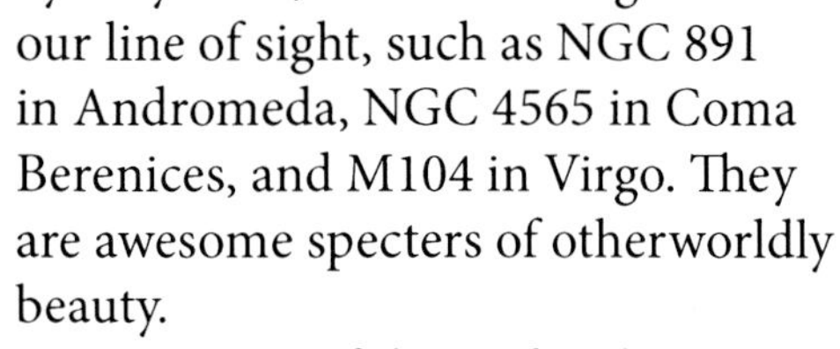

Some of the Molecular Clouds are rather small—like the Taurus Molecular Cloud, at a distance of 450 light years, whose tendril-like dark features reach eerily across one of Barnard's most famous photographs. Others, the Giant Molecular Clouds, are enormous, with breadths of 10 to 100 light years, and containing perhaps 100,000 times the mass of the Sun. The Molecular Clouds may remain quiescent for a long time until, owing to effects of general turbulence in the interstellar medium, the passage of a spiral density wave, a blast from a windy star or sudden concussion due to a nearby supernova explosion, they collapse, and give birth to stars. The smaller clouds, like the one in Taurus, are massive enough only to give birth to smaller stars, of one solar mass or less (and mostly dim and cool M-type stars with masses of only 0.1 to 0.6 that of the Sun). Only the Giant Molecular Clouds are massive enough to form giant stars, such as the brilliant blue supergiant Rigel in the foot of Orion

Fig. 7.19 Molecular clouds such as those shown here are known as Bok globules, for Bart J. Bok, who discussed them in the 1940s. They are on the order of hundreds of light years across. These belong to Pacman Nebula, NGC 281, in Cassiopeia, which was discovered by E. E. Barnard in 1883. He described it as a "large faint nebula, very diffuse." It is located some 9500 light years away in the Perseus Arm of the Milky Way. The globules are silhouetted against HII regions illuminated by the intense radiation of the hot young stars of the IC 1590 cluster. *Courtesy: Hubble Heritage Team (AURA/STSCI).*

Fig. 7.20. Milky Way: 3 x20mm f/3.5 stitched images, by John Drummond in Gisborne, New Zealand. Note the dark dust lane, made up molecular clouds and running the length of the Galaxy. These molecular clouds were captured in detail by E.E. Barnard in his wide-angle photographs. Smaller clouds, such as the nearby one in Taurus, are not large enough to give brith to stars of more than a few solar masses, but in giant molecular clouds (of which some 6,000 have been identified in carbon monoxide surveys across the Galaxy), more massive and luminous stars are able to form. Examples of these giant molecular clouds are found in Orion and Monoceros. *Courtesy: John Drummond.*

and θ^1 Orionis C, in the Trapezium. It is 40 times more massive than the Sun (only 0.00004% of stars are that massive), and its surface, at a temperature of 35,000 degrees K, produces intense winds and staggering amounts of ionizing UV radiation, which excites the gas of the Orion nebula to glowing. The sheer violence of the photoevaporative effects of the UV

Fig. 7.21. NGC 891, an edge-on spiral galaxy in Andromeda. Compare to Fig. 7.20. Image obtained with the Discovery Channel Telescope of Lowell Observatory. *Courtesy: Lowell Observatory.*

radiation and blustery stellar winds of this star produce strong headwinds for the proplyds, young stars preceded by shock waves, featuring bright irradiated heads, and tails pointing directly away from θ^1 Orionis C.

The intense ultraviolet radiation and ferocious stellar winds of these stars are hugely erosive, and produce some of the grandest scenery in the heavens, as in the moth-eaten, ragged, mangled remnants of cosmic dust clouds (IC 2944, Thackeray's globules) in Centaurus, which hang like shipwrecked galleons among wisps of interstellar gas, or the towering "hoodoos" or fairy chimneys such as the famous "Pillars of Creation" in the Eagle Nebula (M16) in Serpens. These, however, grandiose as they are, are only details. The broader cosmic landscape of which they are a part was first captured in the wide-angle photographs of E.E. Barnard taken a century ago, and we can only return to them again and again to reexperience the sheer "Ahhhh" of the scene.

* * *

1. Matsuo Basho, *Narrow Road to the Interior and other Writings*, Sam Hamill, trans. (Boston and London: Shambala, 2000), 29.
2. Lafcadio Hearn, *The Romance of the Milky Way, and other studies and stories* (Boston: Houghton Mifflin & Co., 1905).
3. Thomas Hardy, *Two on a Tower* (Oxford and New York: the World's Classics, Oxford University Press, 1993), 33-34.
4. Sir John Herschel, *Outlines of Astronomy*, 449.
5. Huggins, "New Astronomy," 913.
6. Holden, "Monograph of the Central Parts of the Nebula of Orion," 227.
7. The most complete biography of Barnard is William Sheehan, *The Immortal Fire Within: the life and work of Edward Emerson Barnard.* Cambridge: Cambridge University Press, 1995.
8. Sheehan, *Immortal Fire*, 3.
9. Ibid., 4.
10. Dacher Keltner and Jonathan Haidt, "Approaching awe, a moral, spiritual, and aesthetic emotion," *Cognition and Emotion*, 17(2) (2003), 297-314.
11. Sheehan, *Immortal Fire*, 4.
12. Bruce Catton, *The Civil War* (New York: The Fairfax Press, 1980), 260.
13. Barbara Tuchman, *The Proud Tower*, 119.
14. Sheehan, *Immortal Fire*, 5-7.
15. Ibid., 10.
16. Ibid.
17. Ibid., 11
18. Ibid.
19. Ibid., 12.
20. Ibid., 13.
21. Ibid., 15.
22. As summarized in Donald E. Osterbrock, "The Rise and Fall of Edward S. Holden: Part I," *Journal for the History of Astronomy*, *xv* (1984):81.
23. Sheehan, *Immortal Fire*, 274.
24. Ibid.
25. Ibid., 340-341.
26. Ibid., 343.
27. Ibid., 342.
28. E.E. Barnard, *Atlas of Selected Regions of the Milky Way* (Washington, D.C.: Carnegie Institution, 1927), description of Plate 26.
29. Ibid., description of Plate 19.
30. E.E. Barnard, "Some of the dark markings in the sky and what they suggest," *Astrophysical Journal*, *43* (1916), 1-8:4.
31. E. B. Frost, in ibid., vi.
32. On the other hand, the original is very scarce, and when a copy becomes available, it is extremely costly; one recently sold on e-bay for $12,500, which was actually a steal. A commendable effort is Gerald Orin Dobek, ed., *A Photographic Atlas of Selected Regions of the Milky Way* (Cambridge: Cambridge University Press, 2011), which has the added merit of presenting Barnard's original two volumes (plates in volume 1, diagrams showing the identification of significant objects in volume 2) within one cover.

8. What Stuff Stars Are Made Of

The fires that arch this dusky dot—
Yon myriad-worlded way—
The vast sun-clusters' gather'd blaze,
World-isles in lonely skies,
Whole heavens within themselves, amaze
Our brief humanities.

—*Alfred, Lord Tennyson, "Epilogue"*

Ad Aspera

As we have seen in previous chapters, the Milky Way is a kind of rag-and-bone shop of stars. It contains the gas and dust of molecular clouds, hot young stars forming within them and entering the main sequence, noble stars burning in their prime. But it also consists of the slag and debris of dying or already-dead and depleted stars—the relic exotica of white dwarfs, neutron stars, and black holes. The individual stars evolve; they have their personal histories.

Astronomers have learned a lot about these stars, but at first they had only two features to go on. The first is a star's apparent brightness, or magnitude. A stellar brightness scale was already being used by ancient Greek astronomers such as Hipparchus and Ptolemy, in which the brightest stars visible to the naked eye were called first magnitude, and the faintest sixth magnitude. William Herschel (again!) discovered that a first magnitude star was 100 times brighter than a sixth magnitude star, and this led to the adoption of a precise quantitative scale for magnitude (i.e., since five magnitudes correspond to a factor of 100 in brightness, one magnitude difference corresponds to a factor of $(100)^{1/5} = 2.512$). The other feature is the star's color. Some, such as Betelgeuse and Antares, are noticeably reddish, others, like Capella, yellow, still others, like Sirius and Vega, are white or blue-white.

Another feature that astronomers discovered about the stars is that, though they appear fixed in position over short time periods, in fact (as Edmond Halley discovered in 1718 by comparing modern measures of star positions with those recorded by Hipparchus and Ptolemy) at least some of them have detectable proper motions relative to the other stars. Large proper motions were a clue to a star's proximity. It was no accident that the first two stars to have their parallaxes successfully determined, Alpha Centauri (by Thomas Henderson) and 61 Cygni (by F.W. Bessel), have very large proper motions.

Having measured distances of a few stars was obviously a breakthrough of the greatest importance. If all the stars were of the same actual brightness, as Herschel originally assumed in plotting the cross-sections

Fig. 8.1. Stars within 20 light years of the Sun. As would be expected, most of these are dwarfs. For each star of solar mass, there are more than twenty stars smaller than the Sun—mostly M-type dwarfs. *Image: Wikipedia.*

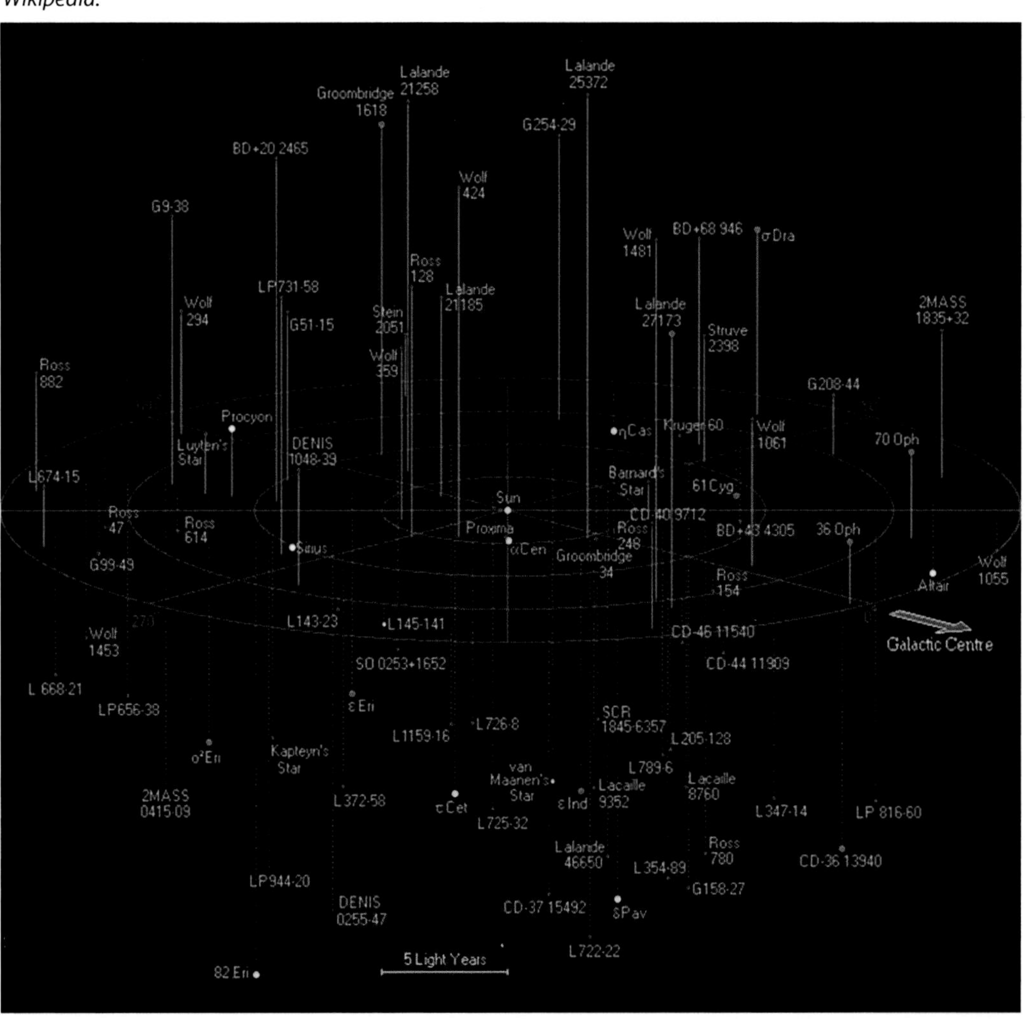

of the Galaxy, then the brighter ones would obviously be closer and the fainter ones farther away. Herschel himself did not believe that this could literally be true, but intuited that the stars might on average be about the same actual brightness as the Sun. Once astronomers had measured distances for stars, the actual brightnesses (or luminosities) could be worked out. A few stars proved to be much more luminous than the Sun. Among stars relatively close by, Sirius, which appears to be the brightest star in the sky, is 25 times more luminous than the Sun; Vega is 40 times more

Fig. 8.2. Stars within 250 light years. Now the brilliant young massive stars dominate the scene. Because of observer selection, they can be seen across great distances—they are truly the stellar beacons. Stars of stellar type A, which include Sirius, Vega, and Altair, comprise about 1% of stars; B stars, like those in the Pleiades and many of those outlining the Scorpion, about 0.1%; while the true stellar beacons—O stars, like Theta 1 Orionis C in the Trapezium, a star 40 times as massive as the Sun, with a surface temperature of 35,000 degrees, producing intense stellar winds and emitting staggering amounts of UV radiation, account for a mere 0.00004% of stars. *Image: Wikipedia*

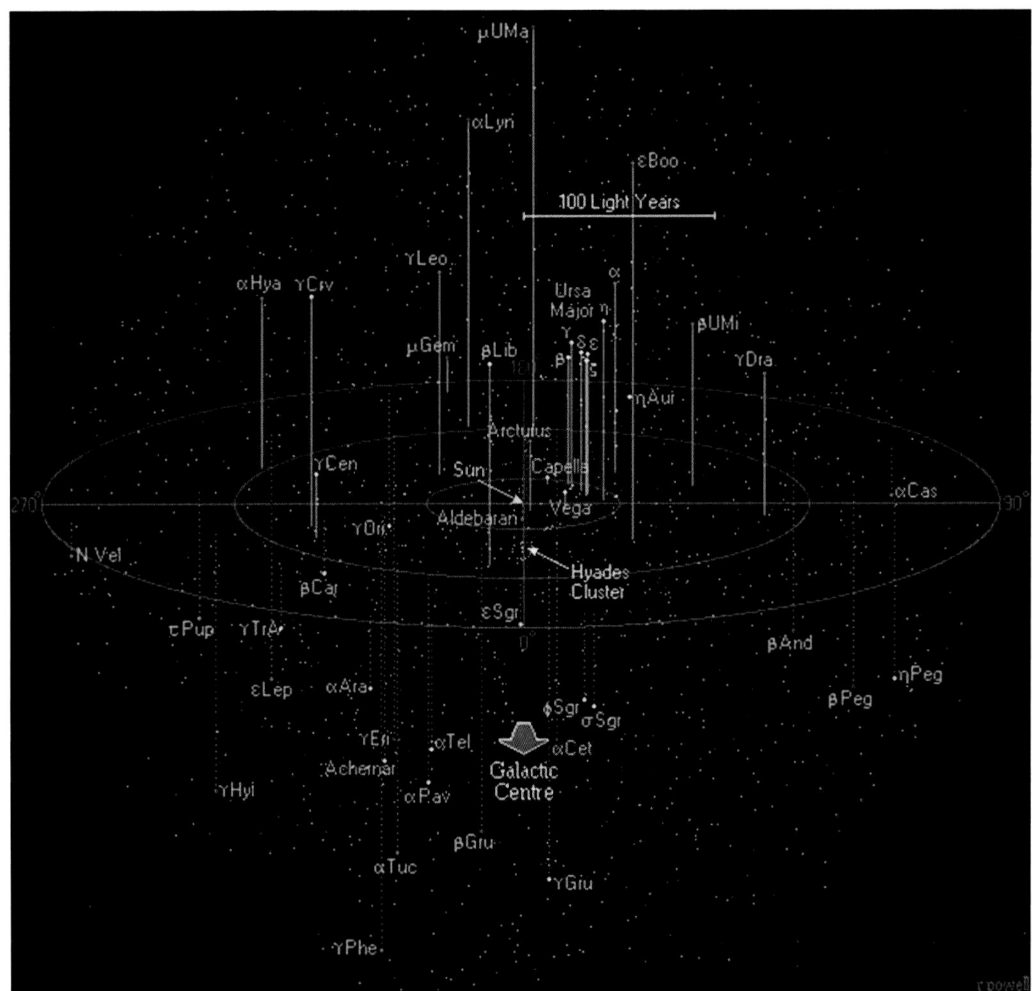

Fig. 8.3. Father Angelo Secchi, pioneer of stellar spectroscopy. *From: Wikipedia Commmons.*

luminous, Arcturus 170 times more luminous. But most of the nearby stars are stellar glow-worms, much feebler than the Sun; 61 Cygni, for instance, a double star, consists of one star with 15% of the luminosity of the Sun, and another with a luminosity of 8.5%; while fast-moving Barnard's Star has a luminosity of only 0.04% of the Sun's—so, despite its proximity to us, it is invisible without a telescope.

Yet another feature of stars of interest to astronomers is their masses. Half of all stars are members of binary systems, and as these stars revolve slowly around a common center-of-mass, they trace out an apparent relative orbit. The center-of-mass travels along a straight line relative to the background stars, and the binary components weave periodically around this line. By using Kepler's laws to work out the relative distance of each star from the center-of-mass, the masses of each of the components can be derived. In 1844, F.W. Bessel, analyzing the proper motion of Sirius, found that the bright component (Sirius A, with a mass of 2.2 times that of the Sun) was wobbling relative to the center-of-mass, and deduced that it was being pulled on by an unseen companion; this faint companion was discovered by Alvan Graham Clark in 1862, when he was testing the lens of the 18-inch refractor built for the Dearborn Observatory, and despite its faintness has a mass equal to that of the Sun. It was later identified as a peculiar degenerate star known as a white dwarf—of which we shall say more in the next chapter.

The Classification of Stellar Spectra

Though the measurement of distances, brightnesses, and masses are fundamental to stellar astronomy, and remain important branches of research right up to the present day, pioneering 19th century astrophysicists gave particular attention to the study of the spectra of stars.[1] When Fraunhofer first noted the dark lines in the spectrum of the Sun, he mapped only the 600 or so most prominent ones. There are actually many more. This spectrum, which as we now know is produced in the Sun's photosphere, obvi-

ously contains an enormous amount of information—if only we knew how to interpret it.

When pioneer spectroscopists first studied the spectra of the other stars, they were confronted with a "bewildering variety."[2] However, after sorting spectra for a while, it became clear that most specimens fell into relatively few classes.

The individual most strongly identified with this realization was an Italian priest-astronomer, Angelo Secchi. Born at Reggio Emilia (west of Bologna), he entered the Jesuit novitiate in Rome at the age of 15, taught physics at Loreto College before being recalled to Rome to study theology, and was ordained a priest in 1847, just before the year of "revolution," and the expulsion of the Jesuits from Rome (then in Austrian hands). He spent his exile at the Jesuits' Stonyhurst College in England and then in Georgetown, near Washington, D.C. Returning to Rome in 1849, a year later Pope Pius IX named him director of the observatory of the Collegio Romano, replacing Francesco de Vico, a well-known discoverer of comets, who had died in exile in England. Secchi upgraded the observatory's instrumentation, acquiring a 9.5-inch Merz refractor with the intention of measuring double stars; but he soon came against structural limitations owing to the observatory's location in the tower of the Jesuit church of San Ignacio which limited his ability to make the very precise measurements of stellar positions that were the mainstay of astronomy at the time. Malleable to circumstance, he turned in 1863 (at the same time as Lewis Rutherfurd in New York and Huggins at Tulse Hill) to investigating the stars with a spectroscope.

It has been said that "classification is one method, probably the simplest method, of introducing order in the world."[3] Secchi made the first sustained attempt to bring order into this field. From 1863, when he began his investigations, until his death in 1878, he carried out visual studies of the spectra of thousands of stars, and developed an influential scheme of spectral classifications. (His friend, Giovanni Schiaparelli at the Brera Observatory in Milan, exclaimed in March 1868: "The spectral classification ... is a very important undertaking! I see this as an opening to new horizons that nobody would have ever dreamed of. I do not know where [you] find the time...")[4]

Secchi's classification system included five groups:

I. White or blue-white stars (Sirius, Vega), whose main spectral features were a few strong absorption lines, attributed by Secchi to hydrogen.
II. Yellow stars (the Sun, Arcturus) with more numerous, narrow absorption lines. Hydrogen lines, though still visible, less intense.
III. Orange or red stars (Betelguese, alpha Herculis), with spectra having wide dark bands and maximum intensity at the red end of the spectrum.

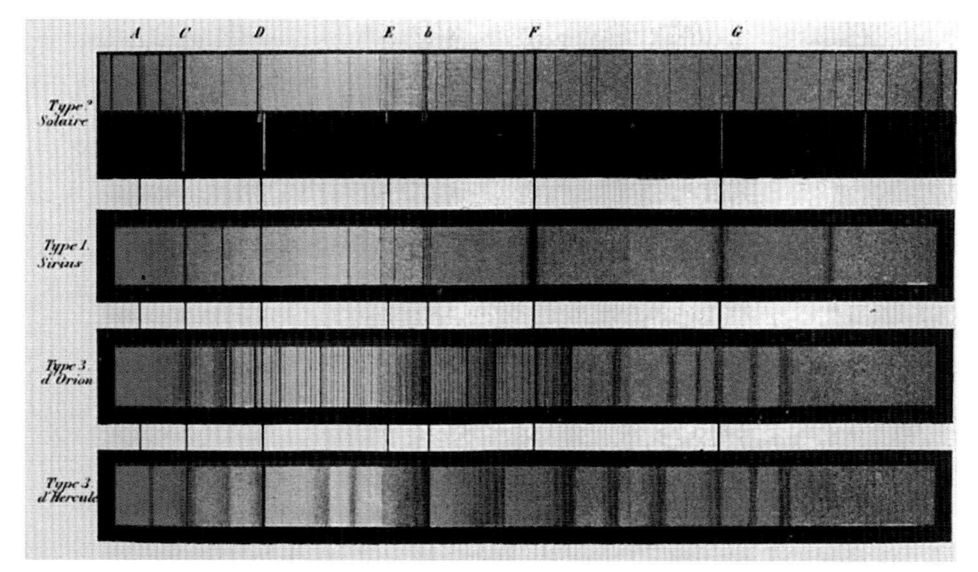

Fig. 8.4. Spectra of typical stars, according to Father Secchi. Included here are: a star of solar type; Sirius; Betelgeuse; and alpha Herculis. In modern terms, these would be classified as G1, A1, M1, and M5 stars. *From: Angelo Secchi, L'Etoiles, 1895.*

IV. Very red stars (19 Piscium), with dark bands but different from those of Type III. Secchi noted that these spectra were similar to the inverted (i.e., flame) spectrum of carbon; he had discovered carbon stars.

V. A rare type (e.g., gamma Cassiopeia) with bright emission lines.

The classification system of Secchi, published as *Le Stelle* (The Stars) in 1877, would remain the standard for fifty years. Unfortunately for Secchi, his later years were sad ones. After the Franco-Prussian war of 1870, Italian nationalist troops occupied Rome and the Vatican. Secchi remained loyal to Pius IX, who spent the last eight years until his death "a prisoner in the Vatican."

The United States: a factory observatory

Up to this point in the history of astronomy, the great breakthroughs had been largely those of the Grand Amateurs, men who made (or inherited) fortunes and who enjoyed independent means. They could, as Henry James put it, "to their hearts' content build their own castles and move by their own motors."[5] Meanwhile, those who toiled at professional observatories, such as the Royal Observatory at Greenwich and the Paris Observatory, were overwhelmingly concentrating on positional astronomy, and often assigned to drudge-like tasks, as human computers poring mindlessly over tables of logarithms. They were more like clerks and accountants than astronomers. So, for instance, when after the death of G.P. Bond in

1865 and the retirement ten years later of his successor, W.C. Winlock, a new director of the Harvard College Observatory was sought by president Charles Eliot, Charles S. Pierce, a sometime assistant at the observatory who was later to attain distinction as a philosopher, wrote to his friend William James:

> Of all situations I know of the one which has the most thankless, utterly mechanical drudgery, together with vexatious interference from … different sources [is] the Directorship of [the observatory].[6]

Simon Newcomb—the eminent astronomer who had sternly rebuffed the aspirations of young E. E. Barnard with his small telescope at the A.A.A.S. meeting in Nashville—was a professor of astronomy at the U.S. Naval Observatory in Washington, D.C. at the time. He was offered the Harvard position, but declined because, as he said in his memoirs, the observatory was

> poor in means, meager in instrumental outfit, and wanting in working assistants; I think the latter did not number more than three or four, with perhaps a few other temporary employees. There seemed little prospect of doing much.[7]

It was certainly poor in means compared to the U.S. Naval Observatory; but, as Bessie Zaban Jones and Lyle Gifford Boyd write in their history of the Harvard College Observatory, Newcomb "might correctly have said the same for nearly all the thirty-odd public and private observatories existing in the United States before 1880. None of them had enough financial support or a large enough staff to operate solely as a research institution. In European countries astronomy was usually supported by the state. In America, however, only the Naval Observatory was so fortunate; astronomy was generally regarded not as a science in its own right but chiefly as a tool to be used in solving practical problems—the determination of latitude and longitude, geographical boundaries, accurate time, and the peculiarities of weather and climate."[8]

After failing to interest an astronomer for the position, President Eliot chose, in 1876, a 30 year-old professor of physics at the recently founded Massachusetts Institute of Technology, Edward C. Pickering. Of course, Eliot was criticized for his unconventional choice, but having known Pickering personally, he was confident of his abilities and his temperament. In the end, Pickering proved a wise choice. The first American Pickering was a Yorkshireman who had settled in Salem, Massachusetts by 1636, and brought with him the family coat of arms (a lion rampant) and motto *Nil desperandum* (never despair). A great-grandfather, Timothy

Pickering, had served during the Revolutionary War as George Washington's quartermaster general, then as Washington's Indian commissioner, postmaster general, secretary of war, and the nation's third secretary of state under Washington and the second President John Adams.

E.C. Pickering, born on Vernon Street on Beacon Hill, attended the Boston Latin School (founded in 1635), the usual prep school for New England's elite; among the exercises he recalled was the memorization of long passages of works such as Xenophon's *Anabasis,* which had the result that he "studied little and learnt less." His interests lay in other directions. He read mathematics books on his own, and abandoned classics for a more congenial course of study at the Lawrence Scientific School at Harvard, where Eliot was an assistant professor of mathematics and chemistry at the school. Pickering received his Bachelor's degree in 1865 on his 19th birthday. With that he was done with his formal education.

After two years as instructor of mathematics at Lawrence Scientific School, Pickering went to M.I.T. (then known as Boston Tech), where he set up the first physical laboratory for student instruction in the United States, and published a two-volume lab manual.

Pickering and his wife Elizabeth, daughter of former Harvard president Jared Sparks, moved to Observatory Hill on February 1, 1877. At once the physicist-technician began nudging the observatory away from the "old" astronomy of measuring star positions toward the burgeoning field of astrophysics, and his characteristic philosophy quickly emerged. The time was not yet ripe for theory, he believed. Instead what was needed was the steady accumulation of astronomical facts. For example, the magnitudes of stars, among the most basic data of stellar astronomy, had not yet been measured accurately. Though the German astronomer Friedrich Argelander at the Bonn Observatory had in 1843 produced a catalog, *Uranimetria nova,* giving the brightnesses of 3250 stars estimated visually by subjective "steps" between standard stars, Pickering wanted something more quantitative. At first he tried an astrophotometer developed by another German astronomer, Johann Zöllner, in the 1860s, in which an observer compared the brightness of stars with the light from a kerosene lamp seen through a pinhole, but it proved difficult to maintain a constant enough light with a kerosene lamp. Finally, with the optician George B. Clark of the firm of Alvan Clark and Sons, he developed photometers that allowed, through suitable arrangements of mirrors and prisms, a direct comparison of stars with Polaris, which he chose as his standard star (he assigned it a magnitude of 2.1; unfortunately, his choice of Polaris was rather unfortunate, since it turns out to be a variable star, with a variation in brightness of 14% over a period of four days. Much later, Vega was chosen as the standard, with a magnitude of 0.0, and continues in the role today).

In the fall of 1879, Pickering and two assistants began a systematic program of remeasuring all the stars in the *Uranometria nova.* According to Jones and Boyd, "the work required physical endurance. Adjusting the images and reading the scales involved a certain amount of bodily contortion. Also ... during the winter months, when the temperatures dipped toward zero, and sometimes below, all the men suffered from the pain of the cold and consequent ailments of the throat and lungs." Despite the arduousness of the work, reminiscent of the harsh conditions under which George Bond worked to the bitter end at his drawing of the Great Nebula of Orion, with practice the measurements became routine, and Pickering and his colleagues were knocking off a star a minute like parts on a moving factory assembly-line. A list of magnitudes for 4,000 stars appeared in 1884, and another, for 45,000 stars, appeared in 1908. Pickering himself seems to have enjoyed making such measures—perhaps they were a form of relaxation for him from his job's other pressures, as the task was sufficiently repetitive that it required no substantial amount of mental concentration once it had been mastered. In any case, Pickering himself made something like 1.5 million photometric measures over a period of a quarter century!

Fig. 8.5. Edward C. Pickering, at the International Solar Union meeting at Mt. Wilson in 1910. *Courtesy: Yerkes Observatory.*

With the photometry program, Pickering had begun the retooling of the Harvard College Observatory as a "factory" observatory. He himself thought in those terms. "A great observatory," he wrote, "should be as carefully organized and administered as a railroad. Every expenditure should be watched, every real improvement introduced, advice from experts welcomed, and if good, followed, and every care taken to secure the greatest possible output for every dollar expenditure. A great savings may be effectuated by employing unskilled and therefore inexpensive labor, of course under careful supervision."[9]

As a member of a prominent Boston family, Pickering had outstanding connections, and was an effective fund-raiser for the observatory. Within a year of his publication of the results of his photometry project, he was in correspondence with Henry Draper's widow. Having in 1872 obtained the first spectrum of a star, Vega, to show definite spectral lines, Draper was at the time of his death planning much more extensive investigations using the new dry plates, and Mrs. Draper was eager to see them carried forward. At first she hoped to continue the work at her late husband's observatory at Dobb's Ferry, New York but realizing the impracticality of the idea, instead turned her late husband's instruments over to Harvard and provided a substantial grant of $1000 to establish the Henry Draper Fund for the study of stellar spectra. It was to be used by Pickering and his colleagues to photograph, measure, and classify the spectra of stars, leading to a catalogue of stellar spectra to be published as a memorial to Henry Draper in the *Annals of the Harvard College Observatory.*

Earlier spectroscopists, such as Secchi and Huggins, had had to carry out their studies visually. This meant many fatiguing hours at the telescope to obtain the spectrum of a single star. The largest catalog of stellar spectra available at the time, listing some 4000 stars, was the work of the German astronomer Hermann Vogel of the Potsdam Observatory, and represented the labors of twenty years. Photography changed everything. By placing a large prism in front of a telescope's objective lens, as many as 200 stellar spectra could be recorded in a few minutes' exposure on a single photographic plate, and though the definition was not, obviously, as good as was attainable using a spectroscopic slit, it was good enough for a practiced eye to classify them at a glance. Though in 1885 Pickering had written to Eliot that in the photometry work, "the fatigue and the exposure to the cold in winter are too great for a lady to undergo," he evidently did not consider the tedious work of sitting at a desk examining and classifying the stars on these plates to be unsuitable for women. It wasn't an early embrace of feminism that led Pickering to this innovation; it was merely an interest in saving the factory observatory money. Astronomy, before the arrival of a few millionaire tycoons who would bequeath large amounts of money to assure its advance, was used to sailing financially close to the wind, and was also at the time very much a man's world. Despite the brilliant contributions of Caroline Herschel, Maria Mitchell (who had discovered a comet in 1847 and was professor of astronomy at the all-women's college of Vassar in New York), and Lady Margaret Huggins, who was her husband's right-hand—woman!—after their marriage in 1875—there were virtually no professional opportunities in astronomy for women. The woman who hoped for such an opportunity could easily imagine that things were still as they were when George Eliot [Mary Ann Evans] wrote in *The Mill on the Floss,*

> I suppose it's all astronomers [who hate women]; because, you know, they live up in high towers, and if the women came there, they might talk and hinder them from looking at the stars.

When work began on the Draper catalog, the classification of spectra was seen as a mindless, repetitive task, with echoes of domestic drudgery. At first, a Miss Nettie A. Farrar did most of the measures of spectra, but by the end of the year she decided to leave to get married. Pickering replaced her with a woman with no particular training in science and one well inured to domestic drudgery. She was his housekeeper!

That woman, Williamina Paton Fleming, had been born and grew up in Scotland, and married James Fleming, an uneducated worker with his hands. They emigrated to Boston when she was 21, but their marriage fell apart soon after that. It was then that Mrs. Fleming—as she was always known—caught on as Pickering's live-in housekeeper. (Over just how many months all this happened tends to be fuzzed over in biographies about her, but it is certain that about this time she had a son, who was named Edward Pickering Fleming, who looked like E.C.P. and whose education E.C.P. later supported.) In any case, Pickering realized that Mrs. Fleming was too intelligent to mop floors all day, and since he was then looking for a "computer" who was willing to work seven hours a day, six days a week, classifying spectra, and was frustrated with the low productivity and lack of interest of his male assistants, he announced, "my Scotch housekeeper could do the work better than you!" With that, he dismissed the male assistants, and hired her for $10.50 a week—about two-thirds what it would have cost to hire a man for the same position.

As it turned out, Mrs. Fleming proved to be a natural at spectral classification. She was promoted to being the straw boss of the other women computers hired at the observatory who became known, unofficially, as "Pickering's harem." She got very good work out of them, and led by example.

It is important to point out that classification is not a "discovery" process. At least in the beginning, it is designed simply for the purpose of bringing order into a huge mass of observational data. When Mrs. Fleming started classifying spectra on those objective-prism plates, she had nothing astrophysical to go on. She published a first catalogue in 1890 in which stars in Secchi's classification I were subdivided into classes A, B, C, D; those in his classification II into E, F, G, H, I, K, L; his III became her class M, IV became class N. She also used O for Wolf-Rayet stars (named for their French discoverers; their spectra had bright emission lines), P for planetary nebulae, and Q for other spectra ("peculiar" stars).

Fig. 8.6. Mrs. Williamina Fleming, E. C. Pickering's former maid, at center of the image, taken about 1890, presides over "Pickering's harem," the women computers at Harvard College Observatory who valiantly cataloged the stellar spectra. They were less expensive than men, earning between 25 and 50 cents an hour—more than a factory worker but less than a clerical worker. *From: Wikipedia Commons.*

Mrs. Fleming's scheme was revised repeatedly as instrumentation and the quality of the spectrograms improved. It had many advantages, but it was not the only way to classify stars. Another Harvard "computer," Antonia Maury, introduced a different scheme. Miss Maury happened to be Henry Draper's niece. She had studied under Maria Mitchell at the all-women's college Vassar, and was the most creative of the women computers employed on the stellar classification project. In 1897, though still using lettered classifications like those used by Mrs. Fleming (though with some of her classes reordered so that, for instance, B preceded A, rather than being intermediate between A and F), she also introduced a second dimension based on the strength of the lines ("a" stood for average spectra with well-defined lines, "b" for those in which the lines were hazy and indistinct, and "c" for those with narrow sharp lines). The significance of this second dimension was not recognized at first. Unfortunately, Pickering did not like the way she classified. She, for her part, refused to change her approach. After her great publication in 1897, she left Harvard and, practically speaking, astronomy. Before she left, Pickering hired Annie Jump Cannon, who had studied astronomy at Wellesley, another all-women's college. Miss Cannon was a cheerful person, easy to get along with (at least compared to the deep and complicated Miss Maury). Most important, she classified stars in the way Pickering wanted them classified, and did so with remarkable dispatch. According to the greatest spectral classifier of a later era, W. W. Morgan, who met her in the 1930s when she was still classifying stars:

> I had quite a talk with her ... and I saw how she classified stars. She had a low-powered eyepiece, and the striking thing was that she had to hold it in the air instead of placing it on the glass [plate], at a constant focus. And that's the way she did the whole Henry Draper Catalogue. Hanging [the eyepiece] up in the air like this. And if you do that, you can't help changing the focus a little. That means your eye jumping back and forth and focusing all the time. That's the way she showed me, she was still doing it that way, still classifying when I was there.[10]

The Henry Draper Catalogue, published in nine volumes between 1918 and 1924, included Miss Cannon's classification of 225,300 stars, with the lettered categories finally reordered in their familiar and final form OBAFGKM. Examples of each type are:

O θ^1 Orionis C (in the trapezium), σ Orionis, AE Aurigae, ζ Ophiuchi
B υ Orionis, η Aurigae, η Ursae Majoris
A Sirius, Vega, Altair, Fomalhaut
F Procyon (A), γ Virginis (A and B)

G The Sun, alpha Centauri A, τ Ceti, 51 Pegasi
K α Centauri B, 61 Cygni (A and B), ε Indi
M Proxima Centauri, Barnard's star

The H-R diagram evolves

Although the classifications had originally been merely descriptive, the classifiers guessed that their order reflected a temperature sequence. They were right, and this temperature sequence was to prove to be one key for understanding the properties and evolution of stars. The other was the absolute luminosity (the true brightnesses) of stars. However obtaining a luminosity is more difficult, as it requires that we know the distance to these stars. At about the same time in the early part of the 20th century, astronomers were finally able to obtain distances to many stars through the measurement of their parallex. Once these distances are computed it then becomes possible to obtain a measure of the luminosity. If one plots, for a large number of stars spectral class (or temperature), on the x-axis and luminosity on the y-axis, as two astronomers working independently, the Danish astronomer Ejnar Hertzsprung and the American astronomer Henry Norris Russell, did by 1913, the resulting graphic—known as the Hertzsprung-Russell, or HR diagram—shows at once that the stars divide into two main groups, giants and dwarfs depending only on their temperature and their luminosity.

Fig. 8.7. Annie Jump Cannon, portrait from 1922. *From: Wikipedia Commons.*

(Hertzsprung realized at once that Miss Maury's spectroscopic criterion "c" identified very bright supergiants, but when he wrote to Pickering to congratulate him, Pickering continued to insist that Miss Maury was wrong.)

The HR diagram is one of the most important diagrams in astronomy—a kind of Rosetta stone for understanding the properties and evolution of stars.

As shown here, stars on the HR diagram divide into two broad swaths. The first swath is called the *main sequence.* It runs diagonally from dim-red stars at the lower right

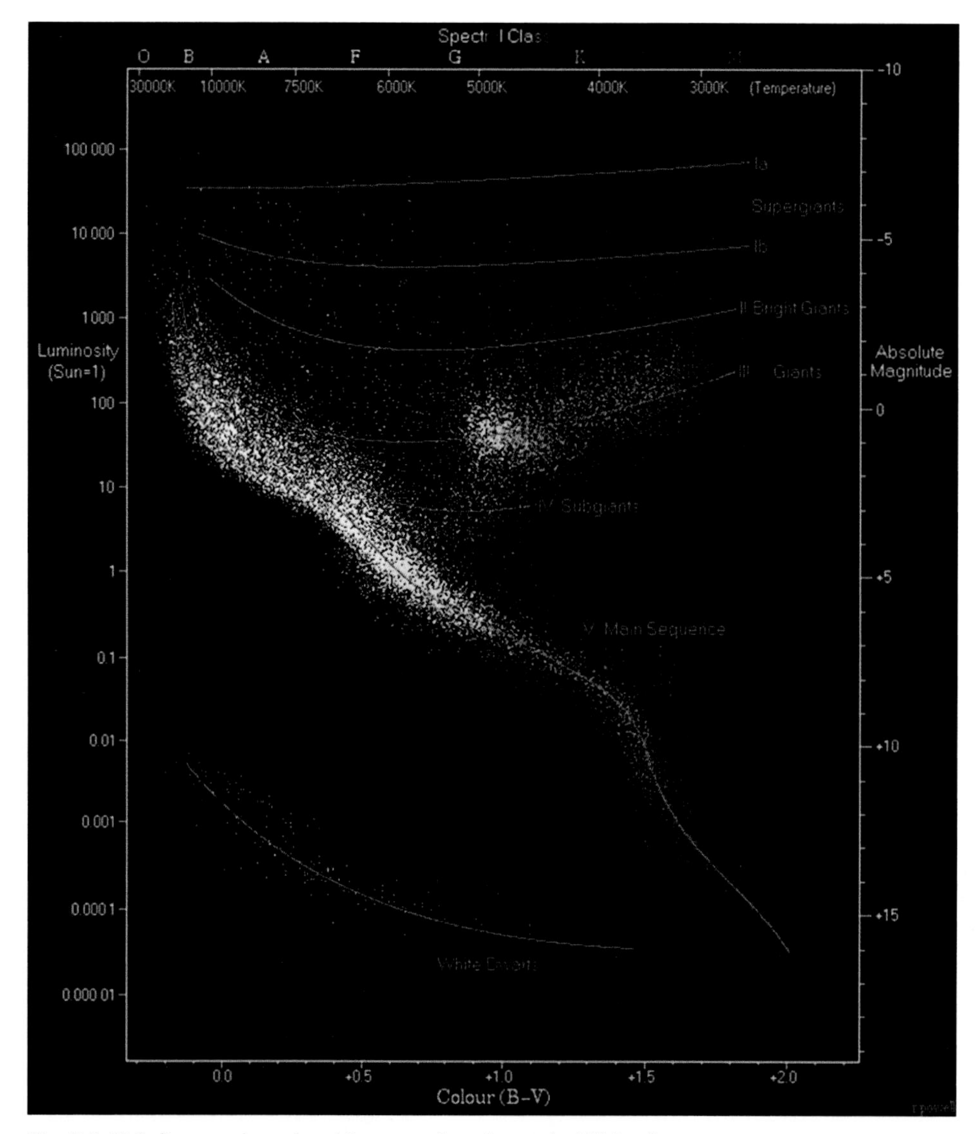

Fig. 8.8. H-R diagram, based on *Hipparcos* data. *Image by Wikipedia.*

to bright blue stars at the upper left. These are dwarf stars, and include the Sun—enormous, of course, by everyday standards, but a fairly typical dwarf star of spectral class G. Stars along this distribution vary somewhat in size (from about 1/10 to 10 times the Sun's radius) but enormously in luminosity (from 1/1000 to 100,000 that of the Sun). This means the difference in luminosity must depend on something other than size, and it does: it depends also on surface temperature. Dim stars of low surface temperature sit on the lower right, while bright stars of high surface temperature sit on the upper left of the HR diagram.

The other swath, consisting of stars in the upper right of the HR diagram, are the much more luminous *giant* and *supergiant* stars. Though very rare, these stars are overrepresented among the brightest naked-eye stars, since their great luminosity makes them celestial beacons visible across great distances of space.

The HR diagram contains a great deal of important data about the way that stars evolve during a significant part of their lifetimes. However, in order to interpret this data, it is necessary to understand the energy sources of stars, and for a long time this was completely inscrutable. Sir John Herschel, addressing the question of the Sun's energy source in 1833, remarked that the "great mystery ... is to conceive how so enormous a conflagration (if such it be) can be kept up. Every discovery in chemical science here leaves us completely at a loss, or rather, seems to remove farther the prospect of probable explanation. If conjecture might be hazarded, we should look rather to the known possibility of an indefinite generation of heat by friction, or to its excitement by the electric discharge...for the origin of the solar radiation."[11] Herschel's conjectures certainly are, as historian Karl Hufbauer notes, "feeble,"[12] and his failure to come up with a plausible explanation was complete.

The energy of stars

By the early 1840s, Julius Robert Mayer, a German physician, and John James Waterston, a Scottish engineer in Bombay, were independently toying with the idea that solar radiation originated in the conversion of gravitational energy to heat—Waterston, in a paper submitted but rejected as too speculative by the Royal Society of London, even tried to connect the idea with the nebular hypothesis, suggesting that as the Sun formed from the nebula, the gradual contraction of its mass might be the source of its heat, and went so far as to calculate that it was massive enough to continue shining for another 9,000 years. He also suggested an alternative source of gravitational energy in the form of meteorites falling into the Sun. Though at first ignored, Waterston's ideas were again presented in 1853 at a meeting of the British Association for the Advancement of Science, and this time they were taken up by the Glasgow physicist William Thomson (later known as Lord Kelvin), who in a presentation a year later before the Royal Society of Edinburgh strongly advocated the meteoric theory of solar energy and went so far as to propose that Kirchhoff's discovery of iron in the solar spectrum in its support. At the same time, the German physicist Herman Helmholtz was enthusiastically arguing that the Sun's gravitational contraction was the source of its heat. By the mid-1860s, the gravitational contraction theory had won out over the meteoric theory, basically because it could be shown that there were not enough meteors to generate enough heat to maintain the Sun's temperature. Thomson himself now became the leading champion of the gravitational contraction theory, and calculated—on the assumption that this was the only source of the Sun's energy—upper limits to the age of the Earth, which went from 400 million years in 1862, to no more than 100 million years in 1868, to only 50 million years in 1876, to nearer 20 million than 40 million years in 1897. As a check to these results, he also tried to calculate the Earth's age based on the theory that it had been slowly cooling since it had formed. Geologists, meanwhile, and supporters of Charles Darwin's theory of natural selection were becoming more and more sure that the Earth's age was considerably older than Kelvin calculated. Joe Z. Burchfield writes: "For a time geologists, convinced or cowed by Kelvin's mathematics, yielded to his results. But as the estimates grew more restrictive they grew more restive. In 1897, for the first time, Kelvin's results from his theory of terrestrial cooling were nearly in agreement with those from his theory of solar heat, but as he was well aware, neither was in close agreement with the hypotheses of the geologists."[13] Needless to say, Kelvin held to his conclusion until his death in 1907.

The *deus ex machina,* which vindicated the geologists, came from a new branch of physics that began with Henri Becquerel's discovery of radioactivity; studying the phenomenon of phosphorescence, and thinking that the recently discovered X-rays might have something to do with it, Becquerel wrapped a glass plate in black paper and then placed various phosphorescent salts on it. Of the salts tried, only those containing uranium blackened the plate. Becquerel made his discovery of what were then known as Becquerel Rays in 1896—ironically, the very year before Kelvin published his final calculations of the age of the Earth. Though X-rays proved to be high-energy electromagnetic radiation, Becquerel Rays were more complicated. By 1899 Ernest Rutherford, a native of New Zealand then working at McGill University in Canada, discovered that Uranium and other radioactive elements gave off two types of particles: alpha particles (helium ions), which were easily absorbed by paper and did not blacken the plate, and much more penetrating beta particles (electrons), which did. In 1909—at just about the exact moment that Hertzsprung was producing the earliest sketch of the HR diagram—Rutherford, who had meanwhile left McGill for Manchester University in England, and his students Hans Geiger and Ernest Marsden, aimed a collimated beam of alpha particles at a thin gold foil and observed scintillations from scattered particles on a zinc sulfide screen. Most of the alpha particles were either undeflected or deflected through very small angles, but surprisingly, some of them were deflected through angles as large as 90 degrees or more. They published the results in 1911. Their experiment showed that, surprisingly, most of the mass of the atom was concentrated in a small positively charged nucleus, which was incredibly dense and small—"like a pinhead in the earthly vastness of St. Paul's," Rutherford himself said.

At the time, the most widely accepted model of the structure of the atom had been that of the British physicist J.J. Thomson, director of the Cavendish Laboratory in Cambridge, who envisaged an atom as a kind of electrically neutral plum pudding—consisting of negatively charged particles, electrons (which he himself had discovered), embedded in some kind of fluid that contained most of the mass of the atom and possessed enough positive charge to make the atom electrically neutral. The gold foil scattering experiment ruled out the Thomson model. Now all the positively charged pudding was stuffed into the dense and massive core at the center, while the electron plums were displaced into distant orbits. But there were problems with the Rutherford atom too. Opposite charges attract and like charges repel, so the electrons in such atoms ought to either undergo a death spiral into the nucleus, or blasted each other out of their orbits as they approached too near each other. Rutherford's atom should have been highly unstable—it was calculated that it ought to last no longer than a millisecond or so. Such atoms ought not to exist.

They don't. Enter Niels Bohr, a young Danish physicist with a rare ability to tolerate dilemmas and contradictions. He started working with Thomson at Cambridge but soon transferred to Rutherford's lab in Manchester, and while there he proposed a model of the atom that combined classical physics and the new quantum theory, according to which energy, including light, could only be emitted and absorbed in discrete packets (later called photons) even though it was known to travel in waves. Bohr's model postulated that electrons could only occupy certain specific orbits around the nucleus, and when in them, they neither emitted or absorbed radiation. They only emitted radiation when they jumped from a higher to a lower orbit, and absorbed it when they jumped from a lower to a higher orbit. The wavelength or frequency of the emitted or absorbed light was proportional to the energy difference between the levels. This was why the lines in the spectrum occupied the specific places they did. Applying this model to the simplest case—hydrogen, which consists of a proton and an electron—Bohr was able to derive the specific wavelengths of the series of Balmer lines, discovered in the late 19th century by German schoolteacher Johann Balmer, and which correspond to discrete energy states of the hydrogen atom. The lines involve emitted wavelengths produced by electrons in higher orbits ($n=3, 4, 5$ etc.) jumping to the second orbit ($n = 2$). Thus the first Balmer line, known as H-alpha (HII), is a beautiful ruby-red line at wavelength of 656.28 nanometers, and is emitted when an electron jumps from the third ($n=3$) to the second ($n=2$) lowest energy levels of the hydrogen atom.

Since the Balmer lines are in the visible part of the spectrum, they were the first recognized. Electron jumps from $n=2, 3, 4$, etc. to the first orbit ($n = 1$) produce emitted wavelengths of higher energy than those in the Balmer series, and are in the ultraviolet; these are known as the Lyman series, after Harvard physicist Theodore Lyman. Electron jumps into the third, fourth, and fifth orbits (making up the Paschen, Brackett, and Pfund series) correspond to lower energies, and thus are found in the infrared. Note that the placement of the lines of these series will be affected by the Doppler shift; thus, for galaxies at cosmological distances, where the red-shifts are very large, even the Lyman series is red-shifted into the infrared.

Higher states of excitation of hydrogen are known as Hβ, Hγ, and so on, and converge at shorter and shorter wavelengths through the blue and violet parts of the spectrum. Though the success of his model in explaining the Balmer lines was, as Bohr himself realized, largely owing to hydrogen's simplicity—it didn't work for more complicated atoms—at least it provided a vindication of his belief that atomic physics could only be understood in terms of quantum theory. The next important step would be the development of modern quantum mechanics by Werner Heisenberg and Erwin Schrödinger in the 1920s, when solutions were found for more complicated atoms than hydrogen.

As electrical forces are weak enough to be rather easily overcome, electrons in the atmospheres of stars are easily stripped away from the nucleus to produce charged nuclei called ions. As shown by Cecilia Payne in 1925 in her Harvard Ph.D. thesis (which Otto Struve called "undoubtedly the most brilliant Ph.D. thesis ever written in astronomy"), temperature controls ionization. The spectral sequence is thus a result almost entirely of temperature progression in the atmospheres of stars. These range from something like 50,000 K in some O-type stars to 2000K in M-type stars. The higher the temperature, the faster the atoms are moving, and the more electrons will be stripped away. Since the number of atoms in one state of ionization versus another depends on competition between collisions of all kinds and radiation causing ionization and rates of electron-ion collisions producing recombination, temperature—not chemical composition—determines which absorption lines appear in a star's spectrum.

This is an important result, and the key to unraveling the details of stellar spectra. In low-temperature stellar atmospheres such as those in M-type stars, metals will be in their neutral states; e.g., iron will be present as Fe I, and even molecules such as TiO will be present. In stars at higher temperature, K-type and G-type (like the Sun), calcium ionizes; this explains why the two strongest lines in Fraunhofer's spectrum, which he called D but are now known as the H and K lines, correspond to once-ionized calcium (Ca^+); at still higher temperatures, as in A-type stars, hydrogen ionizes, and the Balmer lines become prominent.

It follows that metals (by which astronomers mean all elements except hydrogen and helium), despite the prominence of their lines in stellar spectra, are present as mere whiffs; instead, as Payne was first to realize, stars consist preponderantly of hydrogen and helium. This result went profoundly against what was believed by most astronomers at the time, but we now know that Payne was right; hydrogen makes up about 75% (by mass) of a star like the Sun, helium about 23%; all the other elements make up the rest.

Chemical energy—that involving electrons bound to atoms, the source of energy in batteries, for instance—is, it turns out, no more adequate than gravitational energy to sustain the heat of the Sun for a duration consistent with the aeons of geological time. But what about nuclear energy? If an unstable nucleus, like one of Uranium, decays, it gives off high-energy particles such as alpha and beta particles, and highly energetic electromagnetic radiation in the form of gamma rays. This process, which we observe in the case of radioactive decay, involves nuclear fission, and gives some idea of the kind of energies are involved. The reverse process—taking lighter nuclei and compressing them so that they fuse together into heavier nuclei—is known as nuclear fusion. The sequences of nuclear reactions that provided the energy sources for the Sun and other stars engaged in hydrogen fusion (stars on the main sequence) were clarified by the German-American theoretical physicist Hans Bethe in 1939. The details needn't concern us; the important point is that four hydrogen nuclei (protons) fuse to form one helium nucleus (a helium nucleus contains two protons and two more particles of almost the same mass called neutrons). Since the mass of a helium nucleus is 0.007 percent less than that of four hydrogen nuclei (protons), the difference is converted into energy according to the famous equation published by Albert Einstein in 1905: $E=mc^2$. In the case of hydrogen fusing into helium, the energy is released in the form of a gamma ray.

This, then, is the basis of the energetics of stars, and the key, at last, to decoding the HR diagram.

The rest of this chapter discusses the details of stellar evolution. While the below is not required for the rest of our story on how galaxies were discovered and studied in proceeding chapters, we include it for completeness. It can be skipped or skimmed for those interested in continuing the main story in the next chapter.

Stellar evolution: a primer

The Milky Way—and other galaxies—are made up of gas, dust, stars, and radiation, as well as remnants of stars such as white dwarfs, neutron stars, and black holes. Galaxies also include—in large haloes far from the galactic disk—matter which we cannot see and which only interacts with the familiar atomic type of matter through gravitation; this is the famous dark matter, about which we will need to say more later. The galactic environment may be considered an ecosystem in which these components interact with one another, and evolve over time.

One of the great intellectual achievements of the 20th century, accomplished largely in the 1920s through the 1950s, has been to understand the way that stars form and evolve. This is a vast subject, and we can only briefly touch on it here, but it may be useful to at least recall some of the main themes.

The Birth, Evolution, and Deaths of Stars

Consider a star on the main sequence, that is, one that is "burning" hydrogen in its core. Some 90 percent of all bright stars are there, including the star that is brightest of all to our eyes—the Sun. The Sun, of course, is of all stars the nearest and dearest to us, and though it has its cycles and occasional tempests, overall it is exceedingly mild-mannered and mellow as stars go. It is, for the likes of us, like Goldilocks's porridge, just right.

The Sun formed about 5 billion years ago from a collapsing Molecular Cloud, a cold womb containing grains of silicate- and carbon-dust and a breeding ground of stars, like those that are so dramatically silhouetted in Barnard's photographs. The Molecular Cloud from which the Sun formed probably was much like the Taurus Molecular Cloud. At a distance of 450 light years, it is the closest Molecular Cloud to us in which stars are forming, and modest in size, with dimensions of 90 by 70 light years and a mass of about 24,000 Suns. It is also very cold, with a temperature at only a few degrees above absolute zero. Among these cold dusty tendrils— which appear like ribbons of crepe streaming across Barnard's pictures, but are actually swaddling bands—new stars are forming; for example, the Taurus Molecular Cloud is collapsing, by gravity, into dense cloud stars, which will eventually form stars.

A newly formed star is a famous one called T Tauri, which is still entangled in the gas and dust of the enigmatic nebular feature known as Hind's variable nebula—the first "variable" nebula detected in the heavens, which created a sensation when it was first detected in 1852. The nebula itself is a reflection nebula. It is shining as light from the star it envelops is scattered by dust; the nebulosity that swathes the stars of the Pleiades is of the same type. Its variability is due to the fluctuations of light intensity from T Tauri itself, which is still unstable but which will eventually settle down to the business of fusing hydrogen into helium in its core. Once it does so, the tendency to further gravitational collapse is in intricate balance with the heat and pressure pushing outward from within. At this point it is a stellar debutante, a full-fledged main sequence star like the Sun.

Indeed, T Tauri seems destined to produce a star of about one solar mass, so we may see in it a reflection what the Sun and its system of planets were as they formed from a Molecular Cloud almost 5 billion years ago.

The more massive stars form from Giant Molecular Clouds, like those in Orion and Monoceros. They are much larger in scale than that of Taurus, ponderous hulks with breadths of 10 to 100 light years, and containing perhaps 100,000 times the mass of the Sun. It is the star's mass that ultimately determines where it enters the main sequence. As soon as it begins burning hydrogen into helium, it is locked into a well-defined place, shining with a fixed color and brightness for as long as it continues to do so. In the case of a G-type dwarf like the Sun, the period of its main sequence existence is on the order of 10 billion years, for very bright, massive O and B stars, which burn hotter and brighter and more voraciously than lower-mass stars, it may be only a few million. The more massive the star, the hotter and more brilliantly it burns, but its glory—like that of the hero Achilles—is purchased at a price; its very brilliance destines it to a short lifetime. Eventually a main sequence star will 'burn' most of its hydrogen in its center. So what happens when the star has depleted the hydrogen fuel in its core (i.e., reached the point where only 1% of its fuel remains)? At this point it is no longer able to remain on the main sequence. It leaves the main sequence and moves to the upper right of the HR-diagram, becoming colder and more luminous, and joins the realm of "giants." As a giant, it is surrounded by a distended outer envelope, an outward manifestation of internal adjustments occurring unseen within. The core of the star now collapses in a gambit to increase its core temperature to about 300 million degrees K, at which it begins to burn helium into carbon and oxygen by the so-called "triple-alpha" process (the point of ignition is known as the "helium flash").

Like a debtor pursued by creditors, the star is now maintaining its solvency only by relying on ever more desperate high-interest loans. Once it leaves the main sequence, a star of solar mass remains in the red giant, helium-burning, stage for about 1 billion years. As it continues helium burning, it passes through a "zone of instability," to become an RR Lyrae variable star (after the prototype star in Lyra, discovered by Mrs. Fleming in 1901). The helium-burning core remains enwrapped by a shell which is still burning hydrogen; in the boundary between the two, the temperature is just right for helium to alternate between single- and double-ionization states, and these alternations cause the star to oscillate between equilibrium states of slightly different radius and luminosity. During this unstable, manic-depressive period of its existence, the star gives off stellar winds of enormous intensity, a billion times greater than those that characterized it during the halcyon days of its main-sequence prime. The material expelled in this way piles up in the colorful shells of a planetary nebula (and also

Fig. 8.9. (right) The "Mystic Mountain," an H.P. Lovecraft inspired name for a spectacular structure in the Eta Carina Nebula. A three-light-year pillar of hydrogen is being worn away by UV radiation streaming from hot young stars. The blebs are places where radiation from newly formed stars are pushing outward. *Courtesy: M. Livio and the Hubble Heritage Team (STScI/AURA).*

Fig. 8.10. (below left) The Eagle Nebula, a hoodoo of gas and dust located in the constellation Serpens. The pillars are four-light years high—or equal to the distance between the Sun and alpha Centauri (!). Intensely radiating hot young blue stars, OB stars, with their fierce outpouring of UV radiation, are at work here, savagely weathering—photoevaporating—their surroundings. Echoing Clarence Dutton's words on the Grand Canyon, William Waller describes these cosmic structures as bespeaking "tales of wholesale erosion and exquisite sculpting.... The gestation of stars in molecular clouds leads to the remaining nebulosity being reshaped, rekindled, with new star formation—and ultimately destroyed." *Courtesy: Hubble Heritage Team (STScI/AURA).*

Fig. 8.11. (below right) N 180B, an HII region that is the scene of active star formation in the Large Magellanic Cloud. *Courtesy: Y.-H. Chu and Y. Nazé, NASA, ESA, and the Hubble Heritage Team (STScI/AURA).*

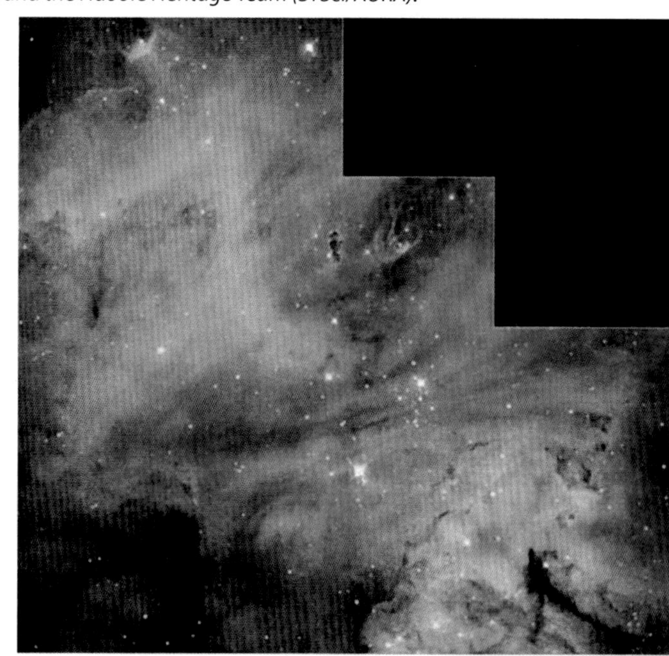

contributes most of the dust in the interstellar medium). Hundreds of these "cosmic butterflies" are known, their brilliant colors owing to fluorescence of gas by the intense star at their centers. Among themost celebrated are the Ring Nebula (M57) in Lyra, the Dumbbell (M27) in Vulpecula, the Owl (M97) in Ursa Major, and the Helix (NGC 7293) in Aquarius.[14] The Sun will eventually pass through such a stage.

Stars from one solar mass up to about eight solar masses puff away most of their material in this way. By the time it is done with all its blowing and wheezing and wasting itself, the star has dwindled to a mere vestige of its former self; unable to sustain thermonuclear burning, what is left of it collapses into a stripped-down hyper-dense remnant known as a white dwarf, in which the mass, typically on the order of 0.6 solar masses, is squeezed down into a volume a million times smaller than the Sun. The matter is so compressed that a cubic centimeter of the stuff weighs a ton, and it resists further collapse because it no longer consists of separate atoms; it now exists as a plasma of protons, neutrons, and electrons called degenerate matter. The Milky Way is littered with these stellar corpses. Among the Sun's closest stellar neighbors, Sirius, Procyon, and 40 Eridani all have white dwarf companions; they are remnants of stars once larger and more luminous than their companions. The companions being more ravenous and spendthrift, they have hurried along their life-courses, branched off the main sequence long ago to become red giants, sloughed off their outer layers, and finally collapsed into these now-inconspicuous embers where they will continue shining feebly till the end of time.

We have so far described the fate of stars of up to about 8 solar masses, or some 99% of all the stars. The more massive stars, the cosmic 1%, follow more complex life-histories. As usual, the details are highly dependent upon their mass. Evolving rapidly, these massive stars use up their hydrogen fuel in only a few tens of millions of years, and become red supergiants. They too pass through a zone of instability, becoming the metronomically regular pulsating stars known as Cepheid variables (after the prototype delta Cephei). Their periods of pulsation are precisely tied in to their luminosities, and because they are also intrinsically very bright—some of them attaining brightnesses equal to 100,000 times that of the Sun—they are important beacons for astronomers measuring the distances to remote star systems. Thus they are important yardsticks across the Galaxy, and indeed form the first rung of the extragalactic distance scale to which we will turn later.

Instead of giving up the ghost at the helium-burning stage, these massive stars are able to stubbornly hang on by pushing through to a series of further internal collapses. Each successive collapse raises the core-

Fig. 8.12. Close-up of Hubble Space Telescope color mosaic (Fig. 1.5), showing the region southwest of the Trapezium stars. Above the center is the star LL Orionis. A bow shock wave surrounds the star and points toward the stream of gas flowing from the neighborhood of the Trapezium stars located to the upper left (just outside the image). The bright star toward the lower left, LP Orionis, is surrounded by a prominent reflection nebula. *Courtesy: M. Robberto and the Hubble Space Telescope Orion Treasury Team; NASA, ESA, and the Hubble Heritage Team (STScI/AURA).*

temperature to the point where a new, more intense stage of thermonuclear burning can occur. Helium burns to carbon, carbon to neon, neon to oxygen; carbon and oxygen are cooked into silicon, oxygen and oxygen into sulfur. Again, the star attempts to shed mass, by blowing off a good part of itself in fierce interstellar winds. A star having thirty times the mass of the Sun may "evaporate" as much as two-thirds of its total mass in this way.

The end, like Hamlet's slaying of his murderous and adulterous uncle Claudius, is postponed, but at last becomes inevitable. Each successive collapse involves the star's ever-more desperate attempt to sustain itself at whatever cost, but at last it plays its trump card. As soon as it attempts to burn silicon to form iron, it has reached the end-game — or at least the beginning of the end. In structure, the star at this stage is onion-layered: the iron core overlain by a sulfur layer, the sulfur wrapped in hides of silicon, oxygen, carbon, and helium, and the whole enveloped in a hugely rarefied atmosphere of hydrogen. When the star attempts to burn silicon to form still heavier elements, it has reached the end of the line. From this stage onwards each reaction uses up more energy than it releases.

It cannot sustain this level of deficit spending for long. Indeed,after maintaining itself for millions of years, the end comes in less than a day. The star's energy finances go into a fateful crash. The long intricately-balanced homeostatic structure between the heat and radiation pressure within the star and the gravitation of its mass gives way. Gravitation must win out. In less than a second, the core itself implodes into a small dense object with the density of pure neutrons — a neutron star – or, if it is massive enough, forms an even denser and more compressed object. In the latter case, it turns the fabric of space-time inside-out like a stuffed stocking to form a black hole. The whole process occurs with such startling rapidity that the rest of the star has absolutely no chance of adjusting: the overlying shells, no longer supported from below, crumple; envelope careens on envelope, and everything ends in a tremendous jumble. The star plunges, as it were, over a Victoria-Falls or Niagara cataract. The shock wave produced by rebound from the core carries off the lion's share — 99 percent — of the energy of collapse in an explosion so energetic that all the elements of the periodic table heavier than iron are formed therein (it is a staggering reflec-

Fig. 8.13. Detail of the above: The famous Trapezium, as imaged by the Hubble Space Telescope. This image shows how examples of proplyds, young stars with circumstellar clouds of gas and dust that are rendered visible because they lie close to an emission nebula. The poplyds have tails pointing away from the brightest and hottest of the Trapezium stars, Theta 1C. With a mass of 40 times that of the Sun, it dominates the radiation field, driving off material from the proplyds, and causing the region behind the proplyds to be in shadow. *Courtesty: John Bally and NASA, ESA, and the Hubble Heritage Team (STScI/AURA).*

Fig. 8.14. Another detail of above: Bow shock wave near LL Orionis. *NASA, ESA, and the Hubble Heritage Team (STScI/AURA).*

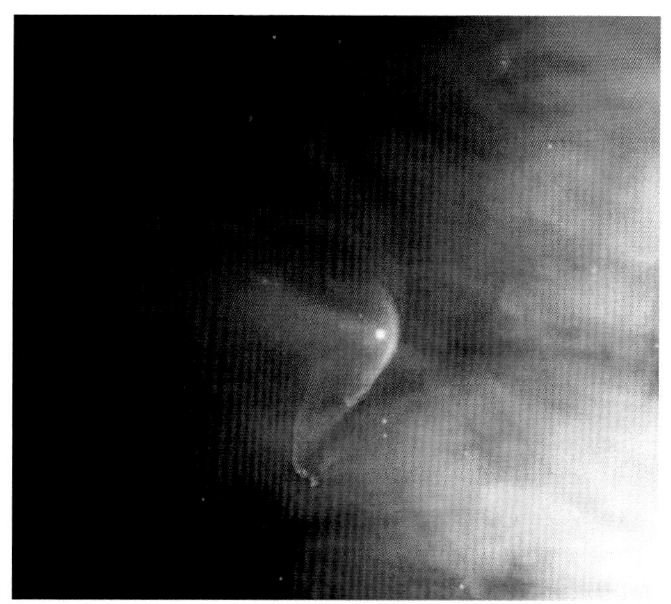

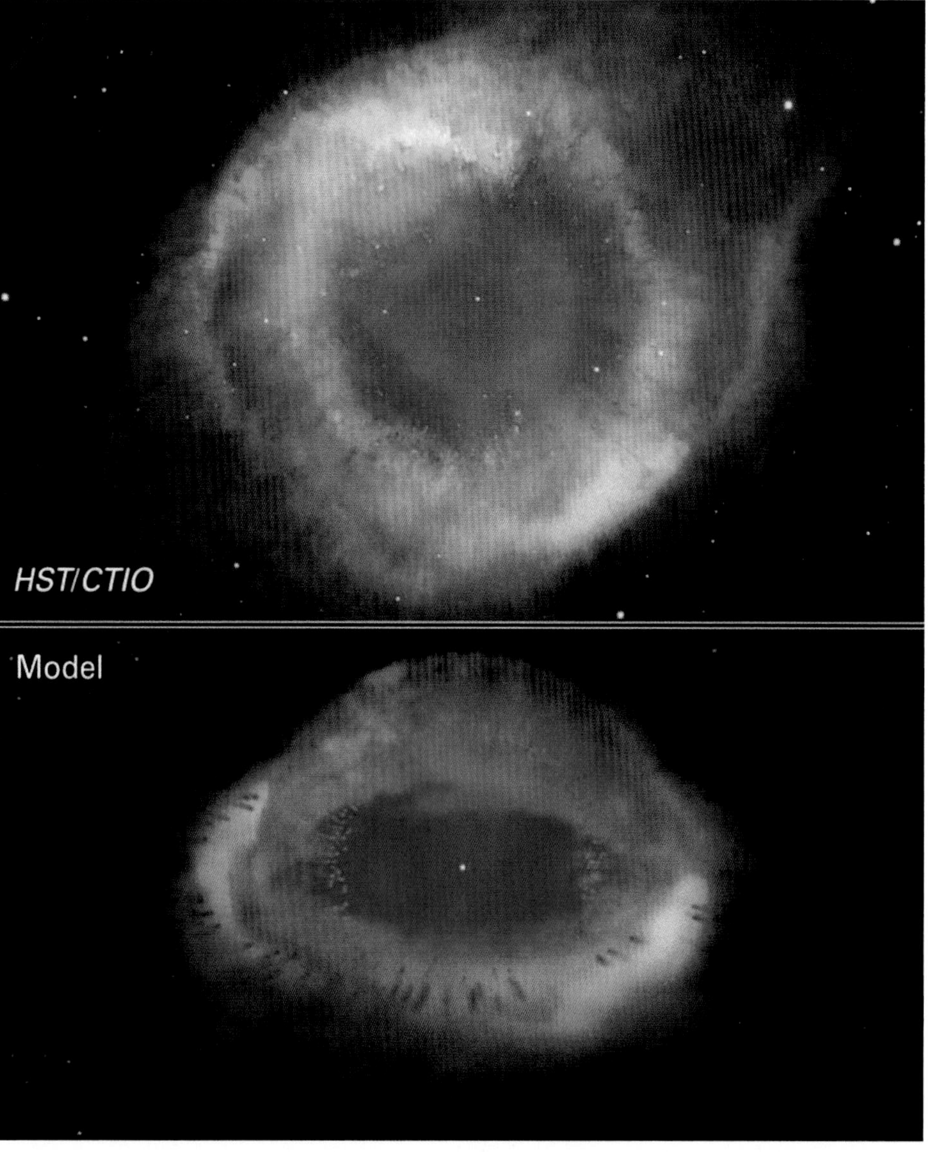

Fig. 8.15. The Helix Nebula in Aquarius (NGC 7293), located at a distance of 650 light years. The top composite image is a view of the colorful Helix Nebula taken with the Advanced Camera for Surveys aboard NASA's Hubble Space Telescope and the 4 meter telescope at the Cerro Tololo Inter-American Observatory in Chile. The object is so large that both telescopes were needed to capture a complete view. The Helix resembles a simple doughnut as seen from Earth. But looks can be deceiving. The bottom illustration is a model constructed from observations from several ground- and space-based observatories, and shows how the Helix would appear if viewed from the side. In this illustration, the Helix consists of two gaseous disks nearly perpendicular to each other. One possible scenario for the Helix's complex structure is that the dying star has a companion star. One disk may be perpendicular to the dying star's spin axis, while the other may lie in the orbital plane of the two stars. The Helix, located 690 light-years away, is one of the closest planetary nebulas to Earth. The Hubble images were taken on November 19, 2002; the Cerro Tololo images on Sept. 17-18, 2003. *Courtesy: NASA, ESA, C.R. O'Dell (Vanderbilt University) and M. Meixner and P. McCullough (Space Telescope Science Institute).*

tion but the energy in all the radioactive elements found on earth, such as thorium and uranium, is a remnant of the energy generated in such massive stellar detonations billions of years ago).

If the star enters this stage we may be privileged to contemplate — hopefully, from a safe distance — one of the most awesome and terrifying spectacles in all nature: a supernova. The outer layers of the destroyed star are flung into space; massively convulsed, they are wrenched into the far-fleeing knots, tendrils, and filaments of an expanding cloud, strewing star-stuff across the interstellar medium. The tortured and twisted material mixes and mingles with new generations of stars in the unfolding cycle of stellar death and rebirth. Since the formation of the Milky Way, at least 11 or 12 billion years ago, there have been tens of millions of these supernovae, spewing oxygen, silicon, magnesium, calcium, titanium, as well as a smattering of the heavy elements, thorium and uranium — through space. The shock-fronts of supernovae from time to time have jostled molecular clouds and triggered new generations of star-birth.

The situation is actually a bit more complicated than the above scenario would suggest. Supernovae like those just described, involving core collapse, are described as Type II and Type 1b. A Type II supernova at its peak may achieve a luminosity equal to 200 million suns. But there is another way for a star to "go supernova," and the resulting explosion, known as a Type Ia supernova, is even more powerful, reaching a luminosity of 5 billion suns. Obviously, such luminous objects can be seen even in remote parts of the universe.

A Type Ia supernova occurs in a binary system in which one component has already reached the white dwarf stage and the other is evolving to become a red giant. An accretion disk may form around the giant star and begin channeling material from the giant to the white dwarf. Small amounts of material may trigger modest outbursts known as "novae," in which hydrogen from giant star is deposited onto the carbon-oxygen "crust" of the white dwarf until thermonuclear reactions occur again; in this case the star may temporarily brighten by a factor of thousands or even millions. Since the white dwarf survives the process, these novae may be "recurrent." But in the special case of a white dwarf in which the flow of material onto its surface is just enough for the hydrogen to push the white dwarf's mass above 1.4 solar masses (Chandraskehar's limit, named for the theoretical astrophysicist who first calculated it in the 1930s), even the pressure of degenerate matter is not able to sustain the white dwarf against further gravitational collapse, and it suddenly detonates in a tremendous explosion that will lead to its utter destruction. Type Ia supernovae within our Galaxy include Tycho's star of 1572 and Kepler's star of 1604; another in our part of the Galaxy is long overdue. Because they are effectively calibrated bombs, with peak luminosities that are very large and in a consistent

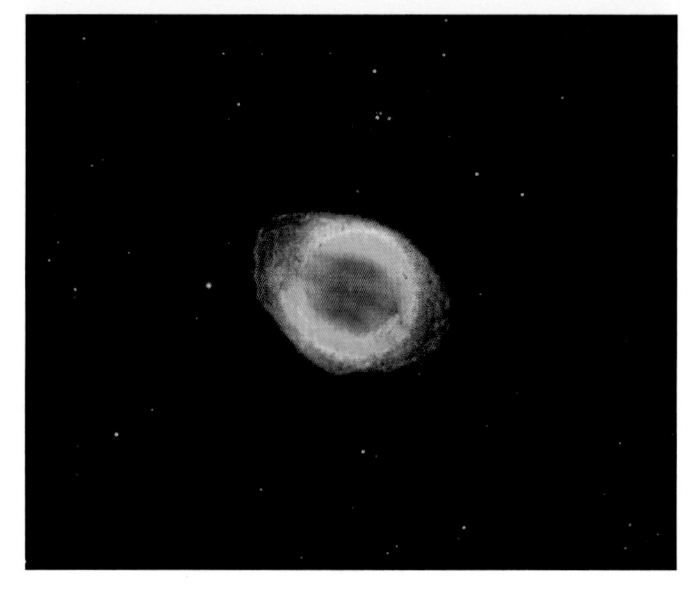

Fig. 8.16. (right) The Ring Nebula, M57. A favorite in small telescopes, here shown in detail in a Hubble Space Telescope image. The vivid colors are produced as atoms are irradiated by ultraviolet light from the hot white dwarf at the center: the deep blue in the center is produced by atoms of helium, the sea-green glow by atoms of hydrogen and oxygen. The red of the main ring comes from atoms of nitrogen and in the outer halo the red comes from molecules of hydrogen. The central blue area is actually shaped like a football, and protrudes through the doughnut of orange gas, while the dark spokes on the inside of the reddish ring are towers of gas denser than their surroundings. *Courtesy: NASA, ESA, C.R. O'Dell (Vanderbilt University) and D. Thompson (Large Binocular Telescope Observatory).*

Fig. 8.17. (below left) The Crab Nebula, the remnant of a Type II supernova, located some 6,500 light years from the Earth (in the Perseus Arm). In this Hubble Space Telescope view taken in 1995, the colorful network of filament constitute material from the outer layers of the doomed star, which went supernova in an explosion observed by Chinese astronomers in 1054 A.D. The collapsed core of the star, which has a mass 1.4 times that of the Sun but a diameter of only about 10 miles, is a neutron star, and spins on its axis 30 times a second. It is the lower of the two moderately bright stars to the upper left of center, and as it heats its surroundings, it creates the blue arc just to the right of the neutron star. *Courtesy: William P. Blair, NASA and the Hubble Heritage Team (STScI/AURA).*

Fig. 8.18. (below right) A supernova remnant in N 63A, a star-forming region in the Large Magellanic Cloud. *Courtesy: Y.-H. Chu and R.M. Williams, NASA, ESA, HEIC, and the Hubble Heritage Team (STScI/AURA).*

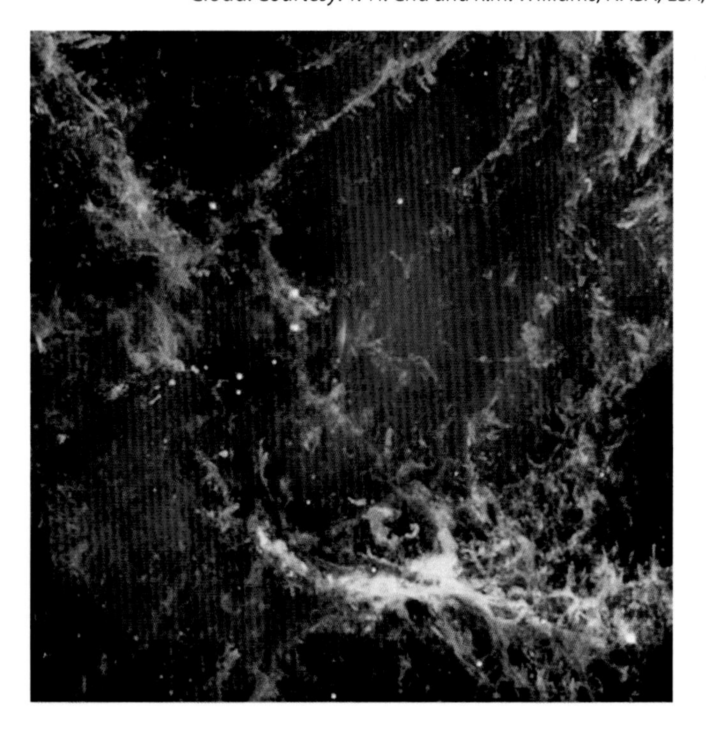

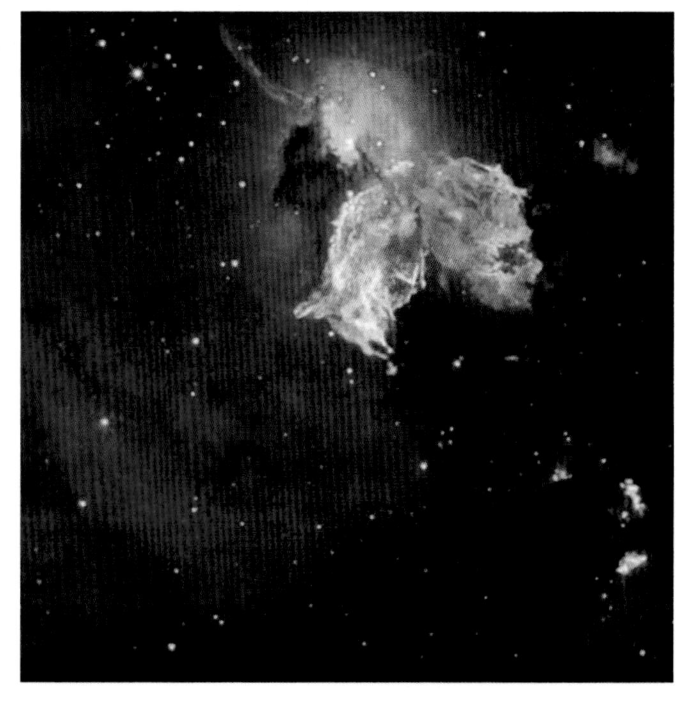

range, Type Ia supernovae are an ideal "standard candle" for measuring distances to galaxies forming in the very early universe. For this reason, we shall return to them again later.

Stars, immortal as they have always seemed from our brief human perspective, age, decline and die. In the phrase of Alfred North Whitehead, "Things fade." Even such things as stars. But ongoing renewal is also part of the story. The dying stages of more modest stars, stars like the Sun that are not massive enough to become supernovae, puff dust and gas over a very long period of time back into the interstellar medium, enriching molecular clouds and forming the substance of new generations of stars. It all begins over again, and will continue to do so just as long as there is cold gas and cold dust to work with.

Dust to dust. Within the soft and muffled dark wombs of the dust clouds of the Galaxy, the stars and their planets are born. They embrace a cycle extending from the softest and most gentle beginnings to the most unimaginably violent ends, a symphony ranging from mute tones and floating whispers to thunderously deafening cacophonies and back.

* * *

1. John B. Hearnshaw, *The Analysis of Starlight: one hundred and fifty years of astronomical spectroscopy* (Cambridge: Cambridge University Press, 1990), is the standard work on the subject.
2. James B. Kaler, *Stars and their Spectra: an introduction to the spectral sequence* (Cambridge: Cambridge University Press, 1989, 62.
3. Abraham Wolf; quoted in Otto Struve, "The Classification of Stellar Spectra," *Sky & Telescope*, May 1953, 184-187:184.
4. Quoted in S. Maffeo, S.J., "G.V. Schiaparelli and Fr. A. Secchi, S.J.," in *Memorie della Società Astronomica Italiana,* 82 (2011):261.
5. Henry James, "Boston," in *The American Scene* (1907), in *Collected Travel Writings: Great Britain and America* (New York: Library of America, 1993), 550.
6. Quoted in Jones and Gifford, *Harvard College Observatory,* 176.
7. Simon Newcomb, *Reminiscences of an Astronomer* (Boston: Houghton-Mifflin, 1903), 67.
8. Jones and Boyd, *Harvard College Observatory,* 177.
9. E.C. Pickering, address to the Harvard Chapter of Phi Beta Kappa, June 28, 1906. Harvard University Archives.
10. W. W. Morgan, Oral interview with David DeVorkin, Niels Bohr Library of Physics.
11. J.F.W. Herschel, *Treatise on Astronomy* (London: Longman, 1833), 212.
12. Karl Hufbauer, *Exploring the Sun: Solar Science since Galileo* (Baltimore and London: The Johns Hopkins University Press, 1991, 55.
13. Joe D. Burchfield, *Lord Kelvin and the Age of the Earth* (Chicago and London: University of Chicago Press, 1990), 43.
14. An accessible account is: Sun Kwok, *Cosmic Butterflies* (Cambridge: Cambridge University Press, 2001).

9. The Nebula Is Leaving the Solar System

Thanks for Percival Lowell. I hadn't realized his errors were so fruitful, I suppose that's the rule in science.

—*Robert Lowell, the poet, to Elizabeth Hardwick, July 2, 1976*

Observational cosmology was born in the early years of the 20th century, and was largely an American enterprise. There were good reasons for this. As Donald E. Osterbrock has pointed out:

> I believe it is because galaxies are faint, and hence to get good observational data required large telescopes at good observing sites. These existed in the United States, partly because of geography and climate (the clear weather, good-seeing sites in California and Arizona), and partly because of a very few American millionaires, such as James Lick and Andrew Carnegie [who could finance the construction of large telescopes].[1]

The first of the great American observatories was Lick, on Mt. Hamilton, a 4200-ft. peak in the Coast Range near San Jose, whose 36-inch refractor was funded by the late eccentric California real estate speculator James Lick. It was there, as we have seen, that E. E. Barnard began the wide-angle photography of the Milky Way. It was no accident, by the way, that California would become not only the premier location for astronomy but also for the nascent motion picture industry: in both cases the *sine qua non* was the climate.

Along with Barnard, the most productive member of the original staff at Lick was James E. Keeler, who was probably the best astronomical spectroscopist of his day. When the staff began to grow restive under the autocratic director Edward S. Holden, Keeler was the first to leave, in 1891, to assume the directorship of Pittsburgh's Allegheny Observatory where he carried out a classic spectroscopic study of Saturn's rings showing them to be composed of meteoritic material. However, Keeler returned to Mt. Hamilton to become director of Lick a year after Holden himself resigned in 1897. One of the first decisions he faced was what to do with a 36-inch reflector that Holden had acquired from a wealthy English industrialist and amateur astronomer, Edward Crossley. It was in such bad shape and so poorly mounted that Barnard, before he had left Mt. Hamilton for Yerkes Observatory in 1895, had told Holden it was a piece of junk he wouldn't have paid five cents for. Keeler rebuilt it so completely that hardly any of the original telescope remained, and as soon as he got it into working order, in 1899, began using it to obtain direct photographs of the nebulae.[2] At first Keeler held, as did most astronomers at the time, that spiral nebulae were nearby masses of gas and dust in rotation, quite possibly solar systems in formation and in any case relatively near. His images with the Crossley revealed that these nebulae existed in staggering perfusion; some of the larger and brighter ones showed subtle detail, but there were many smaller ones, in various orientations but all evidently of the same type. Keeler estimated, conservatively, that 120,000 spiral nebulae might be within reach of the Crossley. There were lots of them, but no one was yet sure what they were.

Keeler had intended from the first not only to take direct photographs of the nebulae but also to investigate their spectra. The brighter stars (and also planetary nebulae, which are of high surface brightness at their emission wavelengths) can be studied with a relatively simple optical set-up. For this reason the pioneers of spectroscopy like Huggins concentrated on studying the spectra of stars and planetary nebulae. The spiral nebulae, on the other hand, are very faint, and have low surface brightness. The feat of obtaining spectrograms of these baffling objects was to prove one of daunting technical difficulty for astronomers at the time.

Fig. 9.1. Pre-dawn snow at Lick Observatory. The large dome in the middle is that of the 120-inch Shane reflector. The Crossley dome is the one on the ridge just to the left of it, while the 36-inch refractor dome is part of the complex of buildings on the far right. *Copyrighted image by Laurie Hatch, used with permission.*

By the 1890s, it was already clear that planetaries and diffuse nebulae have the bright-line spectra characteristic of hot, low-density gas. The spirals appeared to show only faint continuous spectra, like those of stars, though there were inconsistencies in the data. Huggins had earlier reported seeing weak absorption lines in the continuous spectrum of the Great Spiral of Andromeda, but confusingly, he also reported seeing some emission lines, like those in gaseous nebulae. Then, in 1899, a German astronomer Julius Scheiner, using a 12-inch F/3 reflector at the Potsdam Observatory, reported a couple of Fraunhofer absorption lines, so once more it appeared that the Great Spiral consisted of stars.

Even while busy with his direct photography of the spiral nebulae with the Crossley, Keeler was turning over in his mind the design of a spectrograph suitable for studying the nebulae. In contrast to the high-dispersion three-prism stellar spectrograph fitted to the 36-inch refractor, which was meant to obtain highly detailed stellar spectra and to measure radial velocities of the stars by means of their Doppler shifts, he wanted a low-dispersion spectrograph to use on the nebulae. He ordered a quartz prism from instrument-maker John Brashear for a slitless spectrograph that would give very low dispersion spectra. As is usual with the building of new instruments, mishaps delayed its delivery, and it was not finished until the very day Keeler left Mt. Hamilton for the last time. Already ill, he was destined to die unexpectedly soon afterward, at the age of only 42. An "astronomical Adonais," cut short in his prime, one can only imagine what he might have done had he lived a more normal length of years.[3]

Denied by death, Keeler never trained the spectrograph on a nebula. Later, his successor, W. W. Campbell, attempted to use it in its slitless form but found it useless; without lenses, it was impossible to bring it to focus. Campbell mounted quartz lenses in front of and behind the prism, and in this form it proved to be a very efficient instrument—it was especially sensitive to the ultraviolet, which made it useful for studying planetary nebulae, novae, and Wolf-Rayet stars (stars with emission lines). Campbell's graduate student, Harold Palmer, made many studies of such objects, including Nova Persei, in 1901, and

Fig. 9.2. Remington Stone stands by the 36-inch Crossley reflector, one of the most productive telescopes in astronomical history. *Copyrighted image by Laurie Hatch, used with permission.*

after Palmer left, it was used by another graduate student, Joel Stebbins, who took a few spectrograms with it. But by then Campbell was no longer interested in the nebulae. He was almost exclusively involved in making highly precise measures of the radial velocities of stars, which could be determined from the Doppler displacement of Fraunhofer lines in their spectra. A displacement toward the violet end of the spectrum meant that a star was approaching, a displacement toward the red end meant that it was receding. These measures were used to detect spectroscopic binaries and to study the motions of stars through space. Typically, the velocities Campbell found were on the order of 10 or 20 km/sec.

The small spectrograph went unused for a few more years, until Edward Fath, another graduate student, arrived on Mt. Hamilton. Fath had been born in Germany of American parents, and received his education first at Wilton College, Iowa (it no longer exists), then at Carleton College, in Northfield, Minnesota, whose college observatory boasted a 16-inch Brashear refractor (a large instrument at the time) and also contained the editorial offices of the influential journal *Popular Astronomy.* After graduating from Carleton in 1902, Fath taught physics, chemistry and mathematics for three years, then went to the University of Illinois where he studied under Stebbins, who had just completed his Ph.D. under Campbell. Stebbins recommended Fath to Campbell for graduate studies at Lick. For his thesis, Fath wanted to study the spectra of the spiral nebulae. He made a few spectrograms with Keeler's slitless spectrograph, but realized that to learn much about the physical nature of these objects he would need a slit-spectrograph equipped with a fast (short focal-ratio) camera lens. He quickly put together a crude spectrograph of this type, with a box of wood that was shellacked and screwed together. It was just sensitive enough to record absorption lines in the continuous spectra of the brighter spirals, such as M31 in Andromeda, but because the spectrograph had to be completely removed from the telescope in order to add spectral comparison lines to any spectrum, he lacked the ability to make precise measures of radial velocities. That, as we shall see, was to prove a crucial lack.

By the end of September 1908, Fath, with an exposure of 18 hours over three nights, was able to use his nebular spectrograph to obtain a good spectrogram of M31 showing fourteen definite absorption lines of "solar type." There were no bright emission lines such as Huggins had reported. Fath went on to obtain spectrograms of six other spiral nebulae; one, NGC 5194, was too faint to show anything, but the others, though fainter than M31, showed similar absorption lines.

(One or two objects puzzled him: M77 showed the emission lines of a gaseous nebula as well as solar-type absorption lines. We now know that this object is the brightest galaxy with an active galactic nucleus—a so-called Seyfert galaxy, named after Carl Seyfert who studied such objects in the 1940s.

So it really does have emission lines. Another object, NGC 4736, seemed to show one broad emission line, but as later studies were to show, this was only a bright part of the continuum spectrum between absorption features).

Since Fath knew that the nucleus of M31 was an extended object—not a star—he argued that it must be a "star cluster." But why should all the stars in the cluster have the same spectral type? To throw light on this question, he obtained spectrograms of the globular clusters M2, M13, and M15, and found that as with M31 the spectra were continuous, with absorption lines of solar-type stars. In his Ph.D. thesis of 1909, he correctly concluded that the spirals were composed of stars, but neither he nor anyone else at the time had a way of measuring the distance to a spiral, and unfortunately, he watered down his conclusion by suggesting that his hypothesis "must stand or fall on the determination of the parallax" (i.e., a direct distance measurement) to M31. Osterbrock, whose insight into the motives of these astronomers is always keen, suggests that it was quite likely that Campbell—who was very conservative, and who as late as the end of 1916 was still saying of the spirals, "We are not certain how far they are; we are not certain what they are"—might have suggested that Fath include this disclaimer, since "he did not want anyone from his observatory ever to publish and later have to admit a mistake."[4] He may also have been thinking of a recent parallax measure by a Swedish astronomer, Kurt Bohlin, which had indicated that the Andromeda Nebula was only 19 light years away! Whatever the case, Osterbrock concludes, "Fath had brilliantly discovered and proved the nature of the spirals." Unfortunately, the disclaimer weakened his case, so that "Fath's result, so clear to us today, had little impact at the time, and the nature of the spirals remained an open question for at least another decade."[5] Simply put, the paradigm was not yet ready to shift.

Fig. 9.3. Percival Lowell, 1904. *Courtesy: Lowell Observatory.*

Fath published his Ph.D. in 1909, then left for Mt. Wilson Observatory near Pasadena where, with a slower spectrograph than the one he had used at Lick, he continued his studies of the spirals. After three years, he left Mt. Wilson for a teaching position at Beloit College in Wisconsin, and shortly afterwards moved on again to become president of tiny (and now defunct) Redfield College in South Dakota. At last, in 1922, he returned to Carleton, his alma mater, where he spent thirty years as a teacher, editor and textbook writer but never returned to his groundbreaking research on the spiral nebulae.

Before leaving research, however, Fath

attended one of the most important astronomical meetings held in the first quarter of the century, the meeting of the International Solar Union at Mt. Wilson Observatory in 1910 which served as an opportunity for the astronomers there to unveil the recently completed 60-inch reflector (see chapter 10). There he met another young astronomer struggling with the problem of obtaining spectrograms of the nebulae: Vesto Melvin Slipher ("V.M.," as he preferred to be called, of the Lowell Observatory). Slipher was to make the discovery that Fath, owing to his inability to obtain comparison spectra with his crude wooden spectrograph, had missed.

Slipher had spent most of the previous decade working as an assistant astronomer to Percival Lowell, one of the most colorful figures ever to appear in the history of astronomy. Lowell had built a private observatory on a mesa in Flagstaff, Arizona, in 1894 for the purpose of studying Mars and the other planets of the Solar System. His stimulating—but controversial—theory of intelligent life on the red planet, based on his interpretation of linear markings (canals) first reported by the Italian astronomer Giovanni Schiaparelli in 1877, had won him the adulation of the general public as well as the animus of many professional astronomers through promotion of his theory that Mars was inhabited by intelligent beings (Keeler and George Ellery Hale, the leading American astrophysicists of this era, who founded the prestigious *Astrophysical Journal* in 1895, refused to accept his articles as they were not up to their standards).

By any measure, Lowell had followed a rather unorthodox route into astronomy. Born in Boston in 1855 into a prominent New England family, then at the pinnacle of what passed for American aristocracy, he graduated with honors from Harvard in 1876, giving for his part in the graduation exercises an oration on Laplace's nebular hypothesis on the origin of the Solar System. He remained enamored of the nebular hypothesis for the rest of his life—not in Laplace's original formulation so much as in that of the English philosopher Herbert Spencer, whose "System of Synthetic Philosophy" tried to explain how everything in the universe, including man, developed from a simple undifferentiated homogeneous state (i.e., nebula) to a complex, differentiated heterogeneity, through a principle of evolution. Indeed, Spencer's conceptual framework would fit Lowell snugly for the rest of his life, and implied, at least in Lowell's view of the matter (as he expressed it in a talk he gave to the Boston Scientific Society on May 22, 1894, the eve of his departure for his Arizona observatory), the existence of intelligent life on other planets. He said:

> If the nebular hypothesis is correct, and there is good reason for believing in its general truth, then to develop life more or less distinctly resembling our

> own must be the destiny of every member of the solar family which is not prevented by purely physical conditions, size and so forth, from doing so.

As for the Schiaparelli canals—which, remember, he had not even seen as yet—he was already pretty sure what they were:

> The most self-evident explanation from the markings themselves is probably the true one: namely, that in them we are looking upon the result of the work of some sort of intelligent beings.... The amazing blue network on Mars hints that one planet besides our own is actually inhabited now.[6]

With the support of a fortune founded on the textile mills of the city of Lowell, Massachusetts—whose mills were powered by six miles of canals dug by immigrant Irish canal diggers to divert water from the Merrimack River—Percival Lowell was born every inch an aristocrat. After his graduation from Harvard and a Grand Tour of Europe as far as Syria, he briefly tried managing textile mills and trusts in his grandfather's office on State Street, but his interest in business soon flagged and at 28, to the considerable shock of his family, he broke off an engagement (probably to a cousin) and quit the business. He seems to have been suffering from one of the periodic breakdowns that would recur during times of stress over the rest of his life. He soon formulated a plan for his personal rehabilitation, however, having just heard a stimulating lecture on Japan given by the zoologist Edward S. Morse. For Lowell, Morse's lecture served as a siren call to adventure, and as soon as he could, he left Boston, which he called "the most austere society the world has ever known," and set sail for the Far East. He spent the next decade in travel and literary activity in Korea and Japan. Then, as the romance of the Far East began to fade, he switched careers again.

On returning from Tokyo in the latter part of 1893, Lowell's imagination was fired by *La Planète Mars*, a book authored by the French astronomer Camille Flammarion, which Lowell received as a gift from an aunt at Christmas 1893.[7] The book offered detailed descriptions of Schiaparelli's canals. It was the intellectual equivalent of love at first sight. Lowell was captivated by the canals and by the possibility of life on Mars to the point of obsession. Indeed, as Stephen Jay Gould has written, he "fell under the spell of these nonexistent phenomena and spent the rest of his career in increasingly elaborate attempts to map and interpret 'these lines [that] run for thousands of miles in an unswerving direction, as far relatively as from London to Bombay, and as far actually as from Boston to San Francisco.'"[8]

By the time of Mars's next biennial approach to the Earth in October 1894, Lowell was determined to have an observatory up and running.

He commissioned Harvard astronomer (and veteran Mars observer) William H. Pickering, E.C.'s younger brother, and Pickering's assistant Andrew E. Douglass to join him in an expedition to the American Southwest to observe Mars under conditions of clear and steady air. Douglass was sent on ahead to scout sites with a 6-inch refractor; everything being rushed, Lowell somewhat impetuously chose the mesa (which Lowell called "Mars' Hill") above the lumbering and ranching town of Flagstaff; in addition to its purportedly good seeing, the town lay conveniently on a railroad line that connected it to Chicago and Boston, and also had better saloons than other sites considered. Using two borrowed telescopes, Lowell, Pickering, and Douglass gathered on Mars Hill at the beginning of June 1894 to launch the most ambitious and systematic campaign of Mars observations attempted up to that time. Peering at the red planet nightly, and with monomaniacal intensity, for months on end (though Lowell himself was not at his new observatory all the time, observing only through the month of June, briefly late in August, and then most extensively in October and November), they produced hundreds of sketches of Martian surface detail in their observing logbooks, and in a series of books (*Mars,* 1895, *Mars and its Canals,* 1906, *Mars as the Abode of Life,* 1908), articles, and lectures, Lowell proceeded to lay out his theory that the canals on the surface of the planet were nothing less than irrigation channels that had been built by the inhabitants of a world increasingly turning to desert and in the throes of dying of thirst.

Lowell, at age 39, had entered astronomy just as the field was undergoing dramatic transformations. He was a well-educated generalist in an era of increasing specialization in the sciences. But at the very moment he began his career as an astronomer, the initiative had already passed to specialists—astrophysicists, who were combining the forces of spectroscopy and photography to advance the study of stars and nebulae. By contrast, the area of his interest, planetary astronomy, continued a tradition of visual observations going back to Galileo. There were obvious reasons for the difference. The photographic plate is an integrating detector; it builds up a cumulative impression of faint objects. The human eye, on the other hand, is a differentiating detector. It lacks the ability to see faint objects, but is quick and agile, and well-equipped for analyzing objects in motion.

A planet, because of the vicissitudes of the atmosphere ("seeing"), has more in common with a motion picture than a still-life. When we look at a planet, say Mars, through the telescope, what we see is a staccato of fixed images, each immediately erased by the next one. A trick of the brain called flicker-fusion takes these static images and fuses them into what appears to be a seamless continuum—it is this which makes the illusion of "motion pictures" so convincing. The problem in planetary astronomy is not, however, just that of detecting motion: one also wants to "stop action," to freeze the occasional high-definition frame from the series of the blurry ones. Unfortunately, the eye-brain system isn't designed for that.

Indeed, the moving planetary image might be compared to galloping horse. Despite the fact that people have been watching galloping horses with keen interest for thousands of years, it was only in 1878, when Eadweard Muybridge invented stop-action photography, that the sequence of movements involved in the galloping horse was demonstrated. Astronomers in the visual observing era were like the watchers of horses in the pre-Muybridge era. They had photography, but the photographic plate required an appreciable time—from 0.5 to 5 seconds and longer—just to "see" the planet. This meant that detail in photographs of Mars was much more blurred than what the visual observer could make out. The eye remained supreme over the plate for planetary imaging, and was not superseded until the late 1980s when CCD became available, which is even faster and more sensitive than the eye. An 1890s astronomer like Lowell, whose primary interest was planets, had no choice but to rely on the old-fashioned eye-brain-hand technology, in contrast to the astrophysicist who had the newfangled photographic plate and spectrograph in his arsenal.

Also, since making an observation of a planet was something one did in private (in "peepshow" fashion) rather than in a public setting like a movie theater, consensus could only be obtained through indirect means and a process of negotiation (e.g., viewing published drawings and attempting to confirm what one saw in them with detail visible in the eyepiece). Under the circumstances, it is not surprising that the testimony of one authority frequently conflicted with that of another. This would lead to interminable and inconclusive arguments such as those Lowell engaged in about the "canals" of Mars.

Fig. 9.4. Dome of the 24-inch Clark refractor, 2013. *Photograph by William Sheehan.*

After the 1894 opposition of Mars, Lowell returned the larger of the borrowed telescopes to Brashear and as a permanent replacement ordered from the Alvan Clark firm of Cambridgeport, Massachusetts (which had just created the lens for the Lick refractor) a 24-inch refractor. It arrived at the Flagstaff train station on July 22, 1896. At this time, Lowell and Douglass (Pickering having meanwhile sundered his connection with the observatory and returned to Harvard) expected that the Flagstaff site would be temporary, and envisaged the observatory as a peregrinating affair, moving across latitudes and from point to point on the Earth's surface at successive oppositions of Mars so as always to enjoy the best possible locality and atmospheric conditions. In line with this philosophy, the Clark refractor, as soon as it had passed its initial tests, was shipped to Tacubaya, outside Mexico City, for the December 1, 1896 opposition of Mars. Douglass served as front man, arriving in Mexico in November. Lowell himself arrived only in late December, well after the Martian opposition.

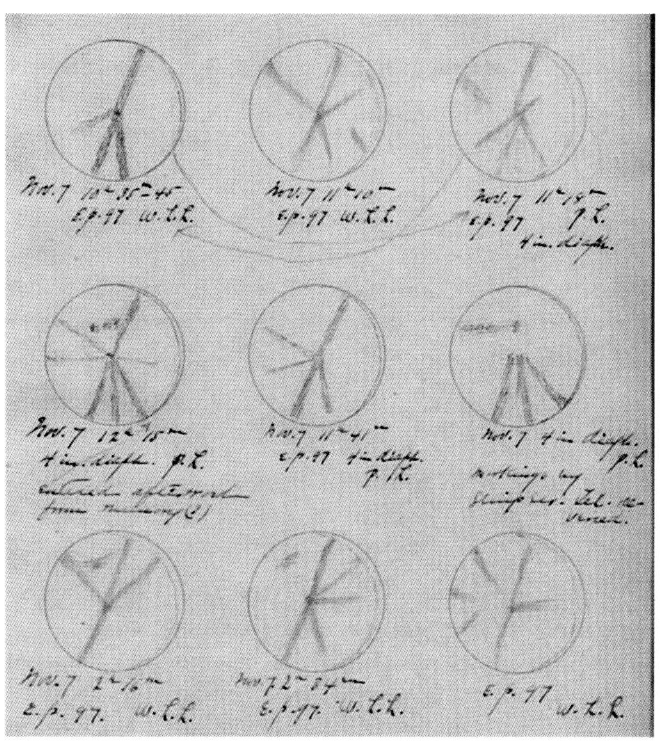

Fig. 9.5. Venus, 1899; observed by Percival Lowell with the 7-inch refractor at the Amherst Observatory. These drawings show the characteristic spokes seen by Lowell and Wrexie Louise Leonard. *Courtesy: Lowell Observatory.*

Several assistants went to Mexico with the expedition. As usual, Lowell's secretary, Wrexie Louise Leonard, was with him (there would be rumors, never substantiated though likely enough, that they had a long-running affair), and several new assistants had been added, of whom the most notable was Thomas Jefferson Jackson See, a Berlin-trained Ph.D. who was on leave from the University of Chicago and then at the height of his fame as an expert on double stars. See's inclusion indicates that, in addition to studying the objects of the Solar System, Lowell had decided to expand the observatory's program of research to stellar studies. Even before the telescope had moved to Mexico, See announced that on August 30-31, 1896, he had rediscovered the notoriously difficult "companion" of Sirius, missing since 1890 in the glare of Sirius itself. It was not observed by other astronomers until November 1896, when it was seen at Lick Observatory—but in a different position from where See had made it out. The Lick director at the time was still Edward S. Holden, and citing a careful series of observations by double-star specialist Robert G. Aitken, Holden concluded, "there is no object in the place reported by the astronomers of the Lowell Observatory."[9] The episode was only one of what would become an endless series of controversies between astronomers at the rival institutions.

Lowell himself provided the best copy from the 1896-97 observing campaign. Ten days before See "rediscovered" the companion of Sirius, Lowell began a study of the inner planets, Mercury and Venus. His observations of Mercury were relatively uncontroversial, and merely seemed to confirm an earlier result from Schiaparelli that the planet's rotation was isosynchronous, i.e., it rotated in the same period as that with which it orbited the Sun. But Lowell shocked astronomers with what he found on Venus—usually a notoriously bland and unrevealing object in the telescope. He made out markings that were, he wrote, "In the matter of contour, perfectly defined throughout, their edge being well marked… a large number of them … radiate like spokes from a certain center."[10] The observations had not even been confirmed before Lowell was advancing interpretations that went against the general consensus that the Earth's sister-planet was shrouded in clouds. The hub and spoke markings, Lowell claimed, had the appearance of "ground and rock," which suggested that Venus had an extensive but transparent atmosphere. Their deadpan stare also suggested that like Mercury, Venus also had an isosynchronous rotation period—one hemisphere was forever sun-lit, the other in night. This confirmed another of Schiaparelli's result, but Lowell went so far as to suggest that the hub and spoke pattern was produced by the "funnel-like indraught of air from the dark side to the bright and then an umbrella-like return of it." This, Lowell added, would "necessarily" lead to the transport of all the water to the night side where it would remain forever as ice!

At the time, there was no agreement about Venus's rotation period—indeed, the matter would not be settled until the 1960s. Some (like Schiaparelli) had proposed a very long rotation, one in which the rotation had been locked to the period of revolution, 225 days, by tidal forces from the Sun. A majority still favored a short Earthlike period, of 24 hours. But no one announced his results with the boldness and self-confidence of Lowell. It is notable that though Lowell's interpretations of the Martian phenomena he observed attracted critics, he was far from the only one to see canals on Mars. In the case of Venus, it was not his inferences but the data themselves that were in doubt.

Seeing him as something of an interloper, rushing in "where angels feared to tread," several astronomers ferociously attacked Lowell and the hub-and-spoke system. To Captain William Noble, first president of the British Astronomical Association, Lowell's map of Venus looked "suspiciously like Mars. I do not know whether Mr. Lowell has been looking at Mars until he has got Mars on the brain, and by some transference … ascribed the markings to Venus." The most respected observer of the day, E. E. Barnard, training the 12-inch refractor of George Ellery Hale's Kenwood Observatory at Venus while he waited the commissioning of the great 40-inch Yerkes refractor in the summer of 1897, entered in his observing log book: "no markings seen—though there were the usual suggestions of large dusky regions. It seems to be fully steady enough to have seen Lowell's narrow markings." In fact Barnard never would succeed in seeing them.

Lowell must at first have experienced these criticisms as a case of *lese majéstè*. But he was clearly unprepared for their ferocity, and arguably the stress of defending them undermined his health. On his return from Mexico, after a brief stay in Boston, he attempted to return to Flagstaff but made it only as far as Chicago before having to turn back. He reached Boston exhausted and a nervous wreck. The doctors prescribed that he take the then-fashionable "rest cure" in his father's Brookline mansion. He was allowed no visitors, no reading material, no distractions, and no intellectual activity. Deciding after a month that the cure was worse than the disease, he set off for Bermuda, and continued his convalescence in places like Virginia, Maine, New York, the French Riviera. It would be nearly four years before he would return to his observatory.

All this may seem like a long digression. In one sense, it is. Lowell's Venus observations were of little real importance, and would prove to be an astronomical dead-end. But his reaction to his critics and his determination to vindicate his observations would lead in fruitful directions, and produce one of the most paradigm-shifting discoveries ever made in observational cosmology.

During his long and enervating illness, Lowell turned over administrative supervision of the observatory's business to his brother-in-law, William Lowell Putnam II, who toward the end of 1898 was forced to dismiss See. That strange and colorful figure, besides being egotistical and self-serving—among other things, he wrote and published grandiose reviews of his own work in which he described himself as a "worthy successor of the Herschels"—behaved in a dastardly manner toward the other assistants (there were even rather explicit accusations of sexual improprieties). The "See affair" completely demoralized the Lowell staff, leading to the departure of several assistants, including Wilbur A. Cogshall who left to take charge of the Kirkwood Observatory at Indiana University (but will surface again briefly at a critical juncture, as we shall see). With See's departure, only Douglass remained. In 1899, on a visit to Boston, Douglass found Lowell still in poor shape. He "wants to work away on the observations of Venus," Douglass noted; "he is not working very hard but is nervous enough to have it worry him if all the material is not at hand."[11] Perhaps stimulated by this visit, Lowell and his secretary Wrexie Leonard set off on a short trip to Amherst, to observe Venus with the 7-inch refractor at the observatory there. These observations continued into 1900. A bit like "the man upon the stairs who wasn't there," the spoke-like markings reasserted themselves.

Fig. 9.6. V.M. Slipher and his family. *Courtesy: Lowell Observatory.*

Douglass, meanwhile, had increasingly begun to doubt the reality of some of the usual markings seen on planets, especially those on Venus. He carried out a series of experiments in which artificial planet disks were observed through the telescope. The results were unsettling, to say the least, as spurious markings were regularly recorded on the artificial planet disks, of the same general type as those recorded in drawings of planets.

Lowell, still in Boston but being kept apprised of developments, was of two minds about all this, and vacillated between ordering Douglass to cease and desist (on the grounds the experiments cast doubt on some observatory publications) and encouraging their continuation. Douglass was still being permitted to continue when, in April 1900, a bombshell landed from Russia. Aristarch Belopolksy, a highly esteemed astronomer at the Pulkova Observatory, published a spectrographic measure of Venus's rotation of 24 hours. But Lowell's hub-and-spoke system was compatible only with a very long, indeed an isosynchronous, rotation. Lowell did not admit defeat, but clearly the testimony of one spectrograph had to be answered by the testimony of another. So Lowell, in December 1900, ordered (on Douglass's recommendation) a powerful spectrograph from John A. Brashear, who had designed and built the Mills stellar spectrograph of the Lick Observatory. In ordering this expensive state-of-the-art instrument, Lowell was explicit about his purpose. It was "to test spectrographically the rotation period of Venus."[12]

Lowell was now well enough to contemplate returning full-time to astronomy. He had missed the Mars opposition of January 1899 completely, and though the next opposition occurred in February 1901, he was in England at the time, where he saw Nova Persei "through a hole in the clouds." He did not finally return to Flagstaff until the end of March, when Mars was already a long way off. Douglass, an ambitious scientist in his own right who had developed a strong independent streak during his boss's nearly four-year absence, was not exactly thrilled. In truth, he had become increasingly dissatisfied with Lowell's whole approach, and confided to his former mentor W. H. Pickering: "It appears to me that Mr. Lowell has a strong literary instinct and no scientific instinct." He also alluded to Lowell's "strong personality, consisting chiefly of immensely strong convictions."[13] He wrote (also in confidence) to Putnam, "His work is not credited among astronomers because he devotes his energy to hunting up a few facts in support of some speculation instead of perseveringly hunting innumerable facts and then limiting himself to publishing the unavoidable conclusions, as all scientists of good standing do, in whatever line of work they may be engaged," and ended, "I fear it will not be possible to turn him into a scientific man."[14]

Putnam kept Douglass's letter confidential for a while, but finally divulged the contents to his brother-in-law in July 1901. Lowell was not amused. He immediately fired Douglass, and fumed for a while in a solipsistic rage in which he resolved to soldier on alone without assistants. At this point Cogshall resurfaced to remind Lowell of an earlier offer he had made to hire on a temporary basis a young Indiana graduate, Vesto Melvin Slipher ("V.M." as he was always known). Lowell shot back: "I shall be happy to have him come when he is ready. I have decided, however, that I shall not want another permanent assistant and take him only because I promised to do so; and for the term suggested. What it was escapes my memory. If, owing to this decision, he prefers not to come, let him please himself."[15] Slipher indeed arrived on Mars Hill that August, and would remain employed at the observatory until his retirement in 1954, at the age of 79.

When Slipher arrived, Lowell was once more pointing his telescope at Venus (then appearing in the evening sky on the way to an evening elongation in December). Apparently, the usual radial markings were recorded (that particular observing logbook seems to have gone missing from the archives). However, though Lowell did not publish anything specifically about Venus that year, he was still evidently thinking a great deal about the problematic planet, for within a month of Slipher's arrival, Lowell, writing from Boston, was advising his new assistant "on the need for short exposures in photographing Venus," adding that experience had taught him that "the markings grew more and more difficult as the phase increased."[16] In the last part of September 1901, the Brashear spectrograph arrived on schedule, and Lowell ordered his assistant to "get the spectroscope [sic.] on and … work as soon as you can. Perhaps—and I trust this may be the case—you have already done so." To which he added, as an afterthought, "How fare the squashes?" (i.e., in Lowell's garden, of which he was inordinately proud).[17]

The query about the squashes was not to be the last in that line. Lowell appears in this correspondence as a fastidious micromanager who did not distinguish among specific roles for those he employed. His assistant astronomers were treated as a rather higher-class type of servant, and having grown up surrounded by servants, Lowell's expectation was that they glide about providing service efficiently without being themselves visible. If they drew attention to themselves, it was inevitably for falling short. And Lowell had a temper. Once, in a fit of rage, he threw a butler down the stairs of his Beacon Hill bachelor's residence, and to add insult to injury, threw the poor man's trunk after him; while in his 1906 book *Mars and Its Canals,* he complained that they didn't make servants like they used to and that even domestics were showing "arrogant independence." It is hardly surprising that whenever Lowell needed something, he turned to the nearest underling, no matter what the task or the employee's title, skills or experience. He never considered taking the time to match the employee to the task at hand. His secretary Wrexie Leonard was often asked to convey his displeasure over seemingly minor problems, like being sent the wrong cigars or a soup ladle with the wrong length of a handle. At other times she was asked to take her place at the eyepiece to verify whatever planetary markings he was studying at the time (and her drawings of Venus alone among those of his assistants capture with the same boldness and assurance the characteristic form of his hub-and-spoke system). In the same way, Slipher was not only expected to look after the squashes, he was also, especially in the early years of his employment, tasked with some very non-astronomer tasks—such as measuring and creating blueprints of all the new rooms in the sprawling Lowell residence known as the "Baronial Mansion," or even purchasing a cow for the observatory (something which the native of Mulberry, Indiana, was able to manage without breaking a sweat). Some of Lowell's requests seem almost humorously trite; for instance, Lowell writes to Slipher from Chicago, "Will you kindly see if shredded wheat biscuits are to be got at Flagstaff? If not please wire me at the Auditorium Annex [now the Congress Plaza Hotel]."[18] But despite the sometimes bucolic nature of his assignments, Slipher always insisted on his dignity, and was never casual. In fact, his manner was formal and rather stiff. He usually dressed in a suit complete with a vest, and "kept his tie perfectly knotted even when alone in the dark of the 24-inch Clark."[19]

Fig. 9.7. V. M. Slipher looking at the plate-holder of the Brashear spectrograph, which he used to obtain spectrograms of spiral nebulae. *Courtesy: Lowell Observatory.*

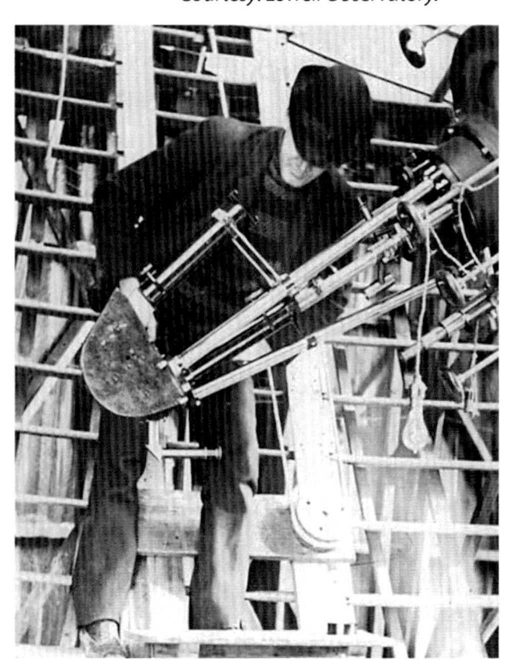

By assigning Slipher the task of getting the spectrograph in perfect running order, Lowell completely discounted the fact that though Slipher's degree had been in astronomy and mechanics he had no specific training in the use of such an instrument. Slipher managed to "get the spectroscope on" the 24-inch Clark refractor within two weeks of Lowell's letter to that effect, but his first attempts to use it left much to be desired; Lowell, with evident disgust, wrote to him that not only did his first spectrograms of Jupiter, with a rapid rotation of 10 hours, show the Doppler-shifted lines of a body in rotation, so did those of the star Capella and Venus (obviously not the desired result!).[20] Lowell corresponded about the problem with an official at the Brashear firm, who suggested that Slipher's disappointing results were due to "lack of adjustment."[21] Slipher can hardly have found the comment particularly illuminating.

The assistant astronomer's perplexity continued until, in December 1901, he suggested to Lowell that he ought to pay a visit to Lick Observatory to learn from the leading expert in astronomical spectroscopy at the time, W. W. Campbell. Campbell's specialty was measuring the radial velocities of stars, but he had also been one of Lowell's most ferocious critics, dating back to a devastating critique he had written of Lowell's book *Mars*, in which he had said, among other things, "Mr. Lowell went direct from the lecture hall to his observatory, and how well his observations established his pre-observational views is to be read in this book."[22] Needless to say, Lowell did not endorse Slipher's request; he told him that it was "inadvisable to go at present,"[23] and lest Slipher fail to catch the hint followed up with a letter that left no room for misunderstanding:

> I am glad the postponing of your desired trip to the Lick commends itself to you. You are quite right in supposing that everybody encounters the same snags; the only difference being the clever ones contrive to get over them themselves whereas the stupid ones have to have recourse to others.[24]

Fortunately, having grown up on a farm, V.M. had great practical sense, and was extremely clever with his hands. By mid-summer 1902 he had come far, and was producing spectrograms of Jupiter and Saturn that Lowell was bragging up to Brashear. Venus was still the main prize. On receiving an appointment as nonresident professor of astronomy at the Massachusetts Institute of Technology, Lowell sent its president, Henry S. Pritchett, prints of Jupiter's spectrum as "the first fruits of the new spectroscope, made by Brashear, in the hands of V.M. Slipher" and added, for emphasis:

> The spectroscope Mr. Brashear considers "the finest and best instrument we have ever made" under the date of Sept. 21, 1901. It was made so that I might be sure of the best possible instrument in the matter of Venus' rotation period.[25]

Lowell seems to have been a bit at sea about Venus at the time. Despite the fact that he registered the hub-and-spoke configuration not only in 1896-97 but also in 1899 (with the Amherst telescope) and in 1901, Douglass's artificial planet work and Belopolsky's spectrograms had clearly shaken him. In 1901, he engaged in his own series of artificial planet observations which led to what was (for him) almost unprecedented: a brief retraction, published as a note in the German journal *Astronomische Nachrichten,* in which he conceded, "the spoke-like markings are probably not upon the surface of the planet but are optical effects of a curious and—astronomically speaking—of a hitherto unobserved kind."[26] Meanwhile, he privately confessed to Slipher, "Belopolsky's sentence is a hard nut to crack."[27] When, in December 1902, he delivered at MIT a series of lectures on the Solar System, later published in book form, he sidestepped the Venus problem altogether. His lectures covered Mercury, Mars, Saturn and Its System, and Jupiter and Its Comets. Venus is noteworthy by its absence.

In 1903, however, Slipher served his master well and brought him apparent vindication. He obtained spectrograms in March of that year which implied, when the tilt of the absorption lines at opposite sides of the visible disk of Venus were compared, practically a null value for rotation. Though in his published report, Slipher claimed nothing more than evidence of a "slow" rotation,[28] Lowell, committing a logical fallacy, went further. Effectively retracting his earlier retraction about the spokes, he wrote to Pritchett at M.I.T. that Slipher's result "which was got without any bias from me and with every precaution of his own to prevent unconscious bias on his part and to eliminate systematic errors, is completely confirmatory of Schiaparelli's period and of my visual work here in 1896-97—the planet undoubtedly rotates in the time it revolves."[29]

Fig. 9.8. Percival Lowell, observing Venus by daylight, October 1914. *Courtesy: Lowell Observatory.*

There was never again to be any doubt in Lowell's mind, at least, about the rotation of Venus, or the planet's place in his Spencerian scheme of planetary evolution (he saw it as a world that had come to the end of its evolutionary career, succumbing to relentless tidal friction from the Sun to become a "planet corpse, circling unchanging, except for libration, around the Sun"). He re-observed the spoke-like markings again in 1903 and 1910, and made his last telescopic foray at the planet in October 1914—at the very moment when, a world away, the bloody standoff at Ypres was beginning in Flanders during the start of the first World War. On a bright sunny day he posed in the Clark dome for a photograph. He was dressed as always in sartorial splendor, his domical bald head covered with a newsboy cap turned back to front as he peered into the eyepiece of the Clark refractor in what has become one of the iconic images of astronomy.

Slipher's spectrograms were not, alas, quite as conclusive as Lowell thought, and the spectrographic analysis of Venus's clouds still failed to yield a specific result as late as the 1960s.[30] Nor has any satisfactory explanation ever been given for Lowell's system of hub and spoke markings; there is not even any consensus as to whether the markings he recorded were on the planet or somehow produced in his own eyeball. When one of us (W.S.) discussed Lowell's Venus map in a lecture to a group of ophthalmologists, he was informed by one of the experts that the Y-shaped marking so prominent in Lowell's drawings and map looked a good deal like a congenital cataract. No better explanation has ever been proposed.

The important thing about all this is that by 1903 Slipher's mastery of the spectrograph had convinced Lowell that his assistant was indispensable. Lowell's preoccupation was always with the planets. He now charged V.M. with investigations of the spectra of the giant planets, Jupiter, Saturn, Uranus, and Neptune, in which the assistant first recorded the dark bands now known to be due to the presence of methane, as well as to the spectroscopic detection of water vapor in the atmosphere of Mars. Slipher believed he had succeeded in detecting Martian water vapor by 1909. His result was vigorously challenged, however, by W. W. Campbell of Lick, who that same year took a spectrograph to the summit of Mt. Whitney, and failed to reproduce the Lowell result. When Mars was not near opposition (at which times Lowell commandeered the 24-inch Clark refractor for his ongoing visual studies of the planet) or when Lowell was on one of his frequent trips to Europe or back in Boston, Slipher had the independence to pursue his own studies. It's clear that the controversies of his employer affected him at times, and that he suffered at times from what might be called the "Lowell Observatory inferiority complex." Though he was fiercely loyal to Percival Lowell, he was also understandably eager to establish a reputation in his own right. Whenever he could, he made spectrographic measures of the radial velocities of standard stars, which led to extensive correspondence with Campbell at Lick and Edwin B. Frost at Yerkes about their radial-velocity programs (the latter long addressed him as "Mr. Victor M. Slipher"). Campbell and Frost were both willing to impart practical advice—though they did so in an often curt and condescending manner, which shows that they regarded the Lowell Observatory astronomer more as a novice in need of tutoring than as an already accomplished peer.

An idea of Slipher's sometimes painful situation can be gleaned from a Lowellian imbroglio in which he briefly and reluctantly became entangled. In 1908 Lowell had written an article for *Outlook* magazine in which he claimed rather provocatively that the Lowell 24-inch was relatively more powerful than the Lick 36-inch, based on the number of stars visible in a given field. Slipher sided with Lowell, which earned him a prompt rebuke from Campbell. Later, Slipher attempted to defend his role in the matter. In a letter marked "Private," he explained, rather plaintively, to Campbell:

> I wanted to write you at the time concerning the Outlook criticism, but was too busy. I deplore controversy much more than the case may suggest and I was induced to make reply solely and wholly because I am selfish enough to wish my own work to receive an unprejudiced hearing, while as you know when discredit falls on Lowell's work, it falls on mine, too.[31]

In yet a further demonstration of his desire for professional standing and respect, Slipher at about the same time privately began negotiations with Frost at the University of Chicago's Yerkes Observatory about doing a Ph.D. thesis, based on his spectrographic work at Flagstaff. (At age 34, he was still "Mr.," rather than "Dr." Slipher.) Frost was tentatively encouraging, and told Slipher: "As I understand it a residence of three quarters is insisted upon for the degree. So that work done at Flagstaff could probably not be counted as residence.... Nevertheless, it might be a good arrangement to secure data there and then bring it here for measurement and discussion. Probably your regular routine duties would interfere with giving as much time to your thesis while at Flagstaff as you would like." He added, "I am assuming of course that any arrangement you propose is endorsed by Director Lowell."[32] Slipher was sure of Lowell's "sanction," but the three quarters residence requirement proved a stumbling block—an insurmountable one unless the University would consider an exception:

> If the residence requirement could be reduced to two quarters it would mean a great deal to me for it would probably make it possible for me to meet the requirements with one less vacation—which would mean a year and a half or two years less time.... Inasmuch as I shall want most of all to do work in the department of physics you might think it worth while to speak of the matter to Professor [Albert] Michelson.[33]

Unfortunately, Frost replied, such an exception as Slipher asked for had never been made at the University, and in any case would have had to be approved by the chairman of the astronomy department, Forest Ray Moulton. Unfortunately, Moulton had just become engaged in a particularly nasty controversy with Lowell about the so-called planetesimal hypothesis of the Solar System's formation, which Thomas Chrowder Chamberlin, a University of Chicago geologist, and Moulton had published in 1905. Since the planetesimal hypothesis played a significant role in Lowell's instructions to Slipher to study the nebulae, there was some irony in this development, as we shall see. But to follow out to its end this particular strand of the story, Slipher, though he continued to correspond with Frost, never mentioned the University of Chicago degree (or his desire to study physics with the likes of Michelson) again. He didn't have to. With some wire-pulling from his employer, he received his Ph.D. from Indiana University later that same year. (Frost, however, who had finally learned to refer to Slipher as "Mr. V.M." rather than "Victor M.," continued nevertheless to refer to him ever after as "Mr.")

If history is not, as Napoleon cynically claimed, "lies agreed upon," it is certainly a matter of quirks and chances. Pondering "what might have been" is seldom useful, but in this case, the temptation is hard to resist. How would the history of astronomy have been different if V.M. had gone to the University of Chicago to do a Ph.D. in 1909? At 34, and already a veteran of practical research with the spectroscope, he might have made a brilliant impression, and been recruited to a career at Yerkes—or even Mt. Wilson. If so, he might have eclipsed the achievements of young Edwin Hubble, who in 1909 was still an undergraduate on the Hyde Park campus and soon to head to Oxford as a Rhodes Scholar, specializing in the study of law and Spanish. On the other hand, it is equally possible that he would not have decided to study the nebulae at all but continued to investigate the spectra of the outer planets or Frost's preferred area of study, the radial velocities of stars.

And this brings us back to Percival Lowell. In truth, Slipher's decision to study the nebulae was not spontaneously arrived at. It was the product of one of many brilliant—if somewhat quirky—brainstorms of Percival Lowell. As so often has happened in the history of discovery, Lowell's directive led not to the expected landfall but to a new and hitherto unsuspected continent, the demonstration of an expanding universe.

Despite his long-standing allegiance to the classical nebular hypothesis of Laplace, by 1905 Lowell had absorbed the *au courant* Chamberlin-Moulton hypothesis of the Solar System's formation, according to which encounters among stars (many of them spent suns, dark stars roaming like "tramps" through space, in Lowell's vivid phrase) might pull filaments of gaseous material and debris from one of the stars to form a spiral nebula. The gases would then condense to form small particles—"planetesimals"—and the planetesimals would then, by accretion, form planets and satellites.

The Chamberlin-Moulton hypothesis had challenged Lowell's Spencerian conviction of the inevitable advance of a nebula to planets in different stages of life and, during at least part of their life cycle, to life-forms including intelligent life as on the Earth and Mars. Now the Solar System was envisaged as beginning with a chance and haphazard event; the inevitability of the origin of life-forms on which Lowell had counted for his Mars theories had been struck out. Nevertheless, Lowell, first and foremost a literary man, at once sensed the inherent drama of the new scenario. While barely hinting at it when, a year after Chamberlin and Moulton published their hypothesis, he gave the Lowell Institute lectures in 1906 (published as *Mars as the Abode of Life*), he developed the idea fully in MIT lectures given in February and March 1909, published as *The Evolution of Worlds.* Lowell pictured the more massive dark star on its approach raising giant tides in the other, tearing it to pieces and spreading the debris in antipodal directions along the line of their relative approach. And this, he claimed, was precisely what appeared in the spiral nebulae:

> The form of the spiral nebulae proclaims their motion, but one of its particular features discloses more. For it implies the past cause which set this motion going. A distinctive detail of these spirals, which so far as we know is shared by all of them, are the two arms which leave the centre from diametrically opposite sides. This indicates that the outward driving force acted only in two places, the one the antipodes of the other. Now what kind of force is capable of this peculiar effect? If we think of the matter we shall realize that tidal action would produce just this result....

Suppose, now, a stranger to approach a body in space near enough; ... if the approach be very close, the tides will be so great as to tear the body in pieces along the line due to their action; that is, parts of the body will be separated from the main mass in two antipodal directions. This is precisely what we see in the spiral nebula. Nor is there any other action that we know of that would thus handle the body.[34]

Ironically, Lowell's deployment of features of the Chamberlin-Moulton hypothesis led Moulton to mount a savage attack in which he practically accused Lowell of plagiarism. Afterwards the two men were hardly on speaking terms. But already Lowell had put Slipher on the quest of vindicating his latest planetological scheme. Thus, on January 29, 1909, Lowell wrote to Slipher: "I think it might be fruitful if you were to make spectrograms with your red plates of a conspicuous green nebula (emission nebula) and then compare the lines as yet unidentified with the known ones of the spectrum of the Major Planets. It occurs that the two may be possibly related." Then, on February 8, 1909, he sent another letter adding:: "I would like to have you take with your red sensitive plates the spectrum of a white nebula—preferably one that has marked centers of condensation." Beneath his signature, he added two asterisked handwritten notes, "continuous spectrum" and "but I want its outer parts." He was thinking that the outer parts might betray the formation of new planets and show dark bands like those Slipher had been recording in the spectra of the giant planets, Jupiter, Saturn, Uranus and Neptune. Among the flurry of letters and requests, this was the really important one. But before Slipher could respond to these letters, there was further correspondence between the two men about many other matters. Slipher asked Lowell to send him and his brother E.C. to South America for the Mars opposition that year (eventually, Lowell declined). Then Lowell sent another letter to Slipher with several numbered requests: "Cannot you manage to take the spectrum of the 3rd satellite of Jupiter now for the red end of its spectrum to bring out any telluric lines it may show? I know it is difficult but I think you can bring it about." Also: "Did you ever publish your spectroscopic determination of Jupiter's rotation showing how close you got to a visible one?" It was the end of February before Slipher could get back to Lowell's letters about the green and white nebulae. A spectrogram of a green nebula would not be difficult, he explained, since the light is concentrated at a few points. The faint white nebulae with their faint continuous spectra would be another matter, however, and here Slipher did not sound encouraging:

> I do not see much hope in getting the spectrum of a white nebula because the high ratio of focal length to aperture of the 24-inch gives a very faint image of a nebula.... This would mean 30 hours with the 24-inch for direct photograph, and as the dispersion of the spectrograph [should] be at least 100 times the slit-width in order to get detail, it would seem the undertaking would have to await the [40-inch] reflector.[35]

The 40-inch reflector referred to here had been ordered by Lowell in order to push his Mars work to a new level, and to combat claims that the "canals," though evident in smaller telescopes, disappeared with larger aperture telescopes. Here was another of Lowell's brainstorms: in order to keep the telescope at uniform temperature, he had it mounted ten feet below ground. Though the telescope did in fact remain at uniform temperature, burying it underground proved disastrous, as a temperature inversion was created above the telescope causing terrible "seeing." On the whole, the telescope was to prove a great disappointment—especially for planetary work. Slipher would never use it for nebular spectrographs. In any case, he realized that he did not need a large reflector to obtain spectrograms of nebulae. The 24-inch Clark would do.

When Fath published his Ph.D. thesis on the spectra of the spiral nebulae in 1909, Slipher acknowledged, it "dulled somewhat the edge of my desire" to do work along that line. But when the two came face to face at the International Solar Union meeting in August 1910, they hit it off at once. It probably helped that both had come from rural backgrounds. Fath was clearly encouraging, and when Slipher returned from Pasadena, he began thinking seriously about Lowell's request that he try to capture the spectrum of a white nebula. By November he had devised a single-prism spectrograph, "from equipment on hand,"[36] which, he told Lowell, "requires only about a hundredth part of the exposure required by the three-prism arrangement."[37] In December, he succeeded in obtaining a spectrogram of the Great Nebula in Andromeda. It seemed to show faintly "peculiarities," which had not been commented upon by earlier workers in the field. This success showed him that he had been wrong in assuming that a large reflector with a short focal length would be needed to make such observations. In fact, he later explained, "no choice of telescopes—as regards aperture or focal-length or ratio of aperture to focus—will increase the brightness of the spectrum of an extended surface."[38] Instead, what he needed now was a fast camera. This insight led to the technical breakthrough that would assure his success. Though, as he later learned from Fath, the same insight had been published by Keeler almost two decades before, it was much to his credit that he had worked it out independently on his own.

Throughout 1911 and 1912 Slipher was often distracted by the mundane details of running the observatory when Lowell was away (and he often was in 1912, since Mars was not available for study and he was mostly in Boston working feverishly, with a team of human computers, trying to calculate a position for a trans-Neptunian planet, which he called "Planet X"; indeed, he worked so hard that at the end of the year, he suffered another mental collapse, but fortunately it wasn't as severe as that of 1897, and he returned to work after seven weeks). Slipher's main project that year was improving the observatory property, which included putting a fence around it, but he pursued his own research whenever he could and in September 1912 had equipped his spectrograph with the fast camera he had realized he needed. With this, a commercial Voigtlander F 2.5 camera lens, he expected to reach "something like 200 times the speed of the usual three-prism spectrograph."

Now, with the faster spectrograph attached to the 24-inch Clark, Slipher, on September 17, 1912, made an exposure on the Andromeda Nebula of more than six hours. Up until now his chief concern had been "the nebular spectra themselves"—he was following along the lines of Fath's dissertation work. But this spectrogram showed enough detail to make him think that a measure of the radial velocity of the nebula might be possible. Since, as noted earlier, Fath's rather primitive wooden-box spectrograph did not allow comparison spectra to be made, but the rigid metal Brashear spectrograph had been designed for such a purpose, he was now pursuing an investigation that was completely new. At the time, the radial velocities of some 1200 bright stars and a few bright planetary nebulae had been measured; all these objects moved at speeds on the order of tens of kilometers per second. Though no radial velocities had been measured for the white nebulae or spirals, no on expected at the time that they would be much different.[39]

Slipher made further attempts to obtain spectrograms of the Great Nebula in Andromeda on November 15-16 and December 3-4. Also in December, he achieved another momentous result—he obtained a good spectrogram of the nebula around the star Merope in the Pleiades, with a 21-hour exposure with the fast spectrograph. Previously, the Pleiades nebulosity, which had made the famous seven stars appear "like a swarm of fireflies tangled in a silver braid," as the poet Tennyson wrote, had been assumed to be gaseous. However, Slipher's spectrogram showed no trace of the emission lines associated with hot gas, but instead showed a continuous spectrum with exactly the same hydrogen and helium absorption lines that occur in the spectra of the hot young B-type stars of the cluster. Slipher correctly concluded that the Merope nebula was a reflection nebula; that it consisted of "disintegrated matter [dust as we should now call it] similar to what we know in the solar system, in the rings of Saturn, comets, etc., and that it shines by reflected starlight."[40]

At first, and naturally enough perhaps, Slipher tried to use the Pleiades reflection nebula as an analogy to what he was seeing in the Andromeda spectrograms. Thus, on December 28, 1912, he wrote to astronomer John C. Duncan, who had been a fellow at Lowell Observatory in 1905-06 and had completed his Ph.D. at Lick under Campbell in 1909, "If this nebula [the Merope nebula] shines by reflected light, why could not the nebula in Andromeda shine in the same way being lighted by a central sun obscured [sic] by the fragmentary material around it?"[41]

That same day, December 28, 1912, he began a heroic spectrogram of the Andromeda Nebula, exposing the plate over three nights and into the pre-dawn hours of January 1, 1913. The spectrum was continuous; i.e., there were absorption lines, which meant that the Great Spiral was star-like not gaseous. He also measured positions of some of these absorption lines—including the celebrated H and K lines of ionized calcium—against those in an iron-vanadium spark comparison spectrum. The first inkling of what he found came in his response to an overnight telegram from Lowell asking Slipher to rush materials, including a transparency of Halley's Comet in 1910, to him for an exhibit back East. "We shall get out the material for the exhibit as soon as possible," he assured his boss, then added the bombshell. "... Since writing you before I have got another spectrogram of the Andromeda Nebula.... I feel safe to say here that the velocity bids fair to come out unusually high. I should be able to send you something definite soon."[42] With classic understatement, Slipher had just announced what would prove to be one of the most momentous findings in astronomical history.

The absorption lines in the spectrum of the Andromeda nebula (M31) were markedly shifted toward the blue relative to those of the comparison spectra. This implied that, if the Doppler shift was a valid indicator of radial velocity (at first there were some doubts), M31 was approaching the Earth with a velocity of 275 km/sec. The blueshift of the lines was only 4 nanometers, but was readily measurable with Slipher's equipment, and gave a velocity that was some three times that of any other object in the universe measured so far.[43] Slipher could hardly believe his eyes! Ever cautious, he spent two more weeks measuring and remeasuring the plates, and for good measure, sent a print to Fath as an independent check. At last he was satisfied that the result was real. He wrote to Lowell giving his final value as 300 km/sec., and it has stood the test of time.[44] Lowell wrote back: "It looks as if you had made a great discovery. Try some other spiral nebulae for confirmation."[45]

Lowell's wish was his command. In April 1913, Slipher obtained a spectrogram of M104 (NGC 4594), the dark-laned spindle-shaped "Sombrero" spiral in Virgo. This time the absorption lines were shifted far to the red, indicating that the nebula was not approaching but receding from the earth, at the almost inconceivable velocity of 1100 km/sec. "This nebula is leaving the solar system," he told Lowell.[46] By the end of the year, Slipher had obtained spectra of two more spirals, M77 (NGC 1068) and NGC 4565. They were also red-shifted, with velocities of recession that were on the order of 1000 km/sec away from the Earth. By the middle of 1914 he had obtained spectrograms of several more spiral nebulae, making 14 in all. He found that M31 and its companion M32 were unique in having a negative velocity; the average velocity for all 14 was 400 km/sec and positive. The high radial velocities for the spirals were presented in a paper read for him at the American Astronomical Society meeting in Atlanta in December 1913. Henry Norris Russell, when he heard these results, was incredulous. Slipher read his own paper in August 1914, at its meeting in Evanston, Illinois. He noted that some of the highest radial velocities had been obtained for NGC 3115, M104, NGC 4594, and 5866, all of which were "spindle nebulae—doubtless spirals seen

Fig. 9.9. Pleiades, "like a swarm of fireflies tangled in a silver braid." The famous open cluster, located in Taurus at a distance of about 380 light years from us. As V.M. Slipher first showed in 1914, the Pleiades nebulosity is an example of a reflection nebula. The light from the bright B-type stars of the cluster is being reflected off a dark cloud of interstellar dust which, in this case, does not represent material from which the stars have recently formed but an independent cloud drifting through the cluster at a relative speed of about 11 km/sec. Imaged with a C-11 by Klaus Brasch.

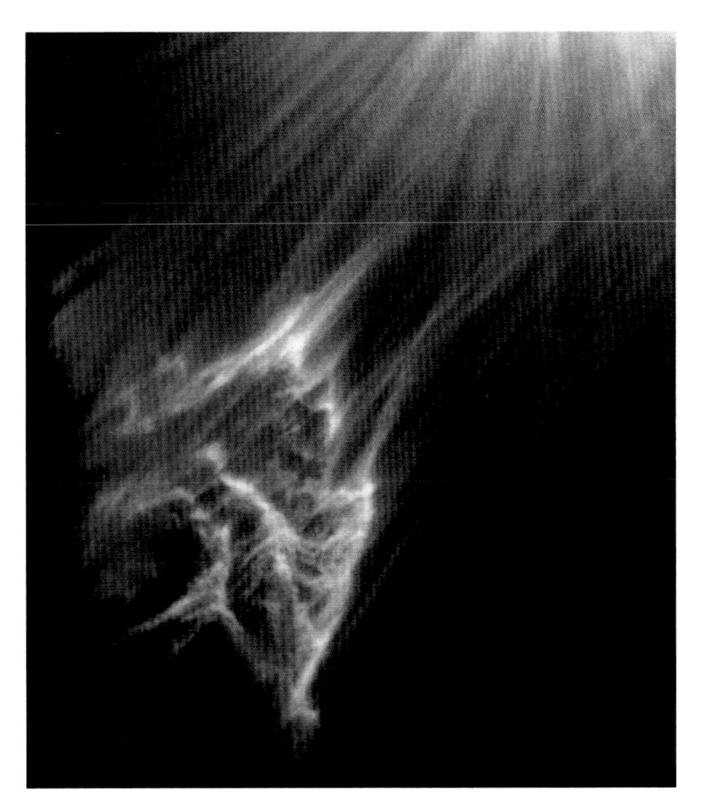

Fig. 9.10. (left) "Barnard's Merope Nebula," IC 349. This small area of exceptionally bright nebulosity was discovered by E. E. Barnard in 1890 with the 36-inch refractor at Lick Observatory. It is a reflection nebula like the other nebulosity of the Pleiades, and appears so bright because it lies extremely close—only about 0.06 light year—from the bright Pleiades star Merope. The cloud is being ravaged by the intense radiation from the star. The parallel wisps are produced as radiation pressure from the star decelerates dust particles. The smallest particles are slowed the most, so there is a sifting of particles by size. The straight lines pointing toward Merope are streams of larger particles that are continuing on toward the star while the smaller decelerated particles are being left behind in the lower part of the picture. *Courtesy: G. Herbig and T. Simon, NASA and the Hubble Heritage Team (STScI/AURA).*

Fig. 9.11. (right) M104, the "Sombrero" Galaxy in Virgo. Image with the Discovery Channel Telescope. *Courtesy: Lowell Observatory.*

edge-on." Their average velocity, 800 km, was greater than for the remaining objects. This suggested that the spirals might be moving "edge forward," rather like plates, through some kind of resisting medium. At the conclusion of his talk, he announced yet another important result: when M104 had been observed with the slit parallel to its "spindle," it had shown inclined spectral lines indicating that a measurable velocity of rotation, and he later showed that in nebulae with spiral arms, the central part of these nebulae were turning into the arms of the spiral as a spring turns in winding up. At the end of his short talk, Slipher received a standing ovation. Astronomers at least at that time were a generally staid and undemonstrative bunch; this was the first time any such spontaneous outbreak of enthusiasm had happened at an AAS meeting, and it would not happen again until 1951 when W. W. Morgan announced the discovery of the spiral arms of the Milky Way (see chapter 12). Despite the fact that he had been the star of the meeting, Slipher was modest and unassuming; in the group photograph of those in attendance meeting, he stands, obscured by others, at the end of the second row from the rear, while the young Edwin P. Hubble, who had not even begun graduate studies in astronomy yet (see chapter 11), stands self-confident and elegantly dressed in the first row,

Fig. 9.12. AAS group portrait, Evanston, Illinois, 1914. E.C. Pickering is the stout man in the front and center, Edwin Hubble is also in the front row, second from right, and V.M. Slipher is in the next to last row, second from left (just below and to the left of the woman with the white hat). *Courtesy: American Astronomical Society.*

having strategically positioned himself near AAS president E.C. Pickering and vice presidents George C. Comstock and Frank Schlesinger. That photo speaks volumes about the personalities of the two men.

Slipher was, as expected, cautious and tentative in interpreting his results, but Lowell, also characteristically, was more flamboyant. According to the *Chicago Tribune* of March 23, 1913, Lowell, stopping en route to New York, had announced that "recent discoveries at the Lowell Observatory, Flagstaff, Ariz., ... tend to confirm the hypothesis of the origin of solar systems... These discoveries ... show that many nebulae shine by reflected light, and that these nebulae consist of clouds of star dust enveloped in gas. 'This is the first step in the evolution of the solar system,' says Prof. Lowell's statement."[47] However, as Slipher's spectrograms accumulated, Lowell changed his mind, and in a talk on "Nebular Motion" given to the Melrose Club in Boston on November 23, 1915, he announced that spiral nebulae were not solar systems after all, but "something larger, and quite different, other galaxies of stars." This address was never published, but Lowell said much the same in a talk, "The Genesis of Planets," given to the Royal Astronomical Society of Canada in Toronto in April 1916 and published that summer:

> The birth of our present solar system was probably catastrophic. Two suns met and we were the outcome.... As to what subsequently occurred, we have no analogy to guide us, for the spiral nebulae indicate themselves to be galaxies, not solar systems in course of construction.[48]

Sadly, within months of the publication of this paper, on November 12, 1916, Lowell—apparently just after exploding in anger at a servant—died unexpectedly, of a massive intracerebral hemorrhage. After his boss's death, Slipher continued to add more spectrograms to the database. In a 1917 paper tabulating the results for 25 spiral nebulae obtained up to that time, he noted that except for M31, M32, and M33 (all Local Group objects) the velocities were all positive. On the basis of this limited material Slipher worked out that the Sun appeared to be moving in the direction

of right-ascension 22 hours and declination -22 degrees, with a velocity of about 700 km/sec. (The perceptive reader will note that this point is diametrically opposite the constellation Virgo, whose abundant spiral nebulae belong to the Virgo Supercluster of galaxies.) "For us to have such motion and the stars not show it means that our whole stellar system [galaxy] moves and carries us with it," he wrote, to which he added this extraordinary conclusion:

> It has for a long time been suggested that the spiral nebulae are stellar systems seen at great distances. This is the so-called "island universe" theory, which regards our stellar system and the Milky Way as a great spiral nebula which we see from within. This theory, it seems to me, gains favor in the present observations.[49]

Slipher's work provided provided strong evidence in favor of the view that spirals were extragalactic systems. He might have received more credit than he did at the time but for a new development. Just as his earlier work had been overshadowed in part by the controversies involving his boss, Slipher's achievement was now obscured by another astronomer's mistake.

Van Maanen's Detour

In 1916 Adriaan van Maanen, a Dutch astronomer at Mt. Wilson who specialized in the precise measurement of star positions on plates, began comparing plates of the spiral nebula M101 taken over intervals of several years, and thought he could detect motions of knots in its spiral arms which suggested that it was in rotation. Subsequently, he published results for another spiral, including M33, in which the knots were more clearly defined than in M101. Though a rotational motion was not unexpected—indeed had been suggested by the spiral form ever since Lord Rosse's pioneering observations—the magnitude of the motions, on the order of 0.02 arcsec/year, implied that if these spirals were indeed extragalactic systems, they must be rotating at speeds at or above the speed of light. Moreover, it was eventually realized (but not until 1924) that contrary to Slipher's spectrographic results, van Maanen's measures implied that the spiral arms were not winding up but unwinding.[50] Though van Maanen's results are now acknowledged to have been as illusory as the canals of Mars,[51] they sowed confusion in the late nineteen-teens and early 'twenties, and slowed the recognition of the extragalactic nature of the spirals for several years after Slipher demonstrated their high radial velocities.

During these years Slipher remained prominent in the study of nebulae, and in 1921 obtained his "personal record," a spectrogram showing that NGC 584 in Cetus was receding at about 1800 km/sec, making it the fastest-moving object yet discovered.[52] He was now reaching the limit of nebulae accessible to his spectrograph and the 24-inch refractor. He played a leading role on the Commission on Nebulae of the International Astronomical Union but was weighed down by worries he faced as director of the Lowell Observatory in the years after Lowell's death. Lowell's litigious widow, Constance, refused to agree to the terms of the founder's will, and started a lawsuit that was not settled until 1925. Of the part that remained after the lawyers were paid off, half went to her and the other half to support the observatory. As a result, for years afterward, the observatory would be famously broke. At times staff had to work without pay, and living conditions for assistant astronomers were appalling. In addition, Lowell's nephew and Sole Trustee of the observatory, Roger Lowell Putnam, was urging Slipher and the other staff astronomers to turn their hand to the rather uncongenial task writing a book on Mars (it was never completed, and they were granted a reprieve by Putnam on Clyde Tombaugh's discovery of Pluto in 1930—the fruit of another Lowellian enthusiasm, the search for "Planet X"—which probably saved the observatory from oblivion).

Perhaps the difficulty of making ends meet and the fact that he had reached the limit of what his instrumentation allowed him to do in the nebular field explains Slipher's increasing neglect of astronomy. He continued to be a diligent, if not very creative, director, while at the same time turning more and more to business, managing rental properties and investing in Flagstaff's Monte Vista Hotel. He was very successful in these pursuits, and ended life as a very wealthy man (though significantly, he did not leave any of his money to the observatory when he died).

Though he remained personally loyal to Percival Lowell and continued to support the idea that conditions on the planet could support lower forms of life, perhaps on the order of lichens, Slipher was not unaware of the fact that the observatory's reputation had suffered because of Lowell's flamboyance and penchant for controversy. He and his colleagues, E. C. Slipher and C.O. Lampland, published sparingly in later years. They arguably overreacted to the criticisms Lowell had faced by becoming overly cautious. But this makes it sound as if Percival Lowell's influence on them was entirely negative. Clearly, V.M. needed Percival Lowell's stimulus. William Putnam III, the current trustee, says, "Uncle Percy was the cattle prod that kept V.M. Slipher productive."[53] Less colorfully, John S. Hall, in his obituary of Slipher in *Sky & Telescope,* put it this way:

> Slipher and Lowell had complementary temperaments. The latter was brilliant, enthusiastic, and a driving personality.... Slipher, on the other hand, was deliberate, fastidious, patient, and showed a high order of technical knowledge.[54]

Each possessed something of what the other lacked. Together, they achieved one of the most paradigm-shifting results in the history of astronomy. The discovery of the large radial velocities of the spiral nebulae was as unexpected as Becquerel's discovery of radioactive, and made in the pursuit of an entirely different result.

That result was the spectrographic validation of Percival Lowell's now long-forgotten (and probably illusory) markings on the surface of Venus. The observatory, devoted to the study of the objects of the solar system, had come a long way from its eccentric frontier beginnings, and had unexpectedly blazed a trail far beyond the solar system in revealing the fundamental result that would underpin the modern concept of the expanding universe.

* * *

1. Donald E. Osterbrock, "The Observational Approach to Cosmology: U.S. observatories pre-World War II," in R. Bertotti, R. Balbinot, S. Bergia, and A. Messina, eds., *Modern Cosmology in Retrospect* (Cambridge, England: Cambridge University Press, 1990), 247-289:289.
2. Remington P.S. Stone, "The Crossley Reflector: a centennial review-I," *Sky & Telescope*, October 1979, 307-311.
3. For a full-length biography, see: Donald E. Osterbrock, *James E. Keeler: Pioneer American astrophysicist.* Cambridge: Cambridge University Press, 1984. For Keeler as astronomical "Adonais," see Donald E. Osterbrock, John R. Gustafson, and W.J. Shiloh Unruh, *Eye on the Sky: Lick Observatory's first century.* Berkeley, Los Angeles, and London: University of California Press, 1988.
4. Osterbrock, "Observational Approach to Cosmology," 255.
5. Ibid.
6. *Boston Commonwealth,* May 26, 1894.
7. C. Flammarion, *La Planète Mars et se conditions d'habitibilitè.* Paris: Gauthier-Villars, 1892.
8. Stephen J. Gould, "War of the Worldviews," *Natural History,* 12/96-1/97, 22-33.
9. William Graves Hoyt, *Lowell and Mars* (Tucson: University of Arizona Press, 1976), 106.
10. Percival Lowell, "Detection of Venus' rotation period and fundamental physical properties of the planet's surface," *Popular Astronomy,* 4 (1896), 281-285; "Determination of the rotation period and surface character of the planet Venus," *Monthly Notices of the Royal Astronomical Society, 57* (1897),148-149; "The rotation period of Venus," *Astronomische Nachrichten* no. 3406 (1897), 361-364; "Venus in the light of recent discoveries," 5 (1897), 327-343.
11. A.E. Douglass to Daniel A. Drew, October 18, 1899; Lowell Observatory Archives.
12. Percival Lowell, 'Venus in 1903." *Popular Astronomy,* 12 (1904), 184-190.
13. A.E. Douglass to W. H. Pickering, March 8, 1901; Lowell Observatory Archives.
14. A.E. Douglass to W. L. Putnam, March 12, 1901; Andrew Ellicott Douglass Papers, Special Collections, University of Arizona Library.
15. Percival Lowell to W.A. Cogshall, July 7, 1901; Lowell Observatory Archives.
16. Percival Lowell to V.M. Slipher, September 20, 1901; Lowell Observatory Archives.
17. Percival Lowell to V.M. Slipher, October 7, 1901; Lowell Observatory Archives.
18. Percival Lowell to V.M. Slipher, January 4, 1903; Lowell Observatory Archives.
19. Gale E. Christianson, *Edwin Hubble: Mariner of the Nebulae* (New York: Farrar, Straus, and Giroux, 1995), 92.
20. Percival Lowell to V.M. Slipher, October 21, 1901; Lowell Observatory Archives.
21. Percival Lowell to V.M. Slipher, October 28, 1901; Lowell Observatory Archives.
22. W.W. Campbell, "Mars" by Percival Lowell; *Publications of the Astronomical Society of the Pacific,* 51 (1896):207.
23. Percival Lowell to V.M. Slipher, December 18, 1901; Lowell Observatory Archives.
24. Percival Lowell to V.M. Slipher, January 3, 1903; Lowell Observatory Archives.
25. Percival Lowell to J.A. Brashear, October 6, 1902; Lowell Observatory Archives.
26. Percival Lowell, "The markings on Venus," *Astronomische Nachrichten,* no. 3823 (1902), 129-132.

27. Percival Lowell to V.M. Slipher, November 5, 1902; Lowell Observatory Archives.
28. V.M. Slipher, "Spectrographic Investigation of the Rotation of Venus," *Lowell Observatory Bulletin 3*, June 1903.
29. Percival Lowell to V.M. Slipher, March 23, 1903; Lowell Observatory Archives.
30. Pol Swings, "Venus through a spectroscope." *Proceedings of the American Philosophical Society*, vol. 113 (1969).
31. V.M. Slipher to W. W. Campbell, February 6, 1909; Lowell Observatory Archives.
32. E.B. Frost to V.M. Slipher, Dec. 24, 1908; Lowell Observatory Archives.
33. V.M. Slipher to E.B. Frost, Dec. 31, 1908; Lowell Observatory Archives.
34. Percival Lowell, *The Evolution of Worlds* (New York: Macmillan, 1909), 24.
35. V.M. Slipher to Percival Lowell, February 26, 1909; Lowell Observatory Archives.
36. V.M. Slipher to E.A. Fath, December 5, 1910; Lowell Observatory Archives.
37. V.M. Slipher to Percival Lowell, November 9, 1910; Lowell Observatory Archives.
38. V.M. Slipher, "Spectrographic observations of Nebulae"; paper presented at AAS meeting in Evanston, Illinois, August 1914.
39. William Graves Hoyt, "Vesto Melvin Slipher, 1875-1969," *Biographical Memoirs of the National Academy of Sciences, 52* (1980) (Washington, D.C.: National Academy Press), 411-449:423.
40. V.M. Slipher, "On the spectrum of the nebula in the Pleiades." *Lowell Observatory Bulletin, 2,* 55 (1913), 26-27.
41. V.M. Slipher to John C. Duncan, December 29, 1912; Lowell Observatory Archives.
42. V.M. Slipher to Percival Lowell, January 2, 1913; Lowell Observatory Archives.
43. V.M. Slipher to E.A. Fath, January 18, 1913; Lowell Observatory Archives.
44. V.M. Slipher to Percival Lowell, February 3, 1913; Lowell Observatory Archives. The result was published as: "The radial velocity of the Andromeda Nebula.: *Lowell Observatory Bulletin, 2,* 58 (1913), 56-57.
45. Percival Lowell to V.M. Slipher, February 18, 1913; Lowell Observatory Archives.
46. V.M. Slipher to Percival Lowell, April 12, 1913; Lowell Observatory Archives.
47. As noted in: Anne Minard, *Pluto and Beyond: a story of discovery, adversity, and ongoing exploration* (Flagstaff, Arizona: Northland Press, 2007), 41.
48. Percival Lowell, "The Genesis of Planets," *The Journal of the Royal Astronomical Society of Canada*, 1916 (July-August), 281-293:281.
49. V.M. Slipher, "Nebulae," *Proceedings of the American Philosophical Society, 56* (1917), 403-410.
50. On van Maanen, see: Norriss S. Hetherington, *Science and Objectivity: episodes in the history of astronomy* (Ames, Iowa: Iowa State University Press, 1988), 83-110.
51. A number of scholarly studies have attempted to resolve the question of how van Maanen could have erred. The most thorough study was carried out by Richard. Hart, "Adriaan van Maanen's Influence on the Island Universe Theory" (Dissertation, Boston University, 1973). This work is summarized in Richard Berendzen, Richard Hart, and Daniel Seeley, *Man Discovers the Galaxies.* New York: Science History Publications, 1976. See, for instance, on p. 143: "the changes he was attempting to measure were at the very limits of precision of his equipment and techniques." Also, on p. 150: "A ... computer analysis was performed in which various types of errors in measurement were assumed and incorporated into the computational scheme actually employed by van Maanen. The analysis showed that a systematic error of only 0.002 mm in measuring the positions of points on photographic plates (i.e., on the order of the accuracy obtainable with the measuring engines) could produce the reported internal motions, provided the direction of the measurement error is such that it is consistent with the spiral features of the plate. That is, if van Maanen had a slight personal bias toward believing that the spirals were in rotation (a bias easily created merely by looking at a picture of a spiral), his results would reflect this bias. This could account for the remarkable consistency of his results because the measurements made at the very limit of perceptibility are extremely sensitive to precisely such errors." Ironically, there are many parallels between this sad episode and that of the canals of Mars.
52. V.M. Slipher, "Two nebulae with unparalleled velocities." *Lowell Observatory Circular,* Jan. 17, 1921.
53. Quoted in Minard, *Pluto and Beyond,* 23.
54. John. S. Hall, "V.M. Slipher's trail-blazing career," *Sky & Telescope,* 39, 2 (1970), 84-86.

10. The "Galactocentric" Revolution

Make no little plans; they have no magic to stir men's blood...
—*Daniel Burnham, Chicago architect*

At the end of the 19th century, the largest telescope in operation was a refractor 40 inches in diameter. It ruled for a few years into the 20th century, when it was surpassed by the 60-inch reflector—called the first "modern" telescope—of Mt. Wilson Observatory.[1] By the century's end, the distinction of the world's largest telescope would be shared by two giant reflectors, Keck 1 and Keck 2, on Mauna Kea in Hawaii, with mirrors 394 inches (10.4 meters) in diameter, while its most productive and awe-inspiring would be the Hubble Space Telescope, a 94.5-inch (2.4 meter) reflector orbiting in outer space above the tumultuous sea of the atmosphere of the Earth.

The man who oversaw the transition from the era of great refractors to that of great reflectors was George Ellery Hale. Undoubtedly the greatest astronomical entrepreneur of all time, Hale always dreamed big. He built the largest refractor of the refractor era, and the first great reflector of the reflector era. He surpassed the latter with an even larger reflector, and at the time of his death, he had dreamed up—and raised funds for—an even larger one. Each telescope he dreamed up and brought into realization would be used by astronomers to push farther and farther into space. By the middle of the 20th century, those telescopes had surveyed the Milky Way and its sister systems, and had set cosmology, hitherto an arena for inconclusive speculation, well on the road to becoming an exact science.

The tale of how all this came about begins with a devoted father's extraordinary gift to an extraordinary son.[2] George Ellery Hale was a child of the "Gilded Age," a scion of Chicago's wealthy aristocracy. He was born in

Chicago in 1868, three years before the Great Fire that destroyed much of the city. His father, William Hale, "with the boundless energy and tenacity his son would inherit," contributed mightily to building a new Chicago. In place of "everlasting sham, veneer, stucco, and putty," a city of massive steel edifices, the forerunners of modern skyscrapers, rose on Lake Michigan. William Hale built the elevators that made the skyscrapers of Chicago possible.

Fig. 10.1. George Ellery Hale; from *The World's Work,* 1912; accessed via Wikipedia Commons.

By contrast, Hale's mother, Mary, lived as a virtual recluse. She was always of delicate constitution. The nature of her illnesses is not clear, but one of George's earliest memories was "of the upstairs bedroom where his mother, a semi-invalid with thin lips, firm chin, and brown eyes deeply set in her gaunt face, spent most of her time."[3]

George inherited his mother's high-strung temperament, Calvinist brooding, and depression, in combination with his father's expansive optimism and superhuman drive for achievement. While his father approved of George's strivings after success, his mother fretted over him and worried about his stomach troubles, backaches, and fainting spells. At the nearby Oakland Public School, where George began his education, he was frequently sick, and after an attack of typhoid, which his doting parents blamed on the school, they decided he must never return. At the age of 12 he was sent to the private Allen Academy, halfway between the Hales' rented home in Kenwood and the center of Chicago.

Already in his pre-adolescence George Ellery Hale became a precocious dabbler in science, and was busy making observations with small microscopes and telescopes. All this activity worried his mother; she feared that "with his intensity and precocity, he would burn himself out early."[4] Nevertheless, when young George decided he couldn't live without his own research laboratory, he persuaded her to turn over to him the small room where she kept her dresses. In this "shop" he set up a Bunsen burner, batteries, and galvanometers, and carried out thrilling chemical experiments such as pouring hydrochloric acid on zinc to form hydrogen gas.

At 13 he installed a lathe in the shop, which he ran with a home-made steam engine known as "the demon." Fed by a second-hand boiler, it was capable of generating an eighth of a horsepower. It made the entire house shake; the steam pressure mounted "toward forbidden heights ... the roar of the exhaust, mingled with the vibration of the speeding engine, brought joy to the excited engineers."[5] Though Hale's mother, rushing on the scene and crying out in horror, was convinced that the speeding engine was about to explode, "for years," as Hale's biographer Helen Wright reflects, "'the demon' continued to run the lathe without mishap."[6]

Hale too seemed to be driven by an inner speeding engine—a "demon," whose workings were sometimes as terrifying and mysterious to others as the noisy workings of this steam-engine. Time and time again, those who knew him in these early days described him in terms that implied boundless energy, enthusiasm, restlessness, and drive. He was a perpetual-motion machine, possessed of "a driven power which was given no rest until it brought his plans and schemes to fruition," according to a later observation by the British physicist James Jeans. "Excited over everything he was doing, he continued to run from one scheme to the next."[7] Friends and colleagues spoke always of "his restless energy and intensity,"[8] his hurry to do things, his inexhaustible curiosity, passion, and enjoyment of everything.

And yet, as with "the demon," there was something unnerving about all this energy. The vibrations and the hissing of the steam engine, the shaking of the building, suggested a machine working near the limits of its capacity. Too presciently, Hale's mother worried that, under similar high pressure, mounting "toward forbidden heights," Hale too would break down or burn out.

While still in his teens, Hale had already started networking with astronomers and instrument-makers near and far. The near included Chicago Court Reporter and amateur astronomer Sherburne W. Burnham, who spent nights searching for double stars with a 6-inch refractor mounted in a dome in South Chicago which his neighbors referred to as the "Cheese-box," and Dearborn Observatory's Jupiter specialist George W. Hough. The far included Pittsburgh instrument-maker John Brashear, whom Hale traveled by train specifically to meet. In 1886, on recommendation of his father's associate, the architect Daniel H. Burnham, he enrolled at the Massachusetts Institute of Technology (then known as Boston Tech), where his classmates recognized him as someone who, on the rare occasions when he left his lab to play tennis, played as he worked, at fever pitch. When he walked he seemed to run, and one of his classmates' most vivid memories was of a small, slight figure tearing through the halls of "Tech," across Copley Square, his dark cape flapping in the wind.[9] However, he saw himself as morose and dissatisfied, and complained to a good friend that "though I slave steadily, I don't accomplish a thing."

Bored with his physics courses at MIT, Hale persuaded E.C. Pickering at the Harvard College Observatory to allow him to spend Saturdays working at the observatory as a volunteer researcher. His family had, by then, moved into a large mansion at 4545 Drexel Boulevard in the Kenwood area of Chicago (it still exists, but has been converted into posh condominiums). By the time Hale was a second year student at MIT, his father had built for him a spectroscopy lab on a vacant lot next to the home. His first surge of scientific creativity occurred during in his last year at MIT. On one of his regular visits to Chicago, he was riding a tram past a picket fence, and in a flash of insight conceived the idea of designing an instrument to draw a slit aperture across the face of the Sun while a photographic plate was being moved synchronously over a corresponding slit at the other end of a spectrograph. This was the kernel of the idea of the spectroheliograph, which he developed for his senior thesis. It marked the beginning of what would be his consuming research interest, the Sun.

Hale graduated from MIT in 1890, and two days after his graduation married Evelina Conklin—their honeymoon included a trip to Lick Observatory to see the 36-inch refractor. Writing forty years later, Hale still recalled "his first sight of that long tube, the largest telescope in the world, pointing up in the darkness of the great round dome toward the slit that seemed to him to be an opening into heaven."[10]

After considering and rejecting solid but unexciting and traditional job offers to teach astronomy at colleges like Beloit College in Wisconsin (and accepting an unpaid position as professor at the latter, just for the title, though he was never in residence and never taught there), Hale decided that the best offer was to stay in Kenwood, where his father agreed to build him a private observatory, equipped with a 12-inch refractor, next to the family mansion. As one of the principal experiments of the new "Kenwood Observatory," Hale hoped to perfect his spectroheliograph, which had the ability to image the Sun in one specific wavelength, such as that of ionized calcium or hot hydrogen gas. He was soon using it to photograph prominences in the light of ionized calcium, and recording disturbed areas (flocculi) all over the solar disk. To follow up these important discoveries would have meant long hours at the telescope, but Hale never had the time or inclination for this dogged type of observing. Instead, he hired a full-time assistant, Ferdinand Ellerman, to do the routine work.

Meanwhile, in Hyde Park just south of the family mansion in Kenwood, the University of Chicago was taking shape with the financial help of John D. Rockefeller, the most golden of the robber barons of the Gilded Age. Hale, the "young man in a hurry," who was only 24, but whose assets included his own spectroscopic laboratory, private observatory, and wealthy father, naturally came to the attention of William Rainey Harper. As president of the university at the age of only 35, Harper was hardly more than a boy himself, and signed Hale on as an associate professor in 1892. Hale made clear that this was only the first—not the last—step of his ambitions. "I would not consider the thing for a moment," he remarked, "were it not for the prospect of some day getting the use of a big telescope to carry out some of my pet schemes." Indeed, he was soon angling for funds to build such a telescope.

Fig. 10.2. Dome of the Yerkes refractor. *Photograph by William Sheehan, 2007.*

Already, he was showing a remarkable ability to raise funds for his projects. Somehow he managed to persuade the notoriously tight-fisted elevated train magnate Charles Tyson Yerkes, nicknamed "the Boodler," to pay for the lens of the largest telescope in the world, and one that would "lick the Lick." (However, Yerkes would not pay for the mount or the lovely neo-Romanesque building in which the telescope would be installed, and John D. Rockefeller had to be "stung" for the remaining funds.)

While it was being built, Hale suffered from a great deal of anxiety over his "great observatory," and during the planning and execution was continually driving himself near the limits of his strength. His main consolation during those stressful days was the Sun. Whenever he could, he hurried back to his beloved Kenwood telescope to observe it, and despite the pressures of setting up the great

Fig. 10.3. Detail of masonry: John D. Rockefeller being "stung" for the money to complete the observatory. The "bee" was later removed. *Photograph by William Sheehan, 2007.*

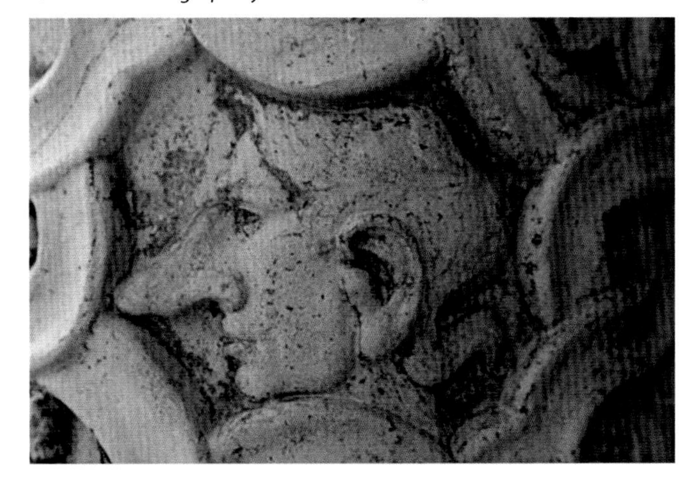

observatory, "in these glorious hours he was excited by everything he did." In 1897, he saw the 40-inch refractor, still the largest telescope of its type, standing in its vast dome at the Yerkes Observatory on the shore of Lake Geneva in Wisconsin.

The official dedication was delayed after the elevator floor of the giant dome collapsed; miraculously, no one was injured, and the telescope survived unscathed, though the floor was completely wrecked, and had to be rebuilt. The dedication took place on October 21, 1897. Trustees, faculty (including Hale) and guests of the University of Chicago, 700 in all, arrived on special trains, and listened as Yerkes expressed his confidence that "in your attempts to pierce the mysteries of the universe which are spread before you by our great Creator, the enthusiasm of your natures will carry you to success." Keeler gave the main address on "The Importance of Astrophysical Research to Other Physical Sciences."[11] "There may be some," he observed, "who view with disfavor the array of chemical, physical, and electrical appliances crowded around the modern telescope, and who look back to the observatory of the past as to a classic temple whose severe beauty had not yet been marred by modern trappings." Whatever the regrets, change was inevitable; mankind could not help "rushing forward with the utmost speed," and astronomy—or rather its increasingly obstreperous stepchild astrophysics—could no longer concern itself entirely with "where the heavenly bodies were," but had begun to consider "what they were." But this was a task which required increasingly technical investigations into the structure of matter and the nature of radiation.

During the first year of the observatory's existence, Hale joined in the work, "with his own boundless enthusiasm for everything he entered into ... unable ... to comprehend the indifferent or perfunctory." On one occasion, a comparatively laid-back European astronomer commented:

> They work too hard at the Yerkes Observatory. Morning, afternoon, and night the work seems to go on continuously.... The center of it all is the 40-inch, which never rests. The whole performance is splendid, and strikes awe into the beholder if he happens to come from lands where folk still retain the mistaken idea that one ought to rest every now and then.[12]

As early as 1898, Hale already realized that the new, largest refractor in the world was really the wrong telescope in the wrong place at the wrong time. Though his personal research interest was the Sun, his energies as a scientist and entrepreneur of astrophysics encompassed the broader question of stellar evolution.

As a youth he had hunted for fossils along the Lake Michigan shore, and absorbed the idea of a slowly evolving Earth together with Charles

Darwin's ideas about the origin of species. He imaginatively projected these principles on to a universe of evolving stars. Even at his private Kenwood Observatory, and more so at Yerkes, he saw the observatory as a special kind of physical laboratory adapted to the solution of this great problem. He also grasped that the future lay not with the refractor but the reflector, and so even before the 40-inch had seen first light, he asked his father to purchase for him a huge disk of high-quality glass for a mirror around which to build a large-aperture reflector, which could feed a large, fixed, flexure-free spectrograph, and which he could erect on a clear-sky mountain-peak. As the last major gift of a doting father to his son, William Hale paid for George to acquire a magnificent 60-inch glass disk from the Saint-Gobain glassworks in Paris. A piece of glass of any size is a far cry from a finished telescope, and Hale did not know where the money would come from to bridge the gap, but he was a gambler with a vision, banking on his certainty that a way would be found.

At least he had some idea of who might fashion the glass disk into the mirror of a great telescope. Some years earlier, he had met George Willis Ritchey at a local astronomical meeting. Only a few years older than Hale, Ritchey at the time was a high-school woodworking teacher with a young wife, an infant daughter, and a second child on the way. In the spare time remaining after work and family he worked in a small home shop he called his astronomical laboratory, quietly reinventing the reflecting telescope. In 1896—the same year Hale acquired the 60-inch glass disk—Hale hired him away from his teaching job to join him at Yerkes. An obsessive perfectionist, just as one would expect of a meticulous instrument-maker, Ritchey brought with him to Yerkes a superb 24-inch mirror he had made in his home workshop, and spent the next four years putting it at the heart of an exceptionally fast (short focus) telescope with a sturdy fork mount. When it became operational, he used it to make a series of astronomical photographs, some of which surpassed anything that had been done before. Clearly, this was a prototype for the 60-inch Hale dreamed of, and just as clearly, Ritchey was the man to help him realize his dream.

When, in 1901, the multimillionaire Andrew Carnegie sold the Carnegie Steel company to J.P. Morgan for $480 million and then decided to give everything away—along with other philanthropies, he proposed to set up a lavishly funded "institution" to finance research and discovery—Hale's die was cast. He now began planning in earnest to found a remote observing station of Yerkes Observatory in California, to be centered on the new 60-inch reflector. In June 1903, he visited Wilson's Peak, near Pasadena, whose exceptional seeing had been known ever since W.H. Pickering had prospected the site for Harvard College Observatory and placed a temporary observing station there in 1889-90. Hale decided that this was where he wanted to build his new observatory.

To others at the time, Hale could seem impetuous and mercurial. Lick Observatory double-star astronomer William J. Hussey wrote to his director, W. W. Campbell, "It appears that Hale's characteristic is to seize one idea and to forget all the rest; to oscillate from one point of view to another. One day he thinks Mt. Wilson the best site in the United States—the next he wants Flagstaff tested; and a day later ... he thinks San Bernardino Mountains fulfill conditions." Hale returned to Chicago in very high spirits indeed, and wrote to John S. Billings: "There is certainly no observatory site at present known which seems to offer the advantages ... at Wilson's Peak."[13] But soon afterwards, the executive Committee on Astronomy of the Carnegie Institution shelved his recommendation for a "solar observatory," as he had called it, at Wilson's Peak—in part because the project had seemed too "grandiose." Hale now became gloomy and depressed, his future plans "far from clear," apart from his certainty that wherever they led, they would one day lead him back to California. In time, his prospects improved again. Carnegie awarded him $10,000— a fraction of what he would need, but it kept his dream alive. On arriving back in California in December 1903, his depression quickly vanished. His wife and children joined him there; henceforth he would not return to Yerkes Observatory except during the summers. "I do not believe it possible that I could ever tire of such glorious conditions," he wrote.[14] He bloomed with the eternal sunshine, the effects of which were exhilarating, and was often in the company of the attractive Los Angeles socialite Alicia Mosgrove. When they were together, he said, the "problems of his life seemed to vanish."

At the Mount Wilson Solar Observatory, a self-contained research center, which Hale eventually succeeded in separating from Yerkes Observatory and the University of Chicago, his career paralleled his father's and other industrial capitalists'—though in science rather than in business. The "second industrial revolution" of the late 19th century, which in America spanned the period between the Civil War and the First World War, "was entwined with the shift from the disorganized entrepreneurial capitalism of the earlier nineteenth century to the organized capitalism of our time." It was also marked by the "rationalization of economic life—the drive for maximum profits through the adoption of the most efficient forms of organization moving into higher gear."[15] Even at Yerkes, he had begun to learn how to become a "master of operating a monopolistic observatory."[16] He recognized that the application of spectroscopy to the study of celestial bodies, the emergence of astronomical photography, and the availability of silvered mirrors for building large reflecting telescopes were creating "boom" conditions for astrophysics—the new blend of astronomy and physics which exhilarated him. It was a period of remarkable opportunity in science analogous to the one his father and other titans of Chicago enterprise had exploited in the post-Civil War era in American business, an opportunity that would respond to some of the same methods of large-scale organization.

In science Hale was, with a few others, such as E. C. Pickering at Harvard and W. W. Campbell at Lick, foremost in recognizing the possibility of creating "a monopolistic observatory, the biggest and most successful in the world, and in organizing combines, in the form of scientific societies and international unions."[17] He would—especially at Mt. Wilson —become the chief architect of this transformation of American science, a transformation from a confused and fragmented activity, in which individuals such as Burnham, Barnard, Henry Draper and Hale's own optician Ritchey, could still play a role, to big science, corporatized science, in which the individual was subordinate to, if not crushed by, large institutions and massed resources. It was this transformation of science—paralleled in industry and social life—in which Hale emerged as a Promethean figure.

At the center of his vision was the 60-inch telescope. After the first $10,000 from Carnegie ran out, Hale supplemented it with an additional $15,000 from his own pocket, betting that the Carnegie Institution, of whose scientific advisory committee Hale was now himself a member, would eventually agree to his plan for a mountaintop astrophysical observatory to the study of the Sun and the problem of stellar evolution. To complement the solar work, it would include a large reflecting telescope for high-resolution spectroscopy of stars. Late in 1904, as his expenses and his worries mounted, he had still not received Carnegie's support. Finally, to his great relief, he learned that Carnegie had granted him $150,000 a year for two years to establish his observatory, and this Carnegie grant opened a floodgate. Hale's California plans, held in abeyance for much of the previous year, now tumbled out. The "Mount Wilson Station of the Yerkes Observatory" became the "Mount Wilson Solar Observatory"; Hale resigned as director of Yerkes (and was succeeded by Edwin B. Frost), and gave up his faculty post at the University of Chicago. He also took with him much of the "first team" at Yerkes: Ritchey, Ferdinand Ellerman, his one-time assistant at the Kenwood Observatory who became the first of a line of legendary observers at Mt. Wilson, and Walter S. Adams, a pioneering spectroscopist. He took a 99-year lease on forty acres of the summit, and purchased the property on Santa Barbara Street in Pasadena where he built a permanent home for the observatory's support facilities. On the mountain, living quarters were improved, and the Snow solar telescope—the first of the mountain's permanent solar instruments—was moved there from Yerkes and put into operation.

Hale used a 5-foot spectroheliograph with the Snow telescope and, with Ellerman, showed that the bright hydrogen filaments known as flocculi, which showed up in H-alpha photographs, often clustered around sunspots like the iron filings around the poles of a magnet. The Snow telescope, however, proved difficult and balky to use, and was superseded by a 60-foot solar tower telescope, completed in 1907 (and still in use). The 60-foot solar tower was used by Hale to make an ingenious series of observations leading to his greatest discovery—the splitting of spectral lines into doublets, known as the Zeeman effect, in sunspots, which suggested that they were the seat of strong magnetic forces.[18] Published in 1908, this result was hailed as the greatest discovery in solar physics, and led to his being nominated for the Nobel Prize in physics (but he never received it; it was the one honor he would never receive).

At the time of the Carnegie grant, the 60-inch mirror—stalled for several years after Ritchey had completed the first fine grinding—remained in the optical shop at Yerkes, but the glass belonged to Hale, who had offered it to the University of Chicago on the condition that it provide a mounting and a dome. The university had plans for neither, and Hale now repossessed the mirror. He settled with the university for costs incurred for any work done on it thus far, and signed it over to the Carnegie Institution. He also instructed Ritchey, who had not yet come west, to send it, post-haste, to Pasadena. Ritchey did so, after designing for it a special case, equipped with double sets of heavy steel springs, to smooth its ride over the Continental divide as it wended its way from chilly Wisconsin to sunny southern California.

The contracts for the large steel and cast-iron parts of the mounting which Ritchey had also designed were awarded to the Union Iron works of San Francisco, which also won the contract for the 60-inch dome. As the large pieces were churned out in Union's factory on the shore of San Francisco Bay, the mounting was assembled for testing. It survived, miraculously undamaged, the devastating 1906 earthquake and fire that destroyed much of the city. The telescope was then disassembled, packed, and shipped to the Santa Barbara Street shops, where the huge parts were machined to a perfect fit, the intricate driving clock assembled, the gearing cut, and the complex mirror support system fabricated, transforming more than twenty tons of castings and steel assemblies into an instrument of exquisite perfection.

When the project had started, the best route up Mt. Wilson was a little more than a trail, in places as narrow as two feet, used by hikers who wished to spend a few days roughing it at one of two camps near the summit (and the occasional itinerant astronomer like E. E. Barnard who assailed it with telescopes in tow). The trail was now widened to 10 feet to make it ready for a specially designed truck, driven by an electric motor at each of its four wheels, to carry the heavy pieces of the 60-inch up the mountain. The truck worked well, though mule teams that continued to carry everyday provisions and the lighter parts of the telescope to the summit were often called in to provide a "power assist." All the while Ritchey, ever the perfectionist, was toiling over the mirror in the optical shop, which had been designed with every precaution against contaminating dust that could insinuate itself into the polishing and scratch the mirror's surface. Windows were double-sealed, the air filtered, walls and ceilings heavily shellacked; canvas baffles hung above the grinding machine, and the floor was kept constantly damp—measures that were the equivalent of the first sterile surgical operating rooms. On a morning in mid-April 1907, just as only finishing touches remained, the mirror was found to be covered with scratches—a mystery that has never been explained—and Ritchey had to rework the mirror to perfection. He was finally satisfied late in 1908 that the mirror was perfect enough to be received in the mounting and dome which were finished and waiting.

The first visual observations were made on December 13; the first plates exposed a few nights later. Ritchey knew at once he had achieved a masterpiece. Unfortunately, he became possessive of it; he was reluctant to open the completed instrument to the scientific staff, who for their part were chomping at the bit to begin the many long-range scientific programs, such as obtaining high-resolution spectrographs of stars, they had planned for it.

Fig. 10.4. Sixty-inch reflector, ready for a night of observing Mars at the Newtonian focus. *Photograph by William Sheehan, 2005.*

Even before the 60-inch was completed, Hale had arranged for the casting of an even larger glass blank, which Ritchey was to grind and polish for a 100-inch telescope. But now strains began to develop between the two men, which would lead to Ritchey's dismissal after he completed work on the 100-inch mirror in 1917. Hale himself, meanwhile, was headed toward a breakdown. For several years, his friends and colleagues had been concerned about the toll that running a scientific corporation had been taking on him. After the 60-inch was placed in working order, his colleagues Georgio Abetti and Harold Babcock noted that work on the mountain seemed to be moving "at an extraordinarily rapid pace."

Only to Hale did the progress seem painfully slow. After leaving Yerkes, he had once again founded the greatest observatory in the world; he was absorbed in research on the magnetic fields of sunspots, was full of plans to revive the National Academy of Sciences in Washington, D.C., and was scheming to transform the Throop Polytechnic Institute in Pasadena into a technological university that would be the West Coast equivalent of MIT (he did; Throop became Caltech). He also dreamed of establishing an organization for astronomy on an international scale (the Solar Union). He had been working at top speed, without relaxing. Now, however, the vibrations and shaking of the engine, "the demon," within him were becoming more evident. He told staff astronomer Babcock that he was having "terribly hard dreams." To the pioneer English spectroscopist William Huggins he confided: "Last summer I had a great deal of trouble from nervousness, and my physician told me I ought to give up my work." He ignored his physician's advice for a time, seemed to recover for a while, but was soon incapacitated again by headaches, bouts of indigestion, and chronic insomnia. He found himself unable to concentrate for any length of time.

Fig. 10.5. Smiling Hale at the International Solar Union meeting at Mt. Wilson, 1910. Soon after this photograph was taken, Hale suffered a mental breakdown. *Courtesy: Yerkes Observatory.*

The crisis came in 1910. Mt. Wilson benefactor Andrew Carnegie and paleontologist Henry Fairfield Osborn paid a visit to the observatory. Osborn observed with Hale till two o'clock in the morning with the 60-inch. During a pause, Osborn turned to Hale and said: "This is grand, but I am worried about one thing." Hale: "What's that?" Osborn: "Why, the most precious instrument, and the one most difficult to replace.... George Ellery Hale." Hale admitted he was tired, but was dismissive of Osborn's concerns; all he needed was a good fishing trip, then he would be rested and recovered for the International Solar Union meeting he planned for August 1910 at Mt. Wilson. But he overestimated his powers of recuperation. At the meeting, which was a virtual "coming out" party for the 60-inch, with astronomers and astrophysicists coming from all over the world to admire the great telescope, Hale appeared for only one day. Unable to face work or people, he hid with the shades drawn in his office, cowering under his desk, before departing precipitately with his wife for a fishing trip to Lake Tahoe where he hoped to pull himself together again. Again, he overestimated his powers of recuperation. In a near-panic, he moved restlessly on to Oregon and more fishing before heading off to Europe with Evelina, in order to "travel about as [they] might choose, to new places and old, avoiding all scientific men and institutions, and renewing youth in a second wedding journey." The mention of "renewing youth" sounds almost pathetic in a man only 42 years old (the same age as Keeler when he died). In fact, he was worse off than he cared to admit; in England, astronomer David Gill found him "looking exceedingly ill and suffering from severe nervous pains and noises in his head, symptoms that demanded rest from all excitement." Having as their eventual destination Egypt, which Hale adored, the couple traveled on to Paris, where Hale worried about the 100-inch disk being cast at the Saint Gobain glass works, and on to Mentone, in southern France, where Hale had a strange experience, intimately tied up with the severe anxious depression he later referred to as his "breakdown." His biographer Helen Wright described the appearance of a "little elf," which Hale connected with a ringing in his ears.[19] This phrase suggests a hallucinated vision, and Miss Wright seemed to believe he really saw an elf. There is, however, little justification for that. What Hale actually had experienced is described in a 1911 letter written in Rome on the way back from Egypt, where he refers to a demon not an elf, and where the term is clearly meant to serve as a metaphor for his depression:

> I have reached a stage where it seems almost impossible to keep interested in anything—except, of course, the forbidden things in my own line forcing themselves upon me. Until I got back from Egypt I was able to read, with pleasure, a great variety of books. But now I can't keep my mind on the subject, as a little demon stands by my side, and every few minutes prods me with the suggestion that, after all, the book is not interesting, and that all my attention belongs to him.... If I could only do a little of my regular work there would be no difficulty. But work excites me and sets the back of my head to aching, and so appears to be out of the question.[20]

On his return from Egypt, Hale finally did as Evelina and his personal physician, James H. McBride, had been encouraging him to do, and checked himself into an asylum in Bethel, Massachusetts run by an acquaintance of McBride's, Dr. John G. Gehring. In the same fanciful vein in which he had personified his depression as a "little demon", Hale wrote a rather tormented letter he wrote to McBride soon after his arrival in Bethel, in which he sounds hardly sane:

Mon Cher (mais, helas!—ci-devant—) Maître

There can be no doubt that I have fallen into the hands of a most dangerous and malevolent Seer and Magician. Whatever sympathy you might have for him—your Rival—will be swept away by the logic of my narrative, and you will be forced to grind your teeth in impotent rage. For willing as I am to admit and to maintain the superiority of the Great Muldoon [i.e., William Muldoon, 1852-1933, professional wrestler and physical culturist] as the Exalted Head of a Kur Ost, I have yet to learn of any pretensions on his part to magic art. Would that he possessed it in trifle measure, that he might fly to my aid and break the shackles that bind me to the Fiend! But I must unfold the tissue of deviltry that has been so about me since the ill-fated day I left the Abode of Peace, now so far away....

Yesterday morning I made my first visit to my future abode [i.e., Gehring's sanitarium]. As I approached the end of the street, I saw two great gate-posts of stone and on them was written "Per me si na villa citta dolente." But through the Necromancer's art the meaning of these words was hidden, and as I entered I seemed to be in Elysian Fields, decked out with four trees and covered with green sward. The grim realities of the Dungeons were also concealed by a goodly Castle, with large and pleasant rooms, well set with chairs and couches and not lacking in many and good books to charm the unwary reader. Anon the Wizardess [Mrs. Gehring] appeared.... Then came the Wizard [Gehring] himself, and his piercing eye transfixed my very soul.[21]

What Hale called his "head troubles" continued the rest of his life, and led to his resignation as director of Mt. Wilson Observatory in 1922. Walter S. Adams, who had served as assistant director for a number of years, succeeded him. In his resignation letter, Hale detailed his medical history and problems from his "preliminary nervous attack" in 1908 through his severe breakdown in 1910, and estimated he had not enjoyed one-third working capacity during the previous sixteen years. He later admitted another reason for stepping down: "I have never been skillful in carrying on steady ... work," he told J. C. Merriam, the director of the Carnegie Institution. "In fact, I am a born adventurer, with a roving disposition that constantly urges me toward new long chances." As soon as he had resigned, he set out for his beloved Egypt, and while he was there, Howard Carter made the sensational discovery of the tomb of Tutankhamen.

As 1924 began, Hale could only endure life by avoiding other people. He did succeed in nerving himself to take part in the dedication of the National Academy of Sciences building, but during the final preparations he was unable to stand the strain; as he had done at the International Solar Union meeting at Mt. Wilson, he fled—this time to New York City.

The doctors there diagnosed him as suffering from high blood pressure and being in a "psychoneurotic state." At the same time, he was fund-raising again—this time for an even larger telescope than the 100-inch. He succeeded in raising a fund of six million dollars for a 200-inch telescope, whose completion, as he must have known, he would not live to see.

Hale desperately tried to find relief in travel and by reliving the triumphs of his youth. He built his own private solar observatory, decorated with Egyptian motifs. It was a new version of the Kenwood Observatory his father had once built for him. Having earlier tried two stints at Dr. John G. Gehring's sanitarium in Bethel, Massachusetts, without much benefit, in 1927 he gave Dr. Austen F. Riggs's sanitarium at Stockbridge, Massachusetts, a try. It was then as now an asylum for the very wealthy. The course of treatment included a strict vegetarian diet, which helped his blood pressure; his head, alas, remained "about the same."

During the last decade of his life, he spent much of his time at his private solar observatory; going to his lab there automatically meant getting away from his wife, and with its pleasant library and solar spectroheliograph in the basement, with which he could indulge his own research, it became a refuge from the anxieties, bouts of depression, and psychoneuroses with which he had so long contended. The grail of his research was trying to detect the existence of the Sun's general magnetic field, which he thought he had discovered in the 1910s. It proved elusive (later, when Harold and Horace Babcock did detect it, using photomultiplier tubes and electronic circuits undreamed of in Hale's scientifically active years, it was found to be much weaker than Hale had supposed; his earlier result had been an illusion). Happiness proved no less elusive than the Sun's general magnetic field. His checkbooks, found in the solar observatory after his death, contained receipts of payments almost all of which went to psychiatrists. The psychiatrists seem to have done him no good. He complained that he had "very little energy," and "was accomplishing almost nothing." After suffering a mild stroke on a train near Toronto in 1936, his wife and brother brought him back to Chicago and then on to Pasadena. Described guardedly by family members as often "disturbed" or "confused," he spent the last year and a half of his life as a patient in Las Encinas Sanitarium. There, in 1938, the entrepreneur of modern astronomy passed quietly away. One of the finest tributes to his almost superhuman achievements was paid to him by Allan Sandage, one of the outstanding observers who used the Hale telescope at Mt. Palomar: "The longer I work in astronomy the more overpowering is the conviction that we owe all to Hale and his dreams and positive actions to put those dreams into glass and steel. Where would world astronomy be today if Hale had not been an 'empire builder.'"[22] But there was a price to be paid. His greatness in organizational matters came at a tremendous cost in mental anguish—or perhaps the greatness and the anguish were two aspects of the same thing, the makeup of his own mind. Certainly he had breakthrough ideas, and none of his contemporaries succeeded as he did in making his dreams of big telescopes and great observatories come true.

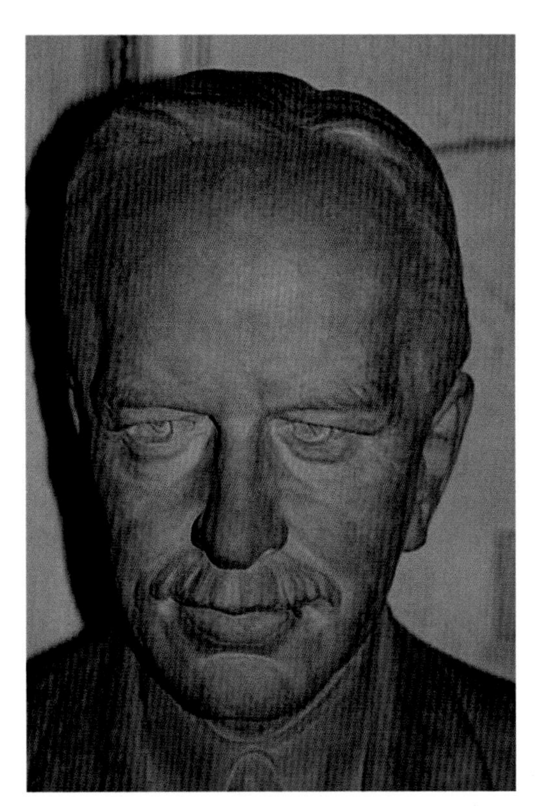

Fig. 10.6. A lovely terra cotta model at Mt. Wilson for Caltech bust of Hale. This model makes clear the massiveness of Hale's frontal lobe, a feature that may help explain why he was so forward-looking and such a visionary. *Photograph by William Sheehan, 2005.*

Until the appearance of the 60-inch, Mt. Wilson had belonged to the daytime astronomers who used two large solar telescopes that comprised the observatory's first scientific apparatus. But the arrival of the big reflector initiated a new era not only in the scientific life of the observatory, but also in its social life which, as long as observing had focused on the Sun, had been regulated by its risings and settings. Walter S. Adams, recounting those early days when work was done by sunset, recalled nightly gatherings for cards and conversation in the "Monastery"—the astronomers' communal living quarters, whose name was inspired by the monasteries of the Levant, and from which even the astronomers' wives were banished on the grounds that their presence would interfere with the great work being done there. He wrote that "Hale's amazing breadth of interests, his great personal charm, and his stories of important figures in science and national affairs make those evenings stand out in memory." But with the arrival of the 60-inch, this convivial routine was largely undone by the introduction of night-work.

Fig. 10.7. Harlow Shapley. Painting at Harvard College Observatory. *Photograph by William Sheehan, 2004.*

Adams recalls the Smithsonian Institution's solar observer Charles Greeley Abbot's "pathetic lament" that the 60-inch telescope had "spoiled" the mountain.

When the 60-inch reflector first went into operation in 1908, the standard view of the Galaxy was that offered at the turn of the century by the Dutch astronomer Jacobus Kapteyn. As William Herschel had done in 1785 in proposing his "grindstone model of the universe," Kapteyn relied on star counts, but because he did not know about the effects of extinction of starlight by interstellar dust, he could still regard the Galaxy as a small disk of stars. He estimated that it measured a mere 4000 parsecs long by 100 parsecs wide, and like Herschel's "grindstone" (and for the same reasons) Kapteyn's universe remained centered on a point at or close to the Solar System—a fact that made even Kapteyn himself uncomfortable, given how poorly geocentric and anthropocentric models had fared in the past (remember Copernicus!). As he wrote in 1909 in an article "On the Absorption of Light in Space":

> We may conclude that if there is a thinning-out of the stars for an increased distance from the Sun, it must be so in whatever direction from the Sun we proceed. This would assign to our Sun a very exceptional place in the stellar system, viz., the place of maximum density.
>
> On the other hand, if we assume that the thinning-out of stars is simply apparent and due to absorption of light, the apparent thinning-out in any arbitrary direction is perfectly natural.[23]

It was Kapteyn's smallish star-system that Hale set out to investigate with the 60-inch reflector. Between 1909 and 1914, Kapteyn himself spent the summers as a research associate on Mt. Wilson, staying in a small residence still known as Kapteyn cottage, and served as Hale's most influential adviser.

The most important work ever done with the 60-inch reflector was that of Harlow Shapley, who came to Mt. Wilson in 1914 as a recently minted Princeton Ph.D. Then aged 29, he had already traveled "by rugged ways to the stars," the title he would later give to his (rather gossipy) autobiography. Born one of a set of fraternal twins in 1885 on a farm outside the small town of Nashville, Missouri, on the edge of the Ozarks, Shapley's formal education was acquired in a one-room schoolhouse whose teacher was his older sister Lillian and, he recalled, from reading the Sunday edition of the *St. Louis Globe-Democrat.* When at 15 he left home to enter business school in Pittsburg, Kansas, he had the equivalent of a fifth grade education. After finishing the course he signed on as a crime reporter for a newspaper in the small town of Chanute, Kansas then in the midst of an oil boom and boasting a Carnegie Library in which he broadened his education. He added a stint as police reporter in Joplin, Missouri, a lead and zinc town, before he decided to do as his sister had urged and get a college education. After two semesters at the Carthage Collegiate Institute, he gained admission to the University of Missouri, in Columbia, intending a career as a newspaperman. Finding, however, that the journalism course in which he wished to enroll had been delayed a year, he became an "accidental astronomer"—opening up the university catalogue to decide on a course of study, he claimed (but surely in jest) that since he didn't even know how to pronounce the first entry, "archaeology," he settled on the course listed on the next page, "astronomy," which he could pronounce.

After a slow start, Shapley's pace quickened, and he was soon making dizzying progress up the ladder of academic success. He completed his B.A. with highest honors in 1910, only three years after arriving on campus, completed his M.A. the following year, and in 1911 went east to Princeton, where he studied on a fellowship under Henry Norris Russell, the co-creator of the HR diagram, who was already emerging as a leader in the study of stellar evolution. With Russell, Shapley began to work on a Ph.D. dissertation on binary stars. About half the stars in the sky belong to binary or multiple-star systems. These include eclipsing binaries, in which the Earth lies nearly in the stars' orbital plane and so one star can pass in front of (or eclipse) the other—the best-known example is Algol, the "Demon," in Perseus. One third of Shapley's work involved making photometric measures of the "light curves" of the eclipses with Princeton's 23-inch refractor, the other two-thirds using a slide rule to calculate the orbits. At first intended to be a rather straightforward project, it soon, Shapley found, "got beyond thesis size."[24] In addition to data such as the orbital period and inclination of the orbit, Shapley and Russell managed to work out the "mean densities" of these star systems. Some of them proved to be only a millionth that of the Sun—these stars, now known as red giants, were, in Shapley's phrase, "enormous gas bags."[25] Their analysis also gave a rough idea of how far away these eclipsing binaries were. "The one thing that surprised us," Shapley noted, "... was that the distances were pretty darned big."[26]

Shortly after completing his Ph.D., Shapley visited the Harvard College Observatory, and in the dome of the 15-inch Merz refractor entered into a discussion with Solon I. Bailey. Bailey had directed the Harvard Boyden Station in Peru since 1892, and had identified on plates taken with its 24-inch refractor a number of variable stars in globular clusters. Shapley would remember Bailey as "pious and kind, a wonderful sort of man, but so New England it made you ache. He said, '... We hear that you are going to Mount Wilson. When you get there, why don't you use the big telescope to make measures of [variable] stars in globular clusters?'"[27]

Some of Bailey's variables were Cepheids. In contrast to eclipsing variables, Cepheids (called after their prototype, delta Cephei, discovered by a deaf amateur astronomer, John Goodricke, in 1784) are indi-

Fig. 10.8. Henrietta Leavitt. AAVSO photograph; accessed via Wikipedia Commons.

vidual stars whose variations in brightness are due to actual pulsations in the stars. These stars undergo regular expansions and contractions, and as they do they undergo changes in temperature and luminosity. Shapley himself was later to suggest, in a paper that was half footnotes, that these stars were actually "throbbing or vibrating" in some way; it was an inspired guess, and it has held up.

However, the Cepheids had a special property that would make them among the foundational objects of the cosmic distance scale. As Bailey's colleague at Harvard, Henrietta Swan Leavitt, had just discovered, the Cepheids with the greatest average brightness also underwent their pulsations most slowly.

In contrast to the flamboyant and self-centered Shapley, Miss Leavitt, who had been born in 1868 of old New England stock on both sides of the family, was one of the quiet, steady workers in astronomy. Bailey wrote of her that she had "inherited … the stern virtues of her puritan ancestors. She took life seriously. Her sense of duty, justice and loyalty was strong. For light amusements she appeared to care little…. [She] was of an especially quiet and retiring nature, and absorbed in her work to an unusual degree."[28]

After graduating from Radcliffe, the all-women's counterpart to Harvard in Cambridge, Massachusetts, in 1892, Miss Leavitt had worked for several years as an advanced student and volunteer research assistant at the Harvard College Observatory before becoming a permanent member of the staff. She specialized in the determination of photographic magnitudes of stars. Among her projects was the careful examination of a large number of plates of the Magellanic Clouds that had been taken between 1893 and 1906 with the Boyden Station's 24-inch refractor and forwarded from Peru to Cambridge for analysis. On these plates, she discovered 1,777 variable stars. By 1908 she had succeeded in working out the periods of several in the Small Magellanic Cloud. They seemed to be just like those Bailey had found in globular clusters, diminishing slowly in brightness,

Fig. 10.9. Small Magellanic Cloud and the globular cluster 47 Tucanae, imaged by John Drummond with a 200mm lens at Gisborne, New Zealand. *Courtesy: John Drummond.*

remaining near minimum for the greater part of the time, and increasing very rapidly to a brief maximum—their brightness curves thus resembling that of delta Cephei and other variables of the Cepheid type. Moreover, she noticed that the ones that appeared brightest on the plates also proved to be those with the longest periods. For instance, the faintest one on her plates varied between magnitudes 14.8 to 16, and had a period of 1.25 days; the brightest, varying between 11.2 and 12.1, had a period of 127 days.

This was important because, as Miss Leavitt knew, the breadth of the Small Magellanic Cloud is small compared to its distance from us. Thus all the stars in the Small Magellanic Cloud are effectively the same distance from us—just as, say, every person in Times Square in New York City is at effectively the same distance from everyone in London. The stars that appeared to be the most luminous really were the most luminous, and what's more, for the Cepheids, their luminosities were keyed to their periods. She first published this result in a 1908 paper. No one paid any attention. In another paper published four years later, she added a plot showing the magnitudes vs. the logarithm of the periods for a somewhat larger sample of Cepheids in the Small Magellanic Cloud. The plot was exactly linear.[29] This meant that the Period-Luminosity relationship for Cepheids was not just a correlation, it was a law of nature.

This was a tremendous breakthrough, and in due course the relationship discovered by this modest unassuming woman would lead to the extension of the distance ladder from the nearest stars—those whose distances had been directly measured through their parallaxes, which at the time included only about 60 stars out to a distance of about 30 light years or so—to the truly cosmic. Ejnar Hertzsprung was the first to realize this. He also realized that in order to use the Cepheids as "standard candles" for determining astronomical distances as great as those to the Small Magellanic Cloud, it was necessary to calibrate the zero-point of Miss Leavitt's plot—i.e., identify Cepheids in the Milky Way whose distances were known so that the apparent brightnesses of the stars on Miss Leavitt's plot could be turned into an absolute brightness scale. The sticking point was that all the Cepheids were very far away. There was none near enough to have a distance determined by the usual parallax method (involving measures using the 2 AU baseline of the Earth's annual orbit around the Sun), so Hertzsprung instead used a clever variation called statistical parallax. Here one takes advantage of the fact (found by William Herschel) that the Sun is moving relative to the other stars with a speed of 13 km/sec in the direction of Hercules, and thus travels about 4 AU a year through space. This furnishes a potentially longer baseline for any stars just near enough to have a detectable proper motion across the line of sight. One assumes that since Galaxy is neither collapsing nor expanding, the random motion of a set of stars relative to the Sun will average to approximately zero, and so by a statistical treatment one can estimate the average distance to these stars. It is not necessary to go into all the details, other than to say that the importance of the method is that some of the groups of stars whose distances were measured in this way included Cepheids. Hertzsprung's sample included only 13 Milky Way Cepheids, meaning his statistical power was not great. Nevertheless, he used this result to estimate a parallax

Fig. 10.10. Period-luminosity relation of Cepheids in the Small Magellanic Cloud. Henrietta Leavitt has plotted the logarithm of the periods in days against the apparent magnitudes of 25 variable stars in the Small Magellanic Cloud. The straight lines have been drawn through the maximum and minimum magnitudes. As is evident from the apparent linearity of the plots, the absolute luminosity is a linear function of the period. *From Harvard Observatory Circular, No. 173 (1912), 1-3.*

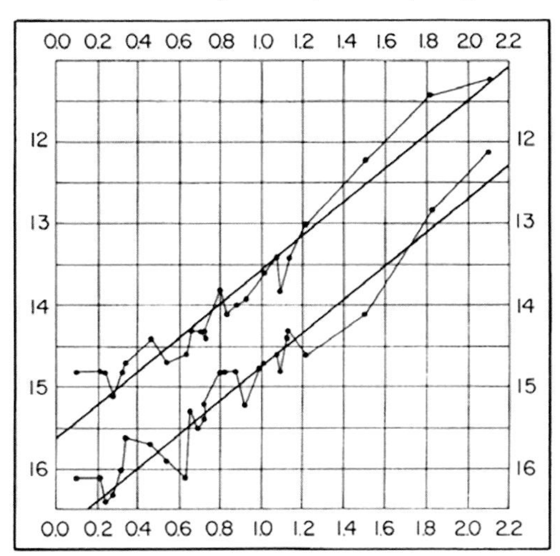

for the Cepheids in the Small Magellanic Cloud. It was 0.0001 seconds of arc, meaning that the Small Magellanic cloud was 30,000 light years from the Earth. This was easily the largest astronomical distance at the time, but remarkably (and perhaps because unconsciously he could not bring himself to believe in such a large distance) Hertzsprung made an arithmetical error, and his paper in the *Astronomische Nachrichten* gave the distance as only 3,000 light years.[30] Subsequently, Russell reworked the same problem, using the same 13 Cepheids, and found the distance to the Small Magellanic Cloud to be about 80,000 light years. He too was diffident about this "enormous distance." The most he would say was that it was "not intrinsically incredible."[31]

In 1918, Shapley would take up the problem yet again. Using 11 of the 13 variables, he significantly revised the calibration, and his calibration seemed so definitive it would be universally accepted for over thirty years, serving as the basis of the estimates of distances not only within the Milky Way, but to extragalactic objects, such as the Magellanic Clouds and the Great Spiral of Andromeda, beyond it. (For instance, it was the basis of Edwin Hubble's estimate, based on his discovery of Cepheids in the Great Spiral, giving its distance as 700,000 light years.)

Mt. Wilson astronomer Walter Baade later realized that Shapley's calibration of the period-luminosity curve for the Cepheids was flawed. Baade showed in 1952 that globular clusters and extragalactic systems such as the Small Magellanic Cloud each have their own pulsating variables, with different period-luminosity relations. These pulsating variables are essentially representatives of two dissimilar types of stars. Shapley did not—and could not—know this, and unwittingly mixed together data from the two types. But though his miscalibration would lead to gross underestimates of the distances to extragalactic objects, his estimates of the distances to globular clusters were essentially correct—and that, at the time, was what Shapley was after.

Shapley's main research project at Mt. Wilson involved the globular clusters, and was along the lines he had discussed with Solon I. Bailey. At the time he began a systematic study, 93 globulars were known, of which many of the brighter ones had been discovered by Messier. Shapley saw that they were evenly distributed in latitude above and below the galactic equator, but almost entirely skewed to one hemisphere in galactic longitude. They were, moreover, the only class of objects that showed this skewed distribution (galactic or "open" clusters like the Pleiades, by contrast, and planetary nebulae were closely concentrated to the galactic circle; they

Fig. 10.11. M80, globular cluster in Scorpio, and one of the densest of the 147 known globular clusters belonging to the Milky Way, as imaged with the WFPC2 camera of the Hubble Space Telescope. This globular is located at a distance of about 28,000 light years from the Earth. Globular clusters can be used to study stellar evolution, since all of the stars in the cluster have the same age (about 15 billion years), but a range of masses. M80 is unusual in having a number of "blue stragglers" at the core—stars that appear unusually young and more massive than the others in the cluster, and thought to occur when two stars merge to produce an unusually massive single star mimicking a normal young star. The number of blue stragglers in M80 is consistent with crowding and a high collision rate. Novae should also occur with high frequency—but though M80 was the site of a nova outburst observed in 1860, Hubble observations have revealed far fewer novae than expected theoretically. *Courtesy: M. Shara, D. Zurek and F. Ferraro and the Hubble Heritage Team (STScI/AURA).*

were immersed in rich star fields in all parts of the Milky Way band). At first he could not account for this distribution. His research project was to photograph the globulars with the 60-inch reflector and scan the thousands of star images on his plates for Cepheids. He later recalled: "I enjoyed observing with the 60-inch telescope, especially the novelty of using such a big instrument. It was wonderful that a beginner could have such machinery.... But observing was always very hard work for me. I 'suffered' quite a bit those long, cold nights."[32]

Before long Shapley was finding variable stars in the nearer globular clusters, such as M3 and M13. By 1918, he had worked out period-luminosity curves for 230 variables in the brighter globulars, with periods ranging from 5 hours to 100 days. Most were short-period variables—so-called "cluster" variables or RR Lyrae stars—whose average period was half a day. But he also found enough of the longer-period Cepheids, seemingly like those Miss Leavitt had discovered in the Small Magellanic Clouds, to use his now-calibrated Period-Luminosity plot to work out the distances. For instance, for M13, the most celebrated of the globulars in the Northern Sky, which had seemed to Messier a *"nébuleuse sans etoiles"* but swarmed with tens of thousands of star images on Shapley's plates, Shapley found a distance of 30,000 light years. It was one of the nearer globulars. The typical globular, he found, was on the order of 50,000 light years away.[33] These were enormous distances, and placed the globulars well out of bounds of the small Kapteyn universe, which was beginning to look very provincial by this time. Thus, Shapley concluded, "There seems to be reasonably clear evidence that M13 and similar globular clusters are very distant systems, distinct from our Galaxy and perhaps not greatly inferior to it in size."[34]

Fig. 10.12. Globular clusters and their aggregation toward the center of the Galaxy. This single patrol-telescope photograph, taken in the direction of the galactic center, covers an area equal to about 2 percent of the entire sky, but shows almost one-third of all globular clusters known at the time. The circles mark the positions of globular clusters for this part of the Milky Way. *Courtesy: Harvard College Observatory.*

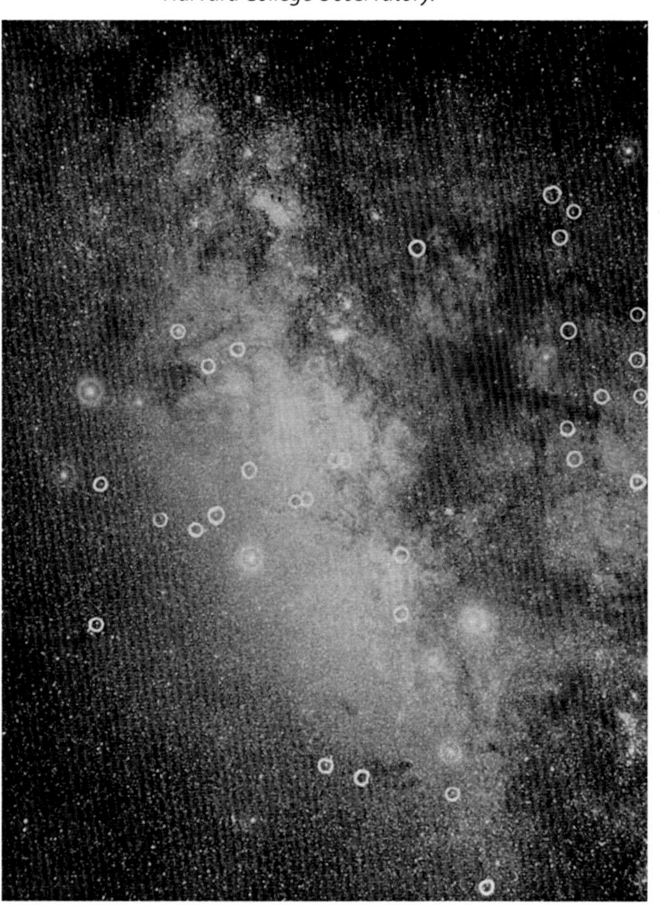

All along, he later confessed, he had been working on the basis of a kind of hunch, "a somewhat bold and premature assumption," to explain the skewed distribution of the globulars; namely,

> the globular clusters are, in a sense, the bony frame of the body of the galactic system. It was argued that the spatial arrangement of fewer than a hundred globular clusters shows the distribution of the billions of galactic stars. It was deduced, therefore, that the center of the ... Galaxy is in the direction of Sagittarius, since there lies the center of the system of globular clusters...[35]

But if this is so, and the distances he was getting for the globulars assured him that it was, then the skewed distribution of the globulars was not due to the globulars' being skewed away from the center of the Galaxy. Instead it meant that we—the observers—were skewed from the center. Instead of being near the center of the sidereal system as in Kapteyn's model, we on the tiny planet Earth were located tens of thousands of light years from the center of the Galaxy. Like the Copernican revolution that had displaced the Earth from the center of the universe as it was known in 1543, Shapley had engineered a "galactocentric" revolution: he had boldly removed man from the central position he had held in the Galaxy.

Shapley wrote to the leading British astronomer Arthur S. Eddington in January 1918 that the globulars had elucidated the "whole sidereal picture" with "startling suddenness and definiteness."[36] In fact, that was true. Eddington himself, just a few years earlier, had estimated the extent of the Galaxy as on the order of 15,000 light years; Shapley recalculated its breadth at 300,000 light years. Because he failed to take account of dimming by obscuring matter, he underestimated the brightnesses—and thus overestimated the distances—of his Cepheids. We now know that the Galaxy is only a third as large as he estimated, but the figure he gave was at least of the right order. Shapley's Galaxy was so large that he could not bring himself to believe that the spiral nebulae could be "island universes," as a number of prominent astronomers such as Heber Doust Curtis of Lick Observatory, believed; instead he thought they were flimsy and insubstantial gas clouds whose high Doppler shifts (as discovered by Slipher) indicated the speed with which they were being driven off by radiation pressure from the Milky Way. They were like puffs of clouds on a windy day. If the Milky Way—the "Big Galaxy," as he called it—was as large as he thought, it was inconceivable to him that there could be other such systems; for him, the Milky Way and the universe were one.

Curtis, who would engage Shapley in the so-called "great debate" (which was actually more like a couple of set speeches with a brief discussion afterward) at the National Academy of Sciences in Washington, D.C., in April 1920, rejected Shapley's "Big Galaxy" idea, and in support of the "island universe" idea pointed to the fact that he, Ritchey, and others, including Shapley himself, had observed in spirals a number of "ordinary" novae, which appeared to be identical to those frequently seen in the Milky Way. (Novae are faint stars that temporarily increase in brightness by 10 magnitudes to become prominent for a brief time. As explained in chapter 9, we now know these outbursts involve white dwarfs in binary systems with red giants; flow of gas from the red giant builds up on the surface of the white dwarf leading to an explosive outburst of nuclear reactions that blasts the material into space and causes the star to increase its brightness 100,000 times in a few days.) Since these ordinary novae occurring in spirals appeared very faint compared with those observed in the Milky Way, Curtis argued that the spirals had to be very remote. Shapley, on the other hand, recalled to mind the exceptional nova in the Andromeda spiral in 1885, which had peaked near the 7th magnitude, as well as another, equally dramatic, in the galaxy NGC 5253 in Centaurus , which had been discovered on a Harvard plate by Mrs. Fleming in 1895. If these "novae" were indeed associated with extragalactic systems, he pointed out they would have single-handedly outshone all the other stars, which seemed inconceivable. (At the time, only Knut Lundmark, a Swedish astronomer, had the courage to suppose there might be two classes of novae, ordinary and "supernovae," which of course proved to be the case). But Shapley's thinking was also influenced by the measures of rotation in the spiral M101 by his friend Adriaan van Maanen, discussed in chapter 9, which strongly implied that the spirals had to be relatively nearby.

In the "great debate," which has been mythologized rather like Thomas Huxley's and Bishop Wilberforce's famous "debate "on evolution at Oxford in 1860, both Shapley and Curtis were partly right and partly wrong. Shapley was right about the Sun's noncentral position in the Galaxy, and Curtis was right about the spirals being extragalactic systems. From today's perspective—but also from the perspective of most of those present—Curtis was clearly the winner.

By now, Shapley's tenure at Mt. Wilson was almost over. His ambitions by then aimed for the directorship of the Harvard College Observatory, vacant since E. C. Pickering died a year before. It would make him, as he liked to point out, the highest paid astronomer in America. Harvard would have preferred to hire Shapley's teacher, Henry Norris Russell, as director, with Shapley—who though possessed of extraordinary assets and capabilities was already widely regarded as rather vain and selfish, and more concerned about his own reputation and advancement than the welfare of the organization or of his colleagues—serving as his second. But Russell decided to stay on at Princeton, and eventually Shapley had his way— he was offered the Harvard position on a probationary basis (for one year) and then named to the position on a permanent basis. All this took time to work out, so that Shapley did not actually take his leave from Mt. Wilson until March 15, 1921. The night-assistant on the 60-inch made the following entry in the observing logbook: "Dr. Harlow Shapley leaves the observatory today to become identified with Harvard College Observatory." Probably most of his colleagues were relieved to see him off. Walter S. Adams, then serving as assistant director to the ailing (and soon to retire) Hale, would recall, simply, "He left Mount Wilson with little regret on the part of staff members."[38]

Milton Humason, mule driver, janitor, night assistant, research astronomer at Mt. Wilson, and raconteur extraordinaire, related a story to Allan Sandage about an incident that is supposed to have occurred before Shapley left the mountain. According to Humason, Shapley one day came in with a handful of plates of M31, the great spiral in Andromeda, taken with the 60-inch in a survey Shapley had begun for novae,

> and asked if he would like to use the Zeiss stereocomparator to blink pairs of them. Humason acceded. After working for several weeks, marking the backs of the plates with ink to indicate promising pairs of images, Humason informed Shapley that many images seemed to differ from one plate to another. "Could these possibly be Cepheid variables," he asked Shapley, "similar to what you have found in globular clusters?[39]"

According to Humason, Shapley was self-assured, and simply lectured him on all the reasons that was impossible, and ended by taking a handkerchief from his pocket and wiping Humason's ink off the plates!

Though Shapley later denied that this had happened (and Humason was rather notorious for telling tall tales), the latter continued to swear by it until his death in 1972. Sandage, who later examined Shapley's plates, concluded that several Cepheids in M31 were indeed visible on Shapley's plates, and that by blinking these plates, Humason might have found at least a few of them. But he adds: "had Shapley stayed at Mount Wilson and continued his survey for novae in M31, he almost certainly would have then found the Cepheids himself. [Edwin] Hubble did not begin his nova survey until the fall of 1923—three years after Shapley had left for Harvard.... By going to Harvard, it is probable that Harlow Shapley—discoverer of the galactocentric eccentricity of the Sun's position—had denied himself his second Copernican-like revolution."[40]

* * *

1. See: Anthony Misch and William Sheehan, "The First Modern Telescope: the Mount Wilson 60-inch Reflector," in *The 2008 Yearbook of Astronomy*, Patrick Moore and John Mason, eds. (London: Macmillan, 2007), 192-221.
2. The discussion of Hale that follows is largely based on William Sheehan and Donald E. Osterbrock, "Hale's 'little elf': the mental breakdowns of George Ellery Hale. *Journal for the History of Astronomy, xxxi* (2000) 93-114.
3. Helen Wright, *Explorer of the Universe: a biography of George Ellery Hale* (New York: E.P. Dutton, 1966), 28.
4. Ibid., 29.
5. Ibid., 30.
6. Ibid., 32.
7. Ibid., 17.
8. Ibid., 51.
9. Ibid.
10. Donald E. Osterbrock, *Pauper and Prince: Ritchey, Hale, and big American telescopes* (Tucson: University of Arizona Press, 1993), 23.
11. Sheehan, *Immortal Fire*, 312.
12. Wright, *Explorer of the Universe*, 135.

13. G.E. Hale to J.S. Billings, July 6, 1903; George Ellery Hale Papers, California Institute of Technology.
14. Wright, *Explorer of the Universe,* p.197.
15. T.J.J. Lears, *No Place of Grace: Antimodernism and the transformation of American culture, 1880-1920* (Chicago: University of Chicago Press, 1994), 9:
16. Osterbrock, *Pauper and Prince,* 284.
17. Ibid.
18. G. E. Hale, "On the probable existence of a magnetic field in sun-spots." *Astrophysical Journal, 28* (1908), 315-343.
19. Wright, *Explorer of the Universe,* 264.
20. G.E.Hale to H.M. Goodwin, March 25, 1911; Henry L. Huntington Library, San Marino, California.
21. G.E. Hale to James H. McBride, July 30, 1911; Henry L. Huntington Library, San Marino, California.
22. Allan Sandage to Helen Wright, 1966; forward to Wright, *Explorer of the Universe.*
23. J. C. Kaptyen, "On the Absorption of Light in Space," *Astrophysical Journal,* 29 (1909), 46-54:47.
24. Harlow Shapley, *Through Rugged Ways to the Stars* (New York: Charles Scribner's Sons, 1969), 34.
25. Ibid., 36.
26. Ibid.
27. Ibid., 41.
28. Solon I. Bailey, "Henrietta Swan Leavitt," *Popular Astronomy,* 30, 4 (1922), 197-199.
29. Henrietta S. Leavitt, "Periods of twenty-five variable stars in the Small Magellanic Cloud," *Harvard College Observatory Circular* No. 173 (1912), 1-3.
30. Owen Gingerich and Barbara Welther, "Harlow Shapley and Cepheids," *Sky & Telescope,* December 1985, 540.
31. Ibid., 541.
32. Shapley, *Through Rugged Ways,* 51.
33. Harlow Shapley, "Globular Clusters and the Structure of the Galactic System, "*Publications of the Astronomical Society of the Pacific, 30* (1918), 42-54.
34. Harlow Shapley, "Outline and Summary of a Study of Magnitudes in the Globular Cluster Messier 13," *Publications of the Astronomical Society of the Pacific,* 28 (1916), 174.
35. Harlow Shapley, *Galaxies* (Philadelphia: Blakiston, 1943),100.
36. Quoted in Robert Smith, *The Expanding Universe: Astronomy's 'Great Debate' 1900-1931* (Cambridge: Cambridge University Press, 1982), 79.
37. Quoted in Christianson, *Edwin Hubble,* 151.
38. Ibid., 306.
39. Allan Sandage, *The Mount Wilson Observatory: Centennial History of the Carnegie Institution,* vol. 1 (Cambridge: Cambridge University Press, 2004), 496.
40. Ibid., 498.

11. From Olympus

Over us, like some great cathedral dome,
The observatory loomed against the sky;
And the dark mountain with its headlong gulfs
Had lost all memory of the world below;
For all those cloudless throngs of glittering stars
And all those glimmerings where the abyss of space
Is powdered with a milky dust, each grain
A burning sun, and every sun the lord
Of its own darkling planets,—all those lights
Met, in a darker deep, the lights of earth,
Lights on the sea, lights of invisible towns,
Trembling and indistinguishable from stars...

—Alfred Noyes, Watchers of the Sky, Prologue: The Observatory

William Herschel was the first astronomer to push his investigation beyond the Solar System and to expressly concern himself with the stars and nebulae—the "construction of the heavens." We now know that it was actually his sister, Caroline, who preceded him and who first showed how much remained to be done. When he began, about a hundred of the nebulae were known, catalogued by Messier and a few others; by the time he finished, he had added nebulae by the thousands, and noted their astonishing variety.

"I have seen," he wrote, "double and treble nebulae, variously arranged; large ones with small, seeming attendants; narrow but much extended, lucid nebulae or bright dashes; some of the shape of a fan, resembling an electric brush, issuing from a lucid point, others of cometic shape, with a seeming nucleus in the center."

Faced with this richness and diversity of material, Herschel compared the nebulae to the plants growing in a "luxuriant garden," and continued into advanced age to stretch his mind — and his data — to organize them into a sequence that showed the influence of the invisible hand of gravity. Gravity acted across space, reached across time, and therefore served as "the cause of every condensation, accumulation, compression, and concentration of the nebulous matter."

His son, who added many more nebulae and listed them in his General Catalogue (the core of the modern catalog, the New General Catalogue or NGC, which is still in use today), rather prosaically sorted them according to their shapes, degree of condensation at the center, and resolvability into stars—thus the nebulae were "circular" or "round" or "oval" or "elongated";" "stellate," "nuclear," "discoid"; "discrete," "granulate," "mottled," "milky." In indicating the characteristics of a particular nebula, a cumbersome system of capital and lower case letters was used; thus, "vBpLvgmbM" meant "very bright, pretty large, very gradual, much brighter in the middle."[1]

The nature of the nebulae remained unknown. And yet—to name a thing is the closest thing to creating it. All of science begins descriptively, with categories. A category is, in the first place, a useful distillation of data. One hopes that—if carefully and thoughtfully chosen—it may represent a well-defined, separate, modally discontinuous group, a true species—carving, as Plato said, nature at its joints.

Classification schemes of all kinds—of snowflakes, seashells, plants, animals, even diseases —involve making judgments. Aristotle — the great classifier of antiquity — approached the taxonomy of organisms by dividing them into large, apparently self-evident groups; he then subdivided these into smaller and smaller groups. His method was grand — aristocratic — top-down; his categories were God-given — once they were ordained, they remained intact. They were absolute, indissoluble, necessary.

But modern taxonomy—the science of classifying nature—begins not with Aristotle but with John Ray, the 17th-century founder of the modern concept of species. Ray, a self-taught English naturalist, took a bottom-up rather than top-down approach, beginning "with awe for the uniqueness of individuals and the wonderful variety of species." "Nature," he wrote in *Methodus Plantarum* (1680), "... makes no jumps and passes from extreme to extreme only through a mean. She always produces species intermediate between higher and lower types, species of doubtful classification linking one type with another and having something in common with both." Extreme positions can be defined; but they merge with intermediates, shade into variations. It is a picture of subtlety. The point is, however, that instead of being dictated from above and eternally unchanging, Ray's species are provisional and open-ended—capable of being "opened and dissolved" in the light of new knowledge.

An important step in the classification of nebulae, which moved beyond mere description to their physical properties, had been taken by William Huggins, whose spectroscope separated the nebular worlds into the two classes of emission ("green") nebulae and continuous-spectrum ("white") nebulae). Later schemes ranged from lumpers—like Heber D. Curtis of the Lick Observatory, who as late as 1919 went on record as saying that all the nebulae in the heavens could be conveniently grouped into three categories, planetaries, diffuse nebulae, and spirals—to splitters, of whom the best example was Max Wolf, a brilliant German astronomer whose pioneering work in astrophotography has been underappreciated, mainly because he worked in Europe at a time when the leading developments in the field were taking place in the United States.

Wolf was born in Heidelberg, Germany, in 1863. At age 21, he discovered an unusual comet, which was at first taken to be an asteroid. In the following year his proud physician-father—in William Hale-like fashion— built him his own private observatory, where he observed for many years. After taking his Ph.D. at Heidelberg University, he succeeded in setting up a new observatory near Heidelberg at Konigstühl, where he worked indefatigably until his death in 1932. He was a friendly competitor of Barnard in the wide-field photography of the Milky Way, and discovered an extremely faint star, Wolf 359, which is the third closest star to the Sun after alpha Centauri and Barnard's star. When Barnard died in 1923, Wolf paid him the tribute of a true kindred spirit, and much of what he said of Barnard could equally well be said (of course) of himself:

> He was a virtuoso in seeing and in measuring, he had the talent to recognize the new, which others carelessly overlooked, and moreover combined indomitable stamina with inexhaustible enthusiasm for research. It is a mighty passion that attracts an observer to his work, and satisfaction with the hard-won results obtained with his instrument provides him his highest pleasure.[2]

That same year, he published an important paper in which, by means of star counts of part of NGC 6960, the "western veil" (part of the "Cygnus Loop," now known to be a supernova remnant) he showed that bright and dark nebulae are associated with another, and that the dark nebulae Barnard had discovered were indeed dust clouds.[3] (The "general absorption of light in space, a more subtle and general result, was not proved until 1930, by Lick Observatory astronomer Robert Trumpler).

Wolf's scheme of nebular classification was as particular as Curtis's was general. In 1908, on the basis of photographic surveys carried out with the 16-inch Bruce photographic telescope at Konigstühl, Wolf devised a highly specific classification of nebulae that included no less than 23 different types, each of which was assigned a lower-case letter.[4] Purely descriptive, and without any pretense of guessing at the physical nature of these objects, it is hardly surprising that Wolf mixed apples and oranges (some of his nebulae are obviously planetaries, while most are variations on the spiral-nebula theme.) It represented little—if any—improvement over Sir John Herschel's classification system. Though it is (justly) all but forgotten today, it does represent the state of the art of nebular classifications in the nineteen-teens when a young astronomer named Edwin P. Hubble began to think seriously about bringing some order into this chaotic field and working toward his own brilliantly elegant solution.

Enter Hubble

Hubble, who "had a romantic, larger-than-life quality that is rare in practicing scientists,"[5] was born in Marshfield, Missouri in 1889, the fifth of the eight children born to John Powell Hubble and Virginia Lee James, who moved soon afterward to Wheaton, Illinois. Details of Hubble's early life are not always reliable; he was not averse to stretching the truth, and in some cases fabricating incidents outright, a trait he probably derived from his father, who trained as a lawyer but spent most of his career working in insurance. Though John later claimed to have served as insurance commissioner for the state of Missouri and prosecutor of Webster County, neither is borne out in the records. Raised in the puritanical values that were fashionable in his day, John became "a stern, hard-bitten blend of moral high-mindedness and relentless ambition," a "demanding taskmaster who 'ruled the roost' in no uncertain terms."[6] His son seems to have absorbed a good deal of this demanding paternal ego, and the relation of father to son may help to account for some of the less lovable aspects of Hubble's own personality, including the careful cultivation of a placid exterior masking "a thorough-driving, competitive nature."[7]

To others, Edwin always seemed to come easily by his success. He never seemed to struggle, never failed at anything. At 16, he went to the University of Chicago, and majored in physics and astronomy, working part-time one year in the physics laboratory of the future Nobel laureate Robert Millikan. Blessed with an awesome physique, and standing over six-feet tall, he also won varsity letters in basketball and track. After graduating in 1910, he applied for a Rhodes scholarship, and won one. He used it to spend three years at Queen's College, Oxford, where—apparently in deference to his father's wishes—he studied law (though again, apparently, not very hard) and Spanish. Though he may not have learned much during his three years at Oxford, he fell in love with everything English—especially the manners of the elite aristocratic class. Raising Anglophilia "to the level

Fig. 11.1. Edwin Powell Hubble. *Courtesy: Huntington Library, San Marino, California.*

of a faith," as one of his colleagues later put it, he developed a persona that included smoking a pipe, wearing tweed jackets and baggy plus-fours, and occasionally even donning a cape. He topped it all off with a fashionable walking stick. Shapley, who was as liberal politically as Hubble was conservative and whose self-consciously "folksy" voice was not always very convincing either, wrote sarcastically, "He was a Rhodes scholar, and he didn't live it down. He spoke with a thick Oxford accent. He was born in Missouri not far from where I was born and probably knew the Missourian tongue. But he spoke 'Oxford.' He would use such phrases as 'to come a cropper.' The ladies he associated with enjoyed that Oxford tone very much. 'Bah Jove!' he would say, and other such expressions. He was quite picturesque."[8] V.M. Slipher found Hubble's Oxfordisms ridiculous; he thought he tried too hard to pass as a "Limey." But Hubble didn't seem to care, and in any case never shed his Oxford mannerisms.

Allan Sandage, a Caltech graduate student who became his assistant (and in some ways, his acolyte) as Hubble approached the end of his life, found his aging idol still wearing Harris Tweeds, using some of the Britishisms from the old days at Oxford, and boasting a cane or walking stick in the affected English manner. In awe, the self-described "hick" from the Midwest saw Hubble as "a noble man," who deported himself as "I imagine a god might."[9] Indeed, Hubble's aloof demeanor fit him as the god of observational cosmology that he was by then. He always sought to keep his distance himself from others. Shapley thought he didn't care for people much. Despite four years of working with him, Sandage admitted that the great man "always remained his formal self, never laughing or cracking a joke, never attempting to put [him] at ease."[10] Even when Sandage was introduced to Hubble's wife Grace some time later, "the same formality prevailed, causing him to wonder about their conduct when alone together."[11]

In the end, even his biographer finds him inscrutable. Comparing Hubble to Galileo, Gale Christianson notes that "unlike [the] renaissance hero, the astronomer revealed almost nothing of his inner universe." Recalling that Hubble's favorite pastime was fly-fishing, Christianson notes that "like the secretive trout of his daydreams, the poetic fly-fisherman chose to remain elusive."[12]

After finishing his Rhodes, Hubble never practiced law, though for some reason he would later claim to have done so (his biography was as carefully groomed as his persona). Instead he taught high school Spanish and coached basketball for a year, and despite being popular with students, and coaching the basketball team to a championship, never mentioned this interlude in his "official" biography. When John Hubble died, Edwin seemed to have felt liberated from the oppression of paternal expectations. Always interested in astronomy, and the recipient of a small telescope when he was eight, he now decided, "even if I were second- or third-rate it was astronomy that mattered." With sudden clarity of purpose he wrote to Forest Ray Moulton, one of his professors at the University of Chicago, and to the director of the University's Yerkes Observatory, Edwin B. Frost, about doing his Ph.D. in astronomy. He had made a good impression as an undergraduate, and (unlike the less traditional student Slipher) had no problem getting accepted. In August 1914, just before he reported to Yerkes to begin his studies, he went to Evanston, Illinois, for the American Astronomical Society meeting at which he heard Slipher talk about the large radial velocities of the spirals, and as we have seen, put himself prominently in the front row in the group photograph, despite not even having started his graduate work. (One thing Hubble never lacked was self-confidence and a knowledge of how to position himself.)

In 1914, the Yerkes Observatory was in its doldrums.[13] The "first team" had left with Hale for Mt. Wilson, and Frost, a well-trained Dartmouth University spectroscopist who took over from Hale as director, was a pedestrian researcher. He also suffered from severe congenital myopia, and was soon to have a retinal detachment in one eye and then in the other. Slowly but surely he was going blind and faced the difficult prospect of being a blind astronomer, with many years remaining before he was eligible for retirement. At Yerkes, the most famous astronomer was still E. E. Barnard, but Barnard didn't take students, and was suffering from diabetes and slowing down. Fortunately, Hubble was a self-starter, and perhaps inspired by Slipher's talk, decided to study the faint nebulae, nominally under Frost's supervision, despite the fact that Frost had never worked on the subject himself. With the uncanny intuition that always served him so well, the 25 year-old graduate student already sensed that, as Fath had told Slipher when they had met at the International Solar Union meeting at Mt. Wilson in 1910, that "there are great opportunities in the nebular field."[14] At the time, it could hardly have been less overworked; the notables included Slipher at Lowell, Heber D. Curtis at Lick, Knut Lundmark in Uppsala,

Fig. 11.2. Yerkes staff, 1915. Hubble stands at the center of the back row (with bow tie); John Mellish, volunteer observer, is next to him at his right. Edwin Frost, the director, is standing in the middle row (with beret) to the left, and E. E. Barnard stands next to the column in the doorway on the extreme right. *Courtesy: Yerkes Observatory.*

Sweden, Wolf in Germany, which was now at war, and a few others.

For his thesis, Hubble wanted to photograph faint nebulae with the 24-inch reflector that Ritchey had built at Yerkes Observatory in 1901, and to find out how the various types of nebulae were distributed around the sky. The telescope had been revolutionary when Ritchey had first built it, and Ritchey had used it to obtain the best photographs of nebulae taken up to that time. But the plate scale was too small for Hubble's purposes. At the Newtonian focus, the scale of the 24-inch reflector was 87 arc seconds per millimeter, so that most of the nebulae Hubble marked on his plates were hardly distinguishable from stars (by contrast, the plate scale of the Mt. Wilson 60-inch was 30 arc seconds per mm, and of the 100-inch would be 16 arc seconds per mm; these larger plate scales were to prove essential for Hubble's eventual success).[15]

While getting underway on his dissertation project, Hubble became intrigued by a comet-shaped nebula, NGC 2261, which envelops the variable star R Monocerotis. The nebula had been discovered by William Herschel in 1783, and the star's variability had been noted by Julius Schmidt, the director of the Athens Observatory, in 1861. Hubble's notice was called to the object by John E. Mellish, who on the basis of his success discovering comets with homebuilt telescopes while working as a farmhand in Cottage Grove, Wisconsin, had been invited to Yerkes as a "volunteer" observer, and had swept it up thinking it a new comet. Now Hubble decided to do a careful study of the object, and obtained sixteen plates of it over six months in 1915 and 1916. When he compared them *inter alia* and with earlier photographs going back to one by the British astrophotographer Isaac Roberts from 1900, he found that not only was the star variable, so was the nebula itself. Frost gave a brief presentation about it to the National Academy of Sciences in Washington, D.C., which led to Hubble's first astronomical publication.[16] Ever afterward, Hubble had a father's pride in his variable nebula, and when, decades later, the 200-inch reflector at Palomar first went into operation, Hubble returned to what he called his "polestar," and captured it on a plate—the very first obtained with the new giant in glass and steel.

In April 1917, the United States entered the European War against Germany. Hubble, ardent Anglophile that he was, rushed through the writing and defense of his thesis so he could volunteer for the U.S. Army. It wasn't a very good thesis, but it anticipates some of the themes of Hubble's mature research. In particular, in trying to address the question of how the various types of nebulae are distributed in the fields taken with the 24-inch reflector, Hubble made a start on the classification question. "Little is known of the nature of nebulae," he began, "and no significant classification has yet been suggested." He decided that Wolf's scheme, "wholly empirical and probably without physical significance," offered probably "the best available system of filing away data," and concluded that it would likely remain of service until "a significant order is established."[17]

The most important of Hubble's findings was that many of the faint nebulae in his plates were not, in fact, spirals, as Keeler and Curtis had believed, but very small round or elliptical nebulae (Wolf's types e and f; the brightest specimen of this type is M60 in Virgo). This realization would later become fundamental to Hubble's own classification system.

Before Hubble left for Europe (he apparently never saw action), Hale hired him for the Mt. Wilson staff. Hale himself was involved in war work, and sympathized with the young Ph.D.s desire to serve overseas; he therefore agreed to keep the position open until Hubble was discharged from the Army. When, in 1919, Hubble, by then Major Hubble, had received his discharge, he arrived at Mt. Wilson by way of Lick Observatory, where his uniform and his affected British accent left an indelible impression on at least one graduate student, C. Donald Shane. He was assigned to the recently unveiled 100-inch reflector, which was then, and would remain for the next 30 years until the 200-inch at Palomar became operational, the most powerful telescope in the world.

Though Hubble's studies of what he came to call "extragalactic nebulae" (Shapley's term "galaxies" did not gain universal acceptance until after Hubble's death in September 1953) would garner by far the most media attention, cosmology was, Allan Sandage notes, "in fact only a small part of the early Mount Wilson accomplishments." Most of the papers issued from Mt. Wilson were concerned with stars; only four per cent with galaxies and the universe. Nevertheless, Sandage adds, "that four percent was the foundation of observational cosmology, and the public spotlight became focused on the Nebular Department and remained there throughout the 1930s and 1940s, providing much of the justification for construction of the Palomar 200-inch reflector—and much of the dismay, jealousy, and dislike

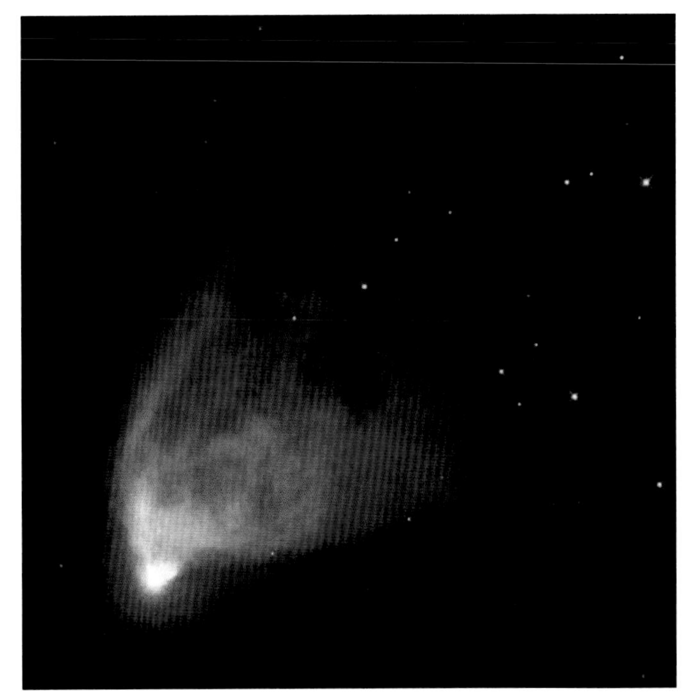

Fig. 11.3. (left) Hubble's Variable Nebula, a fan-shaped cloud of gas and dust illuminated by R Monocerotis (R Mon), the bright star at the bottom end of the nebula. Dense condensations of dust near the star cast shadows out into the nebula, and as they move the illumination changes, giving rise to the variations first noted by Hubble. The star itself, with a mass of about 10 times that of the Sun and only about 300,000 years old, cannot be seen directly, but only through reflected light from dust particles in the nebula that surrounds it. *Courtesy: W. Sparks and S. Baggett, NASA and the Hubble Heritage Team (STScI/AURA).*

Fig. 11.4. (right) 100-inch reflector at Mt. Wilson Observatory. *Photograph by William Sheehan, 1999.*

by the spectroscopists. Part of their animosity toward Hubble was due to the overwhelming media attention he received."[18]

Despite common belief, the 100-inch reflector was not the only telescope in the world at the time powerful enough to resolve extragalactic nebulae into stars. Shapley seems to have been the source of this claim, writing in a note on novae in 1917 that in the 60-inch there were no signs of resolution in nebulae such as M31, NGC 224, and M33. In fact, however, even the earliest plates of M33 taken with the 60-inch by Ritchey in 1909 show images "resolv[ing] into stars beginning at magnitude 16. The images are sharp, and they are fully as stellar as the obvious field stars in the outer areas of the same plates."[19] Even before that, in 1889, A. C. Ranyard had noted that in a celebrated plate of M31 taken the year before by Isaac Roberts, a retired builder who set up a 20-inch silver-on-glass reflector for astrophotography at his private observatory at Crowborough, Sussex, the outer regions appeared resolved into stars.[20] Certainly some of Hubble's discoveries could have been made with the 60-inch, but there is no doubt that the 100-inch gave Hubble's an almost magisterial authority that sorted well with the carefully cultivated grandeur of his personality.

In beginning his observations with the 100-inch as a new staff member of Mt. Wilson Observatory, Hubble returned to his thesis topic, the classification of the nebulae, but with a difference—he now had the plate scale of the 100-inch rather than the inadequate plate scale of the Ritchey 24-inch. After three years of work, in 1922—the same year that F. Scott Fitzgerald published *The Beautiful and Damned,* and one year after Charlie Chaplin appeared in the semi-autobiographical silent film *The Kid*—he published two masterful papers on what he called the "galactic nebulae," those whose provenance was the plane of the Milky Way.

These papers were, apart from his Ph.D. dissertation, Hubble's first publications since 1916-17, when he had written up his work on the variable nebula NGC 2261.[21] In the first of these papers, he subdivided the galactic nebulae into 1) those with emission (bright-line) spectra and 2) those with absorption (dark-line) spectra. In either case, he realized, their visibility depended on the presence of a nearby hot star (of spectral type B or O). The nebula must either shine by reflected light from the star, in which case the reflecting medium is dust as Slipher had shown, and the spectrum contains dark lines, or by fluorescence due to the star's excitation of gas, in which case the spectrum contains bright emission lines. Clearly, gas and dust were mixed together in these nebulae, though as we now know, the mass-density of the dust is minuscule compared with that of the gas. In the second paper, completed only three months later, he beautifully demonstrated that the source of radiation from non-emission nebulae is indeed reflected light from the associated star.

The "galactic" nebulae, the subject of these papers, were in turn distinguished from the "non-galactic" nebulae, those with an aversion to the plane of the Milky Way. So far, his division into "galactic" and "non-galactic" nebulae was neutral on the question of whether the "non-galactic" nebulae were relatively nearby, as Shapley believed, or "island universes" as Curtis believed. Hubble's classification did not depend on the answer to that question.

However, without committing himself to their nature, Hubble was already proposing a tentative classification scheme for the non-galactic nebulae. Building on the result announced in his Ph.D. thesis, he recognized, by inspection of the objects on the deep plates he exposed with the 100-inch, that not all galaxies were spirals. Some, indeed most, were elliptical, and clearly of a different type than spirals. (The same result had already been published in the *Monthly Notices of the Royal Astronomical Society* in 1920 by John Reynolds, an industrialist-turned-amateur astronomer from Birmingham, England, on the basis of photographs taken with a 30-inch reflector at the Helwan Observatory in Egypt; just when Hubble became aware of Reynolds's paper is unclear, but he certainly knew of it, and at some point made notes on it in the Mt. Wilson Observatory library in Pasadena in which he noted the similarity to his own developing scheme.)[22]

Meanwhile, Hubble had been appointed to the International Astronomical Union's Commission 28 on Nebulae, which was to meet in Rome that year. The Commission's chair was a geriatric astrometrist from France, Guillaume Bigourdan; other members included the distinguished, but aged, Danish astronomer Johann Louis Emil Dreyer, who half a century before had drawn up the standard *New General Catalogue* (the basis of the NGC numbers still used for nebulae). The American members included Slipher, Curtis, Barnard (nearing the end) and W. H. Wright of the Lick Observatory. In February 1922 Hubble sent Slipher, who by now was butting up against the limit of the faint nebulae within reach of his spectrograph on the 24-inch refractor, a first ambitious proposal for nebular classifications. "We begin flamboyantly," he rather grandly and self-importantly announced to Slipher.[23] Wright, who also received Hubble's proposal, wrote privately—but presciently—to Slipher, "I must confess that I am rather dazed by [Hubble's] letter.... One can see that the nebulae will have no pivate life when he has his way.... Besides my habit is to think from one plate to the next, and I am afraid I am not much on Empire Building."[24] Slipher agreed. "Hubble's report dazed us too," he acknowledged. "... We know so little about nebulae today that it is no easy task to lay down a line of study that would be so good as not to need very vital alteration in even a few years' time."[25]

By then, Hubble was already tinkering with his own scheme. He now broke the non-galactic nebulae into four groups: 1. spirals, 2. spindles (a class which, he suggested, might be replaced by "elongated," including spindles proper, i.e., those with spiral affinities, and ovates, variants from globular form), 3. globular, and 4. irregular, the last including such relatively amorphous objects as the Magellanic Clouds and Barnard's Galaxy, NGC 6822, soon to be an object of his special attention.

At the Rome meeting that year, which Hubble did not attend, the Europeans, led by Bigourdan, prevailed. They adopted a plan that largely recapitulated the venerable views of John Herschel and proposed making his classification scheme the basis of a new catalog of nebulae (updating the NGC), to be based as far as possible on photographic rather than visual observations. The project was to require the collaboration of multiple observatories. Hubble was not enthusiastic. In 1923, he protested to Slipher, who had now assumed the I.A.U. Commission 28 chair, "It is surely unwise to precipitate a comprehensive program of cataloguing until some classification ... has been devised and generally acceptable. This matter is urgent."[26] Meanwhile, he had continued to tinker with his own scheme, now adding the class "ovate" between "spindle" and "globular. Rather than work up his ideas into an article for publication, however, he told Slipher that he thought it preferable to develop the proposal more informally by circulating his typewritten notes among the members of the Commission, and soliciting their comments in the hopes of eventually gaining for the system the sanction of the I.A.U. That same year, he was stimulated by the presence of the British mathematical physicist James Jeans, who had just arrived at Mt. Wilson as a research associate and who had been working for years on the problem of how masses surrounded by thin atmospheres develop under different conditions of rotation. The problem was related to the Laplace Nebular Hypothesis, for as Jeans tried to demonstrate, as a nebula shrinks it passes through a series of configurations—from almost spherical, it becomes spheroidal, then develops a sharp edge in the equatorial plane that begins to eject matter "symmetrically at two antipodal points on the equator" so as to form two symmetrical streamers or arms. As the process continued, the streamers loping outward from the disk would unwind into ever looser, thinner, and expansive arms. Hubble did not fail to grasp that Jeans's theory was just what he needed to provide a thread of physical significance on which to hang his linear arrangement of the non-galactic nebulae from globulars to the most open (grand-design) spirals. "We seem to be succeeding with the evolutional sequence of the stars," he observed; "we may look forward with some hope to a time when something of the sort can be attempted with nebulae."

Hubble's E, Sa, Sb, Sc and Irr subtypes were similar to Reynolds's classifications I-IV. The E designation included ellipticals ranging from the perfectly circular E0 to the football-shaped E7; his S0 type, introduced later, corresponding to lenticular galaxies, was supposed to be the intermediate form between ellipticals and spirals (it was not noted until half a century later that typical S0 galaxies are typically only half as luminous as typical E and Sa galaxies, thus providing strong evidence against Hubble's own speculation that they constitute an intermediate evolutionary stage between ellipticals and spirals). Then came the spirals, Sa, Sb, and Sc, labeled according to the prominence of the galactic nucleus and the tightness with which the spiral arms were wrapped. The sequence was clearly informed by Jeans's ideas, and Hubble, who clearly thought of them as unwinding as they grew, could not resist referring to them as "early," "middle," and "late" types. Thus the E-type nebulae were "early," the Sa type, with bright condensed nuclei and closely coiled spiral arms, were "middle," the Sc type, which featured "a fainter almost stellar nucleus" and spiral arms "more open—unwound as it were—and conspicuously granular," were "late." The members of Commission 28 in general were qualifiedly positive—Hubble's proposal was "very thorough and complete," "especially interesting," "hard to improve on"—but they would not endorse it at the I.A.U. meeting at Cambridge in 1925 (again, Hubble himself did not attend). A number of members complained specifically about the introduction of the terms "early," "middle," and "late." Slipher complained that this suggested an "evolutional significance that the present state of our knowledge hardly warrants."[27] Indeed, the way in which the spirals were winding was still

moot; van Maanen had had the arms unwinding, and so the evolution corresponded to that in Hubble's sense; but Slipher's spectrograms suggested they were winding up. Moreover, as we now know, galaxy evolution is much more complicated than Jeans or Hubble envisaged, and though it would be better to avoid such terms altogether, the terms "early," "middle," and "late" are now rather conventional (rather like Riccioli's "seas" on the Moon) and are likely here to stay.

On his observing runs on the 100-inch, seated in the observing cage, with the great metal spars of the giant tube rising against the star-powdered night sky like the ribs of a prehistoric beast, Hubble exposed scores of deep plates for the nebular classification project. Though most of the hundreds of thousands of nebulae were mere smudges too small to be useful, several hundred were bright enough and showed enough detail to be classified.

Already, John C. Duncan, an Indiana graduate who had worked at Lowell Observatory before earning his Ph.D. at Lick Observatory and spent the year 1922 at Mount Wilson, had found three variable stars in 100-inch photographs of the spiral M33 (the only other spiral, with M31, visible with the naked-eye). He realized they were variable stars but could not determine their periods. In 1923, Hubble began his own detailed study of several of the non-galactic nebulae, beginning with the irregular NGC 6822 (Barnard's galaxy) in Sagittarius, a system similar to the Magellanic Clouds and serving as something of a stand-in for them as they lay too far south to be seen from Mt. Wilson. He found several small nebulae within it as well as twelve variable stars. That autumn, he began taking deep plates of the Andromeda Spiral itself, looking for novae of the kind Ritchey, Curtis, and others had discovered—by then, over 30 were known. On a night of poor seeing, October 4, he exposed a plate (labeled H331) for 40 minutes, on which he suspected a "nova." The following night, the last of his observing run, he enjoyed much better seeing, and exposed another plate (H335). He confirmed the "nova," and also found two other stars which he suspected were novae and marked them as such on the plate. He then left the mountain and returned to Pasadena. He seems to have put the plates aside for a while, but in due course—exactly when he never recorded—he began a more detailed study of these plates and realized that one of his three "novae" was not a nova after all. It was a variable star. He now crossed out "N" on the plate and replaced it with "V." Moreover, unlike the variables in M33 Duncan had discovered, he was able to determine its period—31.45 days. This made it a Cepheid, like those Shapley had used to estimate the distances to the globulars. However, its apparent magnitude—which ranged from 18.25 to 19.5—meant that the star system to which it belonged had to be very far away indeed. Using the period-luminosity relation for

Cepheids as calibrated by Shapley in 1918, Hubble worked out that its distance was a staggering 930,000 light years. (As we now know, Shapley's calibration, which was used between 1918 and 1952, was incorrect, and Hubble greatly underestimated the distance to the Great Spiral M33; it is now known to be 2.51 million light years.).

Hubble capped a triumphal year in February 1924, when he married Grace Leib, the widow of a geologist who had been killed in a mining accident, and set out with her on a honeymoon to Europe, which included a visit to his beloved Oxford. Before he left, Hubble penned a note to Shapley, "You will be interested to hear that I have found a Cepheid variable in the Andromeda Nebula (M31). I have followed the nebula this season as closely as the weather permitted and in the last five months have netted nine novae and two variables."[29] He included a rough diagram of his first variable's light curve, which showed it to behave in the usual Cepheid fashion. Shapley was stunned. Supposedly—according to his graduate student Cecilia Payne, who happened to be in the office when Hubble's letter arrived—exclaimed, "Here is the letter that has destroyed my universe!"[30] That certainly sounds like the kind of thing Shapley would have said.

Shapley was not the only one to learn of Hubble's results. He shared them informally with Heber D. Curtis, Henry Norris Russell and others, and while on his honeymoon, discussed them at the Royal Astronomical Society in London, and in November, they even figured in a report in the *New York Times*. But he was holding back. As he confided later to Russell: "the real reason for my reluctance ... was, as you may have guessed, the flat contradiction to van Maanen's rotations."[31] Only on January 1, 1925 did Hubble allow Russell to read a short paper, "Cepheids in Spiral Nebulae," to a meeting of the National Academy of Sciences in Washington, D.C.

For all practical purposes, the "great debate" was over. The spirals were "island universes," true galaxies. Admittedly, van Maanen continued to stand firmly behind his rotations, and never would retract, but at least everywhere except on Mt. Wilson, the matter was settled. Even van Maanen's close friend Shapley publicly conceded when Hubble published a paper giving light curves for twelve variables, all of the normal Cepheid type, in NGC 6822—thus it was this rather inconspicuous object in Sagittarius, discovered by E. E. Barnard while hunting for comets in Nashville, and not the Andromeda Spiral deserves to be remembered as "the first object definitely assigned to a region outside the galactic system."[32] It is a dwarf galaxy, a minor member of the Local Group, whose distance has now been set at 1.63 million light years.

Hubble published his paper on NGC 6822 and another, similar one, on Cepheids in M33 in 1926. That year, also, he published his classification

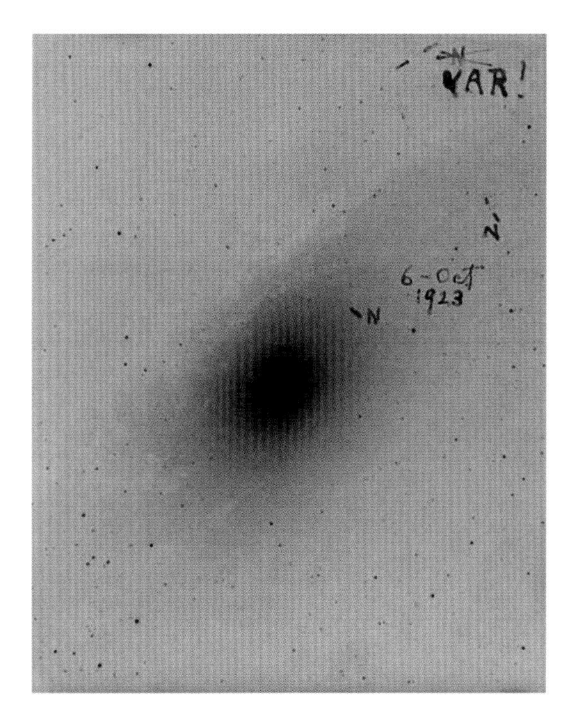

Fig. 11.5. (left) Hubble's negative plate, showing the star he at first thought was a nova but later realized was a Cepheid variable (VAR!). Courtesy of The Carnegie Observatories (the copyright holder and doing business as "*The Carnegie Observatories*")

Fig. 11.6. (right) Cepheid variable V1, indicated by red arrow. *Courtesy: Hubble Heritage Team (STScI/AURA).*

system—though without the I.A.U.'s imprimatur. Having been criticized by his colleagues on Commission 28 for introducing unjustified evolutionary terms, such as "early," "middle" and "late," he tacked with the winds, and took pains to emphasize that what he proposing was meant to be "descriptive and entirely independent of any theory."[33] The published classification was, in fact, very similar to the draft he had circulated in 1923, featuring ellipticals, spirals, and irregulars, but he had waited so long to publish that in the event he was forestalled. A similar classification scheme was published a few months before his by the Swedish astronomer Knut Lundmark, who had been working on the problem since at least 1922 and knew nothing of Hubble's private correspondence with Slipher or the other members of the I.A.U. Committee. Lundmark likewise divided his nebulae into ellipticals, spirals, and the small group of irregulars which did not fit into the other categories and which he termed "magellanic."[34] By this time, Hubble was clearly coming to regard the "non-galactic" nebulae—henceforth referred to as "extragalactic"—as his exclusive province, and went so far as to accuse Lundmark of plagiarism; he was always intensely competitive, and jealous of his prerogatives. He could well have afforded to be more magnanimous, since by 1926, with his discovery of the Cepheids in the spiral nebulae, he was like

Zeus on Olympus towering over his rivals. In Hubble's hands, the 100-inch telescope was achieving the "Manifest Destiny" that Hale, with his dream of a monopolistic and dominant factory observatory, had always planned for it.

Hubble would never repeat the plagiarism charge, though in future publications he did relegate mention of Lundmark's work to brief mention in a footnote, and said no more about it. A footnote it would remain. But it is at least worth noting that Walter Baade, the greatest observer on Mt. Wilson after Hubble, said long afterward, "In the early 1920s, two classification schemes were proposed almost simultaneously, one by Lundmark, one of by Hubble. Both were based on the large collections of plates that had been made at the Lick and Mt. Wilson Observatories. There was not much difference between the two systems."[35]

In Hubble's 1926 version of his classification scheme, there were ellipticals (E-type galaxies) as before, but now the spirals were separated into two branches, "normal" spirals (S-type) and "barred" spirals (SB-type). Barred spirals, to which attention had been first called by Curtis, were a subfamily of galaxies of the basic spiral type in which the nucleus appeared to be surrounded by a ring, across which a more or less prominent bar extended. Curtis, who had majored in classics before taking up astronomy, had compared the form to the Greek letter φ; to Hubble they resembled the Greek letter θ.

The final flourish, the famous tuning-fork diagram, appeared only in 1936. Again, Hubble was borrowing from Jeans, who had already published a similar Y-shaped diagram in his 1929 book *Astronomy and Cosmogony* to show the different forms assumed by his rotating masses. Hubble's diagram was basically a musical tuning-fork laid onto its side. From the base of the tuning fork to the beginning of the prongs are the ellipticals, ranged from globular to increasingly flattened. Along one prong lie the normal spirals opening up from Sa to Sb to Sc, along the other the barred spirals, opening up from SBa to SBb to SBc. Influenced by Jeans—and also, perhaps unconsciously, even by his arch-enemy van Maanen, whose rotations had the spirals unwinding rather than winding up as Slipher had them—he seems to have believed the direction of galactic evolution was from the base of the tuning fork to the ends of the prongs. As we now know, both ways are possible, though not in the exact sense of an evolution along the Hubble sequence. Spirals merging to form ellipticals is not the generally accepted idea anymore, and ellipticals can sometimes form into spirals, not in the way that Jeans and Hubble thought, but by accreting new gas.

Though evolutionary schemes tied to it have been abandoned, the tuning-fork diagram certainly proves the adage that a picture is worth a thousand words, and rescued Hubble's classification system from the danger of being buried (and forgotten) in the back issues of the *Astrophysical Journal.* It served, as Sandage notes, as a device from which generations of astronomers would learn to classify galaxies:

Fig. 11.7. Barnard's Galaxy, NGC 6822, in Sagittarius. This is a typical dwarf galaxy, and a member of the Local Group. Imaged by Stephen Leshin, Sedona, Arizona, for the "Little Things Survey" at Lowell Observatory. *Courtesy of Stephen Leshin.*

> Why did the diagram become so overwhelmingly important? Despite the excellence of Hubble's 1926 word descriptions of the classification, the diagram is much easier to understand and remember. It became the visual mnemonic. Indeed, we all learned to classify from it. Only later did we read the verbal descriptions in the 1926 fundamental paper. That was true in my generation. It is true now.[36]

Despite imperfections, Hubble's classification has stood the test of time and remains as a monument to his genius. A number of astronomers—including Reynolds, Shapley and others at the time, and much later, in the 1960s, Vorontsov-Velyaminov and his colleagues—complained that Hubble's system was too simple to accommodate the wide diversity of spiral forms. The fact is, however, to quote Sandage again, "their work failed to advance any ideas of the physics of galaxies." Of Hubble's system, he writes appreciatively:

> The great merit of Hubble's system is the bin size of the classification boxes. It is so large that it can accommodate all the general forms of the spirals in a continuous sequence using only ... three parameters isolated by Hubble. These are (1) the relative sizes of the nuclear bulge

and the disk for the spirals; (2) the openness of the spiral pattern, or the "pitch angle" of the spiral arms; and (3) the degree of resolution of the arms into stars and gaseous nebulae. Of course, these parameters focus on gross features only, ignoring the fine detail that Reynolds desired.[37]

Another of Hubble's Mt. Wilson colleagues, the legendary observer Walter Baade, paid a similar tribute in a lecture at Harvard in 1958. Having used Hubble's classification system for 30 years and searched "obstinately" for extragalactic systems that did not fit it, Baade confessed, "the number of such systems that I finally found—systems that really present difficulties to Hubble's classification—is so small that I can count it on the fingers of my hand."[38]

In anticipation of later discussions, we ought to point out here that the limitations of Hubble sequence have become much more apparent since the 1950s. In particular, the Hubble sequence does not reproduce well the vastly different structures seen in galaxies in the distant universe (which abounds in "peculiar" types). On the other hand, for local galaxies, Hubble's system still does quite a good job, and is likely here to stay.

Although Hubble no doubt assimilated ideas from others in devising his classification system, its very success is a tribute to his judiciousness and forceful advocacy of his ideas, the discovery of the Cepheids in NGC 6822 and M31 and M33 were furthermore a tribute to the power of the 100-inch reflector, and guaranteed him astronomical immortality. Today, however, Hubble is best remembered for his velocity-distance relation for galaxies, known as Hubble's law. It was declared, by Otto Struve and Velta Zebergs in 1962, "the most spectacular discovery that has been made in the past 60 years,"[39] and has become inextricably entwined with the idea of the expanding universe. But here also matters become a bit more complicated. As historians of astronomy Hilmar Duerbeck and Waltraut Seitter point out,

Fig. 11.8. Hubble Tuning Fork Diagram. *Image by: Wikipedia.*

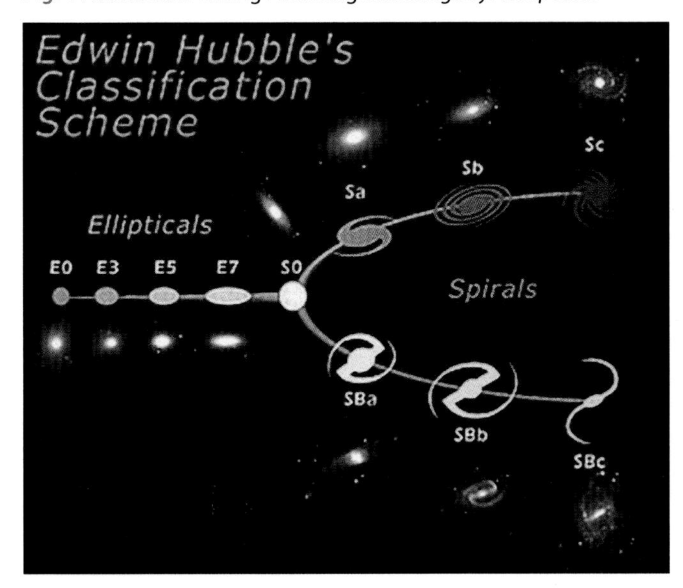

> Hubble has been put into the foreground by a glorifying historiography in such a way that astronomers who had in their hands equally large parts of the truth (or better, confirmation of a model by measurements) now live a shadowy existence [in the historians' consciousness].[40]

This story begins, of course, with Slipher's discovery of the large radial velocities of spiral nebulae. Already by 1914, when Hubble heard the wild applause for his paper at the A.A.S. meeting in Evanston, Illinois, Slipher had measured radial velocities of 13 spiral nebulae, most of which, he found, showed large recessional velocities. Up to that time, his greatest redshift was for M104 in Virgo, and corresponded to a recessional velocity of 1180 km/sec. In 1921 he obtained a spectrogram of the spiral NGC 584 in Cetus that required an exposure of 28 hours spread over the clear and

Fig. 11.9. The giant elliptical galaxy ESO 325-G004, at the center of the cluster of galaxies known as Abell S0740, located at a distance of 450 million light years in the direction of the constellation Centaurus. This giant elliptical has a mass of 100 billion solar masses. In this Hubble Space Telescope image thousands of globular clusters are noted, which appear as pinpoints of light contained within the diffuse halo. Other fuzzy elliptical galaxies dot the image, and several spiral galaxies are also present. *Credit: J. Blakelee and NASA, ESA, and the Hubble Heritage Team.*

Fig. 11.10. M101, in Ursa Major, at a distance of 22 million light years is the classic example of Hubble's open Sc-type spiral. M101 is a large system, nearly twice the size of the Milky Way. The bright knots of glowing gas (HII regions) highlight regions of active star formation (marked by the hot, brilliant blue stars of Population I) along M101's spiral arms. The dark dust lanes are also visible in this image, and are colder and denser regions where interstellar clouds may collapse to form new stars, while the softer, less-bright areas near the core and between the spiral arms consist mainly of older Population II stars. *Courtesy: K.D. Kuntz, NASA, ESA, and the Hubble Heritage Team (STScI/AURA).*

moonless parts of all the nights from December 31 to January 14; the redshift was half again as large as for M104, and gave a recessional velocity of 1800 km/sec. Slipher announced his result, of which he seems to have been justifiably proud, in an article in the *New York Times* for January 21, 1921, registering his satisfaction with it by clipping it and carefully preserving it in a personal scrapbook, which is something he always did rather sparingly. In this article he noted that if NGC 584 had started out in the region of the Sun when the Earth had formed, it would have traveled to a distance of "many millions of light years." This was, as Donald E. Osterbrock notices, "one of the first published statements by an observational astronomer on the expansion of the universe or the separation of the galaxies."[41]

At the same time the observers were making such headway, another chapter in the evolution of cosmology was being written by the theoreticians. As historian Norriss S. Hetherington writes, "Cosmology, for centuries consisting of speculation based on a minimum of observational evidence and a maximum of philosophical predilection," was now reaching a point where "its theories [were] subject to verification or refutation to a degree previously unimaginable."[42] The observational part of this grand project arguably began with the 1914 observations of V.M. Slipher, the theoretical part with the 1916 publication by Albert Einstein, then at the Friedrich-Wilhelm University in Berlin, of his General Theory of Relativity, which featured a profound re-interpretation of gravitation in terms of the curvature of space-time. Within only a year, Einstein himself and Willem de Sitter, a Dutch astronomer who had studied under Kapteyn at Groningen and was now on the faculty at the University of Leiden, published pioneering papers on the cosmological implications of Einstein's theory.[43] Their solutions were based on assumptions that the universe was homogeneous and isotropic. Einstein, on encountering problems finding meaningful "boundary conditions" at infinity, had the remarkable idea to circumvent these problems by envisioning space as curved and closed. De Sitter, on the other hand, took time as completely equivalent to space, and assumed a constant curvature of space-time.

Both solutions were "static," since it was assumed (according to the understanding of the time) that the universe was constant in size and

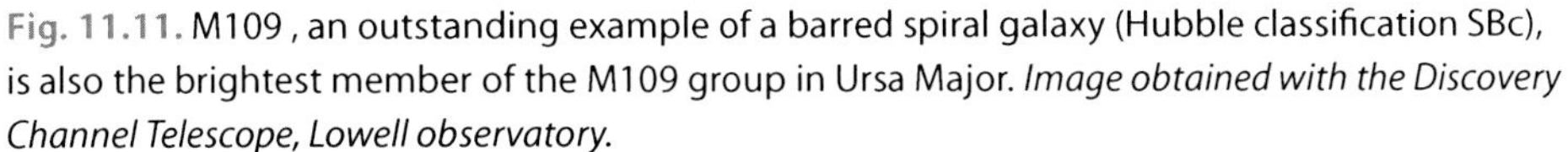

Fig. 11.11. M109 , an outstanding example of a barred spiral galaxy (Hubble classification SBc), is also the brightest member of the M109 group in Ursa Major. *Image obtained with the Discovery Channel Telescope, Lowell observatory.*

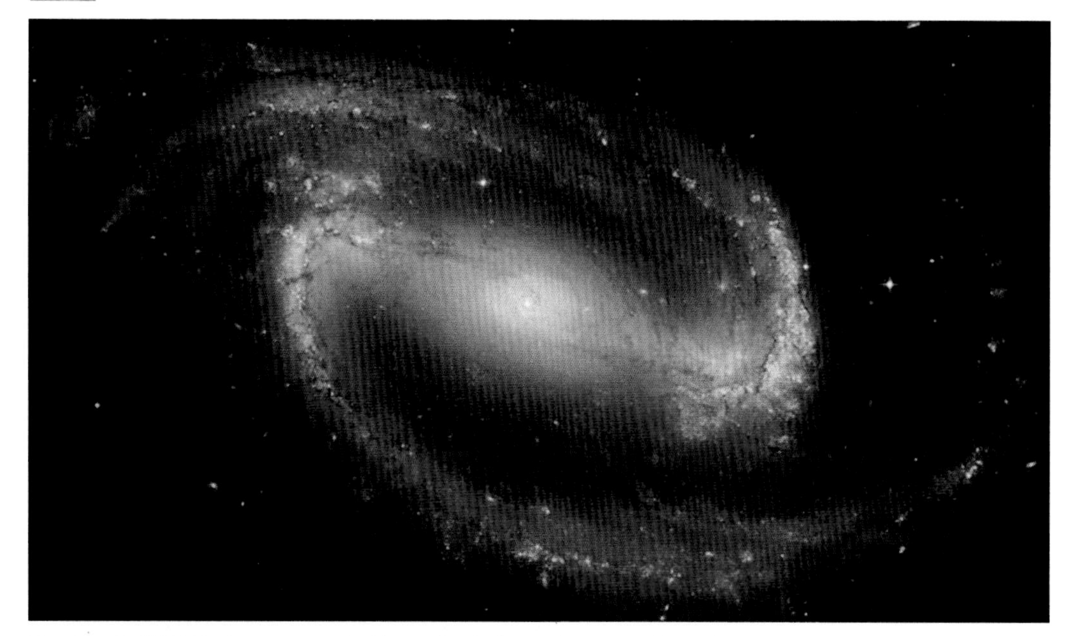

Fig. 11.12. NGC 1300, a prototypical barred spiral, at a distance of 69 million light years away in the constellation Eridanus. Barred spirals differ from normal spirals in that the arms of the galaxy do not spiral all the way into the center but are connected to the two ends of a straight bar of stars containing the nucleus at the center. The nucleus of NGC 1300 shows an extraordinary "grand-design" spiral structure that is about 3,300 light years in diameter. Only galaxies with large-scale bars appear to have these grand-design inner spiral disks. Models suggest that the gas in a bar can be funneled inwards, then spiral into the center through the grand-design disk, where it can potentially fuel a central black hole. However, NGC 1300 is not known to have an active nucleus; either there is no central black hole, or it is not accreting matter at the present time. *Courtesy: P. Knezek, NASA, ESA, and the Hubble Heritage Team (STScI/AURA).*

did not vary systematically over time. When Einstein first began working on the problem, he found that his solutions caused the universe either to collapse or expand; he couldn't get it to stand still without introducing a cosmological constant (lambda) that was not present in the original General Relativity. If he had simply accepted the result he was getting, he might have predicted that the universe was expanding and—as one wag has put it—he might have been famous!

In Einstein's matter-filled universe, the cosmological constant counteracted the gravitational attraction of masses, and was needed to avoid a gravitational collapse (or indefinite expansion) of the universe. In de Sitter's model, the matter-density was taken to be zero, and only the cosmological constant—which de Sitter in 1917 interpreted as the "vacuum energy density," something not connected with matter but a property of the vacuum itself—was present. In de Sitter's universe, the vacuum energy density produced the curvature. These solutions were each special cases of the general solution; the Einstein universe contained matter and no motion, the de Sitter universe contained motion and no matter. They represented the extreme cases, and obviously

Fig. 11.13. Albert Einstein in 1921. *From: Wikipedia Commons/F. Schmutzer.*

would have to be abandoned once more realistic models were found.

At a time when the extra-galactic nature of the spirals was still in doubt, de Sitter far-sightedly suggested that they might serve as "probe particles" whose masses were too small to influence the curvature of space-time. This idea was to become especially suggestive in light of Slipher's radial velocity data, which neither Einstein nor de Sitter knew about in 1917. The Einstein universe could not explain the redshifts at all, but the de Sitter universe, as it turned out, with positive cosmological constant and negligible matter, possessed an overwhelming tendency to expand. The galaxy "particles" in the de Sitter universe would appear to be moving away from each other and from the observer—i.e., they would show accelerated motion away from the observer (large redshifts). As soon as he learned about Slipher's velocities, de Sitter concluded:

> If ... continued observation should confirm the fact that the spiral nebulae have systematically positive radial velocities, this would certainly be an indication to adopt the hypothesis B (de Sitter-universe) in preference to A (Einstein-universe).

The first models to avoid the arbitrary and oversimplistic assumptions of the Einstein and de Sitter universes were proposed by the Russian mathematician Alexander Friedmann in 1922 and the Belgian priest-astronomer Georges Lemaître in 1925. However, de Sitter's model remained very influential during much of the 1920s. Several astronomers who "now live a shadowy existence," in Duerbeck's and Seitter's phrase, including the German astronomer Carl Wirtz at the Strasbourg Observatory and the Swedes Knut Lundmark and Gustav Strömberg (Lundmark always did seem to just miss out), attempted to confirm what was called the "de Sitter effect" (large redshifts at cosmologically great distances) by mapping distance (mainly based on apparent brightnesses of galaxies) against Slipher's radial-velocity measures. Because of the limitations of their data, their results were suggestive rather than definitive. By then, moreover, Hubble had decided to

Fig. 11.14. Willem de Sitter. *Courtesy: Yerkes Observatory.*

pile Pelorus on the Etna of his earlier achievements. On one of his regular trips to Europe, in 1928, he met de Sitter in Leiden, and (at least according to his wife Grace's notes made during the trip), the Dutch astronomer persuaded him to extend Slipher's scale of velocity measurements out to greater depths in space.[44] Slipher had by then reached the limit of the capability of his instruments. More data were clearly needed, specifically radial velocities of more distant galaxies, and this required both a larger telescope than the 24-inch Clark and a faster spectrograph. Hubble then had his hands full with direct photographic work on galaxies with the 100-inch, but he was able to assign the spectrographic work he needed to Milton L. Humason, who had had only an eighth grade education, and was probably the last person with an eighth grade education to become a professional astronomer. Humason had been a one-time mule driver on the Sierra Madre trail up Mt. Wilson but was eager to be involved in the work being done amidst the pines on the peak, and in 1917 caught on as janitor. He was then promoted to night assistant on the 100-inch, where he proved to be "careful, conscientious, and skillful with the mechanical tasks of observing."[45] In 1921 became a full-fledged member of the research staff. Hubble and Humason together made an excellent team. "Almost the exact opposite of Hubble," Osterbrock, who knew him well, has written, "the quiet, careful, self-effacing Humason was outstanding as a spectroscopic observer who studied and understood the mechanical details of his instrument and techniques completely, and could sit for endless hours in the dark, quiet telescope dome, concentrating his entire attention on getting all the light of faint 'nebulae' through the slit and onto the plate in perfect focus."[46]

In 1929, Hubble published his first paper on the velocity-distance relation, the important result that came to be known as Hubble's law. Hubble estimated distances on the basis the assumption that each type of galaxy in his classification scheme obeyed a simple relationship—that of constant surface brightness—so that fainter objects were physically similar to brighter ones, but at larger distances. He further assumed that for each type the difference in apparent magnitude between the entire "nebula" and it brightest stars was a constant. For radial velocities, almost all that he used had already been published by Slipher (whom he notoriously did not credit). There was only one measurement by Humason using the 100-inch, of the nebula NGC 7619, a member of a cluster of galaxies in Pegasus, but it was an important one, as the redshift (relative change in wavelength of spectral lines, denoted by z) was twice anything Slipher could get and yielded a radial velocity of +3779 km/sec. It was this data point that allowed the linear velocity-distance relation of Hubble to emerge from the scatter-plots of his data-deficient rivals. Hubble thus came up with a linear plot

$$V = HD,$$

where V is the velocity of the galaxy, D its distance, and H the slope of the line, known as the Hubble constant, which Hubble gave as 500 km/sec per megaparsecs of distance (where 1 million parsecs = 1 megaparsec).

At the time, it was still believed that the redshifts of galaxies were velocities due to their motion through space. We now know that they are actually due to the general expansion of the universe, and that Hubble's law states that redshift is proportional to distance. Since Hubble's constant is the ratio of a galaxy's distance to its redshift "velocity," it implies that a galaxy (using Hubble's value for the moment) shows a redshift velocity of 500 km/sec for every Megaparsec (million parsecs) of distance from us.

In fact, it turns out that Hubble's own value for this constant was much too large, partly because his "bright stars in distant galaxies" are not stars at all but HII regions of high absolute magnitude but mostly because he used, as everyone did until 1952, Shapley's incorrectly calibrated period-luminosity relation for the Cepheids. In 1952, Hubble's Mt. Wilson colleague Walter Baade showed that Hubble's distance scale was in error by a factor of 2, while later in the 1950s, using the 200-inch reflector at Palomar, his successor Allan Sandage revised the value still further, and found a value for Hubble's constant that made the universe not twice but more like five to ten times larger than Hubble had thought it.

Note that the inverse of the Hubble constant, 1/H, has the units of time. Indeed, this Hubble time, gives the expansion age of the universe, the approximate time since the beginning for everything to spread out to its current state. Since Hubble's value for H was too large, his value for 1/H was too small; in fact, his work implied a Hubble time of only 2 billion years, which made the universe much younger than the Earth as determined from radioactivity studies like those of Rutherford. (The currently accepted age of the Earth is 4.6 billion years.) This was a something of a paradox, since obviously the Earth cannot be older than the universe, so astronomers were long doubtful that the Hubble time could have anything to do with the age of the universe. Only with Sandage's revisions, which made the universe five or ten times older than Hubble had made it—in other words, 10 to 20 billion years—was the paradox resolved, by making the universe comfortably older than the Earth. Cosmologist Steven Weinberg has gone so far to say, "The removal of the age paradox by the tenfold expansion of the extra-galactic distance scale in the 1950s was the essential precondition for the emergence of the big bang cosmology as a standard theory."[47]

The latest value of the Hubble constant is in the range of 67-74 km per sec per Megaparsec, which means that a galaxy shows a redshift velocity of around 70 km/sec for each Megaparsec of distance from us. The expansion age of the universe now becomes 13.8 billion years, and the inconsistencies involving the age of the Earth and the ages of everything else (including the oldest stars in globular clusters, at 12 billion years) are reconciled.

In the 1930s, Hubble and Humason were busy pressing the 100-inch across greater and greater distances of space. First Hubble would find a faint cluster of galaxies on a direct plate, then Humason would obtain a spectrogram of one of its members. "Until lately," Hubble wrote in 1936 in the Silliman lectures at Yale (published as the classic book *The Realm of the Nebulae*),

> the explorations of space had been confined to relatively short distances and small volumes—in a cosmic sense, to comparatively microscopic phenomena. Now, in the realm of the nebulae, large-scale, macroscopic phenomena of matter and radiation could be examined. Expectations ran high. There was a feeling that almost anything might happen and, in fact, the velocity-distance relation did emerge as the mists receded. This was of the first importance for, if it could be fully interpreted, the relation would probably contribute an essential clew [sic.] to the problem of the structure of the universe.[48]

Humason, meanwhile, was measuring larger and larger redshifts. One for a galaxy in the Coma cluster gave a redshift velocity of +7500 km/sec; another in the Gemini cluster gave +23,000 km/sec; finally, for galaxies in clusters in in Boötes and Ursa Major, he obtained velocities of 39,200 km/sec and 41,600 km/sec, respectively. The latter were, as he noted, a seventh of the velocity of light itself.

Humason's spectrograms were now reaching distances 35 times greater than the distance of Virgo cluster—some of the galaxies measured had magnitudes of 17.5, making them, entire galaxies, almost as faint as the individual Cepheids in M31 that had first betrayed the extragalactic nature of the spirals.

By leapfrogging their way to redshifts of galaxies in clusters at greater and greater distances, Hubble and Humason by 1936 had pushed the 100-inch was reaching the reflector to the limits of its capability and had extended the linear relation of the Hubble diagram out to distances thirteen times that of the original diagram of 1929.

Hubble's superb biographer Gale Christianson refers to Hubble as the "Mariner of the Nebulae." The reference to a mariner is obvious, and certainly apt; commanding a great telescope and tacking (no doubt on the starboard side!) for the shoals of distant space is like skippering one of the great vessels of the Age of Exploration. But perhaps Hubble himself would have preferred a different analogy, one to his favorite pastime of fly-fishing. Fly-fishing is indeed the "high church" of fishing, and requires great patience and detailed skill and knowledge of the habits and habitats of one's particular quarry. Hubble was the fly-fisherman of the nebulae, and cast his line across hundreds of millions of light years.

By the 1930s, Hubble was a celebrity, and with little more to do on the 100-inch, was more interested in hob knobbing with Hollywood's elite (for a time, Charlie Chaplin was especially in vogue) than with his fellow astronomers, who either envied him his celebrity or regarded him as enigma. He was eager for the completion of the 200-inch Palomar reflector, which would allow him to push even farther into space, but it was delayed by the war. Hubble volunteered, as he had in World War I, and left Mt. Wilson for the Aberdeen Proving Ground in Maryland, where he was put in charge of ballistics and wind tunnels. (According to L.S. Dederick, who shared an office with him, he was "mildly ineffective" as well as poorly informed when it came to ballistics).[49] Gradually, over 1948-49, the 200-inch "Hale" telescope at Palomar became operational, and on his first run with the nebular spectrograph on the telescope, in 1950-51, Humason obtained a spectrogram for a galaxy belonging to a cluster in Hydra which was so redshifted that the cosmological velocity came out as nearly 20% the speed of light. By then, however, Hubble's health was rapidly failing. In July 1949, while fly-fishing his favorite stream at Rio Blanco Ranch in northern Colorado (west of Rocky Mountain National Park), he suffered a major heart attack. He gradually recovered to the point where he could return to work, though no longer able to smoke his signature pipe on doctor's orders. Prematurely aged,

Fig. 11.15. A diagram depicting Hubble's law, the linear Hubble Law relation between galaxy distance and redshift. The diagram shown here is a modern version of Hubble's original one, which was based almost entirely on Slipher's spectrograms and included, as shown here, a large number of galaxies belonging to the Virgo cluster. *From Wikipdeia Commons/W.C. Keel.*

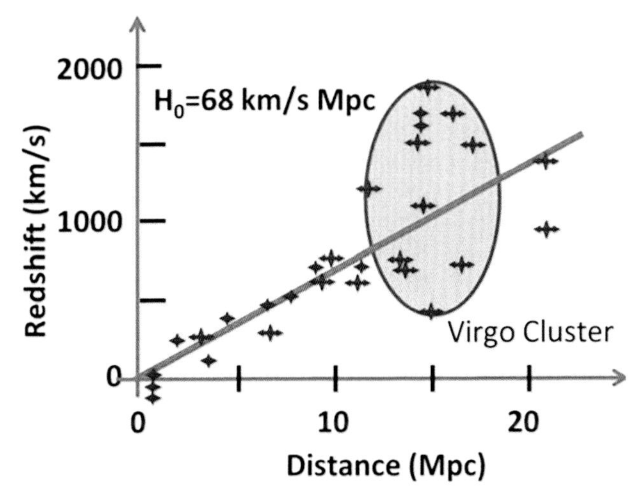

clearly slowing down, he began turning over his cosmological program of extending the velocity-distance relation farther and farther out into space largely to Sandage, who became his scientific heir (and was the only person who could in any sense claim to have been his student). The "mariner of the nebulae" died suddenly of a cerebral thrombosis while riding in the car with Grace in September 1953. Afterwards, his wife Grace destroyed most of his personal letters and papers.

His biographer has done his best to invest him in a human quality. But it is not entirely doable. Similarly, even his portraits have an "official" quality—there is none like the one of Einstein sticking his tongue out. In fact, they rather resemble the Greek portrait busts of the sort that art historian William M. Ivins, Jr., has described as

> "idealized" or "ennobled" ... i.e., abstractions with only the faintest personal character and no psychological value. ... As by definition, the hallmark of "idealized" and "ennobled" representation is vacuity. This is the common "stuffed shirt" quality of most official portraits, ancient and modern. That it should be an outstanding quality of by far the greater part of classical sculpture should be and probably is indicative of many things, as is also the fact that it is a quality that is most frequently singled out for enthusiastic approbation.... [And yet] I, personally, in spite of the portrait busts, cannot believe that Pericles had a vacant face.[50]

Fig. 11.16. Coma Cluster galaxies. *Courtesy: Hubble Heritage Team (STScI/AURA).*

Acquaintance with Hubble—or with Isaac Newton, to give another example, whose portraits have the same "stuffed shirt" quality —makes one wonder whether, perhaps, Pericles did, after all.

From his first result with the 200-inch for the cluster in Hydra, Humason, now in collaboration with Sandage, tried to get redshifts to still more distant clusters, only to encounter what appeared, at the time, to be an insuperable barrier: on long-exposure plates the sky brightness swamped the faint H and K lines of the distant galaxies' spectra on which the distance measures depended. New techniques—electronic spectrographs, which allowed astronomers to perform sky subtraction techniques—allowed astronomers, including Sandage himself, to surpass each barrier in turn—we have already referenced Sandage's five to ten times expansion of the extragalactic distance scale in the 1950s. But even then, the end was not yet in sight. Before he died in 1972, Humason noted that "there is apparently no horizon, at least as far as the 200-inch goes—it just continues... the nebulae go on, they get fainter and fainter."[51]

The latest iterations of the Hubble diagram combine data not from a couple dozen redshifts like Hubble's first in 1929 but from millions of redshifts. His graph has now been extended deeper and deeper into space (and its extension has brought some surprises, described in chapter 15). We are now almost reaching the horizon Humason referred to.

The progress of astronomy in the past sixty years can be read from the successive records for redshifts, which went from the z = 0.2 of the Hydra cluster in the early 1950s, to z=0.461 for the radio galaxy 3C295 by the early 1960s, to z=2 for the quasar 3C9 in 1965. By the late 1980s, quasars were being regularly found with z greater than 4. The current record, for the galaxy UDFj-39546284, imaged in 2009 in the Ultra Deep Field obtained with the Wide Field Camera 3 (WFC3) on the Hubble Space Telescope, is claimed to be at z = 11.9.

Fig. 11.17. From left to right in this 1986 image: Olin Eggen, Donald Lynden-Bell, and Allan Sandage. Their influential 1962 paper suggested that the Milky Way formed by collapse of a large proto-galactic nebula. Sandage was the closest thing to a student that Hubble ever had. *Courtesy: Professor Kenneth C. Freeman.*

We hasten to add that z=2 or z=4 does not mean that the quasar or galaxy is flying away from us at a velocity twice or four times the speed of light. Velocities of greater than about 0.3 cannot be thought of as velocities in the ordinary sense; instead the relative wavelength redshifts are almost entirely due to the cosmological expansion, and the calculations of expansion velocities now require a correction involving the special theory of relativity, according to which nothing can travel faster than c, the speed of light. For a relative wavelength redshift of 2, the corresponding cosmological redshift is z=0.8 (the expansion velocity is 80 percent the speed of light), for a relative wavelength redshift of 4, z = 0.9 (expansion velocity is 90 percent the speed of light), and so on. For these distant objects, the light is shifted from the optical into the near-infrared—and even, in the case of the remnant radiation from the Big Bang known as the Cosmic Microwave Background (see chapter 14), for which the relative wavelength redshift is 1100, with the peak emission all the way into the microwave.[52]

In some ways, it is true, we are now entering the post-Hubble era. The Hubble classification system fails for galaxies in the distant universe, and the resolutely straight-line of the Hubble diagram has now begun to bend for galaxies with cosmological redshifts. Einstein's cosmological constant is non-zero after all. Moreover, as discussed later, we now have not only ordinary baryonic matter to contend with, but dark matter and dark energy as well.

Nevertheless, the basic framework remains, and will always remain, Hubble's. Observational cosmology is now opening up exciting frontiers of which Hubble never dreamed, but it still rests majestically on the shoulders of the Olympian.

* * *

1. Richard Berendzen, Richard Hart, and Daniel Seeley, *Man Discovers the Galaxies* (New York: Science History Publications, 1976), 145-146.
2. Quoted in Raymond S. Dugan, "Max Wolf," *Popular Astronomy, 41*, 5 (1933):239.
3. Max Wolf, "On the dark nebula NGC 6960," *Astronomische Nachrichten,* 219 (1923), 109-116; translated by Brian Doyle and Owen Gingerich in Kenneth R. Lang and Owen Gingerich, eds., *A Source Book in Astronomy and Astrophysics,* 1900-1975 (Cambridge, Mass. and London, England: Harvard University Press, 1979), 568-570. The region chosen by Wolf for analysis is the so-called Cygnus Loop, now recognized as a supernova remnant. Wolf correctly shows by star counts that the obscuration here is caused by dust rather than by gas.
4. M. Wolf, "Die Klassifizierung der Kleinen Nebelflecken," *Publ. Astrophys. Inst. Konigstühl-Heidelberg,* 3 (1909), 109-112.
5. Osterbrock, "Observational Approach to Cosmology," 270.
6. Christianson, *Edwin Hubble,* 14.
7. Ibid., 331.
8. Shapley, *Through Rugged Ways,* 57.
9. Quoted in Christianson, *Edwin Hubble,* 326-327.
10. Ibid., 326.
11. Ibid.
12. Ibid., 248.
13. Donald E. Osterbrock, *Yerkes Observatory, 1892-1950: the birth, near death, and resurrection of a scientific research institution.* Chicago: University of Chicago, 1997.

14. E.A. Fath to V.M. Slipher, Feb. 16, 1917; Lowell Observatory Archives.
15. Sandage, *Mt. Wilson Observatory,* 483.
16. E. P. Hubble, "Changes in the Form of the Nebula NGC 2261," *Proceedings of the National Academy of Sciences, 2* (1916), 230.
17. E. P. Hubble, *Photographic Investigations of Faint Nebulae* (Chicago: University of Chicago Press, 1920) (Hubble's Ph.D. dissertation).
18. Sandage, *Mt. Wilson Observatory,* 481.
19. Ibid., 493.
20. On Roberts, see: Lee T. Macdonald, "Isaac Roberts, E.E. Barnard and the Nebulae," *Journal for the History of Astronomy, xli* (2010), 239-259.
21. E.P. Hubble, "A General Study of Diffuse Galactic Nebulae," *Astrophysical Journal,* 56 (1922), 162-199; and "The Source of Luminosity in Galactic Nebulae," *Astrophysical Journal,* 56 (1922), 400-435.
22. For details about Reynolds, see: David L. Block and Kenneth Freeman, *Shrouds of the Night.* New York: Springer, 2008.
23. E. P. Hubble to V.M. Slipher, Feb. 23, 1922; Lowell Observatory Archives.
24. W.H. Wright to V.M. Slipher, March 7, 1922; Lowell Observatory Archives.
25. V.M. Slipher to W.H. Wright, March 14, 1922; Lowell Observatory Archives.
26. E. P. Hubble to V.M. Slipher, April 4, 1923; Lowell Observatory Archives.
27. V.M. Slipher, note on Hubble manuscript, "The Classification of Nebulae," Lowell Observatory Archives.
28. J. C. Duncan, "Three Variable Stars and a Suspected Nova in the Spiral Nebula M33 Trianguli," *Publications of the Astronomical Society of the Pacific, 34* (1922), 290-291
29. E.P. Hubble to Harlow Shapley, Feb. 19, 1924; Hubble manuscript collection, Henry L. Huntington Library, San Marino, California.
30. Cecilia Payne-Gaposchkin, *An Autobiography and Other recollections* (ed. Katherine Haramundanis (Cambridge: Cambridge University Press, 1984), 209.
31. Richard Berenzden and Michael Hoskin, "Hubble's Announcement of Cepheids in Spiral Nebulae," *Astronomical Society of the Pacific Leaflet No. 504* (June 1971), 11.
32. E.P. Hubble, "NGC 6822, a remote stellar system," *Astrophysical Journal, 62* (1925), 409.
33. E.P. Hubble, "Extra-galactic nebulae," *Astrophysical Journal, 64* (1926), 321-369.
34. Knut Lundmark, "A Preliminary Classification of Nebulae," *Arkiv for Matematik Astronomi och Fysik,* Band 19B, no. 8 (1926).
35. Walter Baade, *Evolution of Stars and Galaxies* (Cambridge, Mass.: Harvard University Press, 1963), 12.
36. Quoted in Block and Freeman, *Shrouds of Night,* 204.
37. Sandage, *Mt. Wilson Observatory,* 489.
38. Baade, *Evolution of Stars and Galaxies,* 18.
39. Otto Struve and Velta Zebergs, *Astronomy of the 20th century* (New York: Macmillan, 1962), 469.
40. Hilmar W. Duerbeck and Waltraut C. Seitter, "In Hubble's Shadow: early research on the expansion of the universe," in C. Sterker and J. Hearnshaw, eds., *100 Years of Observational Astronomy and Astrophysics* (Brussels: C. Sterker, 2001), 238.
41. Osterbrock, "Observational Approach to Cosmology," 275.
42. Norriss S. Hetherington, "Hubble's Cosmology," in Norriss S. Hetherington, ed., *Cosmology: historical, literary, philosophical, religious, and scientific perspectives.* (New York and London: Garland Publishing, 1993), 366.
43. A. Einstein, "Kosmologische Betrachtungen zur allgemeinen Relativitätstheorie," SB Prussian Akademie Wissenschaften, Berlin, phys.-math. Kl. No. 6; and W. de Sitter, "On Einstein's theory of gravitation, and its astronomical consequences." *Monthly Notices of the Royal Astronomical Society, 78* (1917), 3.
44. Christianson, *Edwin Hubble,* 198.
45. Osterbrock, "Observational Approach to Cosmology," 276.
46. Ibid., 276-277.
47. Steven Weinberg, *The First Three Minutes: a modern view of the origin of the universe* (New York: Bantam Books, 1977), 26.
48. E.P. Hubble, *The Realm of the Nebulae* (New Haven, CT: Yale University Press, 1936), 120.
49. Christianson, *Edwin Hubble,* 305.
50. William M. Ivins, *Art and Geometry: a study in space intuitions* (Cambridge, Mass.: Harvard University Press, 1946), 26-27.
51. Christianson, *Edwin Hubble,* 347.
52. A very clear exposition of this is found in Maria Luiza Bedran, "A comparison between the Doppler and cosmological redshifts, *American Journal of Physics, 70,* 4 (April 2002), 406-408.

12. W.W. Morgan and the Discovery of the Spiral Arms of the Milky Way

Dear Book, what a strange thing the unbridled mind is. A sequence of thoughts can develop—move rapidly from stage to stage, and end in a conclusion (a definite, unique conclusion) in a few eye-closings. And what is the "unique conclusion" worth? Perhaps absolutely nothing. Conclusion may not result from premise; there may be spaces—infinities wide—between successive steps.

—W.W. Morgan, personal notebook; December 29, 1956

It is now common knowledge that the Milky Way is a vast spiral star system which we view edgewise from just outside one of the spiral arms. The first clear demonstration of the fact, by Yerkes Observatory astronomer William Wilson Morgan, occurred only in 1951. This was one of the grandest discoveries in the history of astronomy, and when Morgan presented it, in a fifteen minute talk at the American Astronomical Society Society in East Cleveland, Ohio, the day after Christmas 1951, he received a resounding ovation, that included not only clapping but stomping of feet. (It was the first time that had happened at an A.A.S. meeting since V.M. Slipher announced the large radial velocities of the spirals in 1914.) But for various reasons—not least that Morgan suffered a nervous breakdown that led to hospitalization only a few months later—no definitive account of his discovery appeared at the time.[1]

Morgan had used optical methods to detect the nearer spiral arms. When he left Billings Hospital at the University of Chicago, Morgan was determined to reconstitute himself and reorganize his psyche through a systematic program of self-help and psychoanalysis which he would document in a remarkable series of personal notebooks he kept for the rest of his life (first studied in detail by one of us, W.S., in preparation for a full-scale biography of the great astronomer). By the time he returned to Yerkes, the Dutch astronomer Jan Oort and his Dutch and Australian collaborators had independently announced the discovery of the spiral-arm structure of the Milky Way on the basis of radio telescope observations. At the time their results seemed more far-reaching, since whereas Morgan had only identified the nearby spiral arms, radio astronomers were able to detect structures on the optically hidden far side of the Galaxy.

At first, the discoveries of Oort and his collaborators overshadowed Morgan's work. Only later, in about 1970, was it realized that the radio distances were not as accurate as had been assumed because of large-scale systematic deviations from circular orbits of the hydrogen gas clouds on which they had relied for their measurements. Thus, the radio maps were not very reliable, and the uniqueness of Morgan's achievement began once more to be fully appreciated.[2]

Recall that at the beginning of the 20th century, the standard model of the Galaxy was that offered by the Dutch astronomer J.C. Kapteyn who, much as William Herschel had done in 1785, regarded the Galaxy as a small disk of stars. Since Kapteyn ignored the effects on starlight of interstellar dust, his model, like Herschel's, included only the nearer stars; as a result, his disk measured a mere 4000 parsecs long by 1000 parsecs wide, and remained centered on or near the Solar System. As we have seen, Kapteyn's model was eventually undermined by Harlow Shapley's work on globular clusters, while it was also becoming clear, from Barnard's wide-angle photographs of the Milky Way, that there were dust clouds scattered along the Galactic Plane that would redden and obscure more distant stars. Meanwhile, Heber D. Curtis, then at Lick Observatory, making a careful study of spiral nebulae on images with the Crossley reflector, noted their family resemblance—they appeared to him to form a class of similar objects distributed at different angles and at different distances. Moreover, where they were seen edge-on, dark rifts divided them, which Curtis recognized as similar to the dark dust clouds of the Milky Way that Barnard had photographed. Though Curtis was sure that the spirals were "island universes," it was only with Edwin Hubble's discovery of the Cepheid variables in the Andromeda Nebula in 1923-24 that this result was accepted as proved by most astronomers. Thus, within only a few years Kaptyen's rather quaint model of the Galaxy had been discredited, and the Milky Way itself now seemed one of countless millions of "star systems" strewn throughout the Universe. It might well be a majestic spiral in its own right (as had been suggested as early as 1900 by a Dutch amateur astronomer, Cornelis Easton), though it might also be a flattened elliptical. Determining its actual structure would prove to be one of the most daunting problems of 20th century astronomy.

For many years, optical astronomers continued to tackle the problem by means of brute-force star counts and methods of statistical analysis such as those used by Kapteyn. These were basically refinements of the methods introduced by William Herschel when he carried out his star-gages at the end of the 18th century. The basic idea was simple: as one counted stars, the number of stars was expected to rise in the vicinity of a spiral arm and then drop off beyond it. Unfortunately, though a number of astronomers made enormous efforts in these directions, it became increasingly apparent that brute-star counts wouldn't lead to the desired result.

Kapteyn died in 1921. Largely as a result of his influence, Dutch astronomers continued his work on the problems of galactic structure. The most brilliant of them was Jan Oort, who during the 1920s refined knowledge of galactic rotation and modified Kapteyn's model by introducing the notion that there must be much more interstellar absorption by dust than had been realized. He noted that because of galactic rotation, "there was a well-defined relationship between radial velocities, distances and angles, which meant that measured systematic radial velocities could be converted to approximate distances in a straightforward way."[3] Unfortunately, Oort did not have the telescopes to provide the kinds of data he needed – after all, the Netherlands, whose mean elevation is below sea-level, is one of the worst imaginable places on Earth for optical astronomy! On the other hand, the electromagnetic spectrum includes far more than just the optical window. Beyond the red, there are the infrared, microwave and radio regions of the spectrum; beyond the violet, the ultraviolet, X-ray, and gamma-ray regions. Radio was the first area outside the optical to be exploited by astronomers in their studies of the Galaxy. In 1931, radio emissions from the center of the Galaxy were detected serendipitously by the Bell Telephone Labs engineer Karl Jansky using an antenna on a "merry-go-round" track at Holmdel, New Jersey, while just before the war, detailed radio maps, which included such powerful radio sources as Cygnus X-1 and Cassiopeia A for the first time, were made by an amateur

Fig. 12.1. W. W. Morgan inspecting M31 plate. *Courtesy: Yerkes Observatory.*

radio operator and amateur astronomer, Grote Reber at Wheaton, Illinois, using a homebuilt steerable radio dish he set up in his own backyard. Oort, however, was the professional astronomer to fully realize the great potential of a new and powerful technique, and grasp that it would potentially allow penetration even of the interstellar dust clouds.

Unfortunately, Oort's research was disrupted by the war, during which Holland came under German occupation. Nevertheless, there a conceptual breakthrough occurred after a wartime seminar in which he posed to a brilliant Utrecht student, Henrik C. van de Hulst, the question of whether there was any spectral line at radio frequencies that could in principle be detected with a radio receiver. After a few months of studying the question, van de Hulst announced that indeed there was—the 21-cm line of the ground state of hydrogen (neutral hydrogen, HI). Given the vast abundance of hydrogen in the gas in the Galactic Plane, Oort at once realized that mapping interstellar hydrogen would likely lead to the discovery of the Galaxy's spiral arms. However, there were delays in getting the proper equipment—not least the tragic loss of one of Oort's receivers, which was destroyed in a fire.

Meanwhile, optical astronomers were unfolding their own strategy. Noting the failure of the brute star-counts approach, historian Owen Gingerich has noted, "the solution to this puzzle lay elsewhere, with the observational analysis of the Andromeda Nebula and other nearby galaxies."[4] This observational analysis was emerging at almost the very same time that Oort was first exploring the use of radio astronomy to unravel galactic structure. Though it would add a touch of drama to suggest there was a race, in fact radio and optical astronomers pursued their research quite independently; neither was particularly concerned with the other's methods or results.

An Artist, and a Philosopher of the Stars

Among the optical astronomers researching the structure of the Galaxy, the key figure was William Wilson Morgan.

Like Barnard, Morgan was a native of Tennessee. He was born on January 3, 1906 in the tiny hamlet of Bethesda, which no longer exists. With his parents, who were home missionaries in the Southern Methodist Church, he was constantly on the move. From Bethesda, he moved to Crystal River, Florida; then to Starke, Florida, where he claimed to have seen Halley's Comet in 1910 (though it may actually have been the much brighter Great Comet of 1910); to Punta Gorda; to Key West; to a farm near Punta Gorda; to Perry, Florida; to Colorado Springs, Colorado; to Poplar Bluff, Missouri; to Spartanburg, South Carolina; and finally, at age 12, to Washington, D.C., with his mother (his father, a rather domineering, volatile, and sometimes violent man, was away much of the time working as a kind of itinerant inspirational speaker). He hardly spent two successive years in the same place, and what education he received was through homeschooling by his mother. As with others who have suffered from unhappy and abusive childhoods, Morgan found a refuge in the stars. In an interview, he recalled:

> The stars gave me something that I felt I could stay alive with. The stars and the constellations were with me, in the sense that on walks in the evening, I was part of a landscape which was the stars themselves. It helped me to survive.[5]

Morgan's first formal encounter with astronomy occurred in the winter of 1918-19, when the deadly influenza epidemic was raging. His father was as usual gone for an extended period and Morgan, with his mother and sister, moved to Frederickstown, Missouri, where a Methodist junior college (Marvin College) and an attached high school were located in a cow pasture. Morgan entered high school there in the fall of 1919. He received his first astronomy book (a collection of star maps) from his Latin teacher, Alice Witherspoon, and she arranged his first view through a telescope: of the Moon. At the same time, Morgan was discovering his father's set of the Harvard Classics—"Dr. Eliot's six-foot shelf of books," so-called because Harvard president Charles W. Eliot (the same man who had hired E. C. Pickering to direct the Harvard College Observatory) had selected them. One of these books included the Elizabethan play *Doctor Faustus* by Christopher Marlowe, and 67 years later Morgan recalled the electrifying effect the play had on him:

> The picture of the partially legendary Faustus, the man who longed to press outward toward the horizons of knowledge—and beyond to the stars—has been the ruling passion of my life.[6]

Morgan finished his last two years of high school at Central High School in Washington, D.C. His father, who had apparently started out as a fire-and-brimstone preacher who took a rigid and literal-minded interpretation of the Scriptures, viewed the prohibition against working on the Sabbath literally, so that when Morgan was in school he was strictly forbidden to do any work on Sundays at all. As a result, he was always falling behind:

> I remember late in high school, in Washington, D.C., I always dreaded Sunday night because I never was prepared for Monday. So it was a question of just survival. Just passing was all. And that's what it was like through these years.[7]

In the fall of 1923 Morgan enrolled at Washington and Lee University in Lexington, Virginia. Although he was interested in astronomy, he had no idea that it would become his profession. Instead, he decided to specialize in English, in preparation for becoming a teacher. However, he performed well in mathematics, physics, and chemistry, and even talked his physics teacher, Benjamin Wooten, into acquiring a small astronomical telescope, so he could observe sunspots.

Then in the summer of 1926, a year before Morgan was to finish his degree, Wooten, during a vacation to the Midwest, paid a visit to Yerkes Observatory. When he went up the stairs and rang the doorbell, the director, Edwin B. Frost, happened to be just inside. Wooten told Frost about the student who had pressed him to buy a telescope. Frost, meanwhile, was just then looking for an assistant to operate the Observatory's spectroheliograph and use it for the routine work of obtaining daily images of the Sun (the previous incumbent had just left to take a job as chair at Northwestern University in Evanston, Illinois). Morgan was to be offered the job, but there were still difficulties, not least the fact that Morgan's father was violently opposed, thinking that he would "end up just in a laboratory working for somebody else, [and] that's nothing."[8] That was to be the last conversation Morgan had with his father about anything; his father, who had been absent for long periods previously, now decided to abandon the family completely—Morgan would never see him again, and could not even find out what happened to him or where or when he died.

The rage and disappointment Morgan felt toward his father was reflected in a symbolic act. After his father left, Morgan appropriated the set of Harvard Classics, and savagely ripped his father's name plates out of all the books save one (where he missed the name plate because it had been accidentally affixed to the back rather than to the inside front cover of the book, like all the rest). One has a sense that with this act Morgan was symbolically attempting to tear his father out of his life.

At Yerkes, Morgan lived at first in the basement of the Observatory, but soon afterward married Helen Barrett, the daughter of Yerkes astronomer and librarian Storrs Barrett (who had also been E. E. Barnard's closest friend on the staff), and the couple moved into a small house 30 meters east of the Observatory. Morgan would remain there for the rest of his life.

In 1931, Frost finally retired, and was replaced by Otto Struve, a Russian immigrant descended from a very distinguished family of Russian astronomers. Struve was an imposing, hulking man whose eyes were not quite congruent—you could never quite tell if he was looking at you. He had a gruff, bearish manner, but he was also possessed of enormous drive and an incredibly hard worker, and he became something of a father-figure to young Morgan. Struve was an astrophysicist who was interested in using stellar spectra as a tool to understand what was going on physically in stars, while Morgan was from the first temperamentally drawn to problems of stellar classification. Struve once remarked that he had never looked at the spectrum of a star, any star, where he did not find something important to work on. The remark made a lasting impression on Morgan, and as early as 1935, he produced his first paper on stellar classification, "A descriptive study of the spectra of A Stars."[9]

A year later, Bart Bok, a Dutch-American astronomer who had studied with Oort in Leiden but did most of his work at Harvard, gave a series of lectures at Yerkes. Bok had been making heroic attempts to solve the problem of galactic structure by means of brute-force star counts and methods of statistical analysis but, after ten years of hard slogging, had begun to think the task of tracing out the spiral arms of the Milky Way was almost hopeless. "In public lectures during that period I often said it was unlikely the problem would be solved in my lifetime," he afterward recalled.[10] In his lectures at Yerkes, Bok inspired Morgan "… to improve the distances of the hotter stars and to investigate the structure of the Milky Way with the help of these distances." These hotter stars included stars of spectral type B and their even brighter, much rarer cousins, the O stars. Unfortunately, none of the closer stars are B stars, and there are only a few within a distance of 300 light years. Nevertheless, these stars, which include such splendid suns as Rigel, Achernar, beta Centauri, Spica, alpha and beta Crucis, and Regulus, are so intrinsically bright that they dominate the naked-eye sky all out of proportion to their numbers.

The B stars are young hot stars, and are very bright in ultraviolet. They are rapidly rotating stars, and in some of them the velocities of rotation can be as high as 200 km/sec, so that they eject matter into equatorial rings that radiate bright emission lines. They had been grouped together in a spectral classification in the Henry Draper Catalog. The Draper Catalog's familiar categories—OBAFGKM—was a one-dimensional classification which, as suspected by the Harvard classifiers themselves but proved at Mt. Wilson in 1908—was keyed to temperature. (In the hotter stars—those of classes A, B, and O—the spectral type is determined largely by the strength of the hydrogen lines (Balmer lines; see chapter 8) and the increasingly dominant presences of lines of singly or doubly ionized helium. In the even hotter O stars, these include not only absorption but emission lines.)

As early as 1897, Antonia Maury had realized that there were distinctly different spectra for stars of a given temperature. In some stars of a given type—her "c" stars—the hydrogen lines were sharper; in others they appeared broadened and more diffuse (showing "wings"). Between 1905 and 1907, Ejnar Hertsprung showed that the stars in which the lines were sharp were much more luminous than the corresponding main sequence stars. They were supergiants. The stars with lines broadened into wings were main sequence stars—dwarfs; the wings being produced by effects of surface gravity and pressure.

The Henry Draper Catalog was based on the way spectra appeared to Annie Jump Cannon, and were based on low-resolution objective-prism spectra. When good high-resolution spectra of stars began to be obtained with the Mt. Wilson 60-inch reflector, there was much more fine structure visible, and beginning in 1914, Mt. Wilson spectroscopists W. S. Adams and Arnold Kohlschütter began to document in great detail the effects of luminosity on line strengths and line ratios in the spectra of these stars. In contrast to the one-dimensional (temperature-keyed) Harvard classification system, the Mt. Wilson astronomers were developing a two-dimensional classification system combining temperature and luminosity criteria. Inevitably, the classes they assigned often differed markedly from Annie Jump Cannon's, and this produced some tension between the Harvard and Mt. Wilson groups.

It turns out that line strengths and line ratios in stellar spectra are very sensitive to the precise physical conditions in stars and their atmospheres. Since they are luminosity-dependent, the spectral classifications will be very sensitive to calibration effects. As with Shapley's Cepheids, the determination of absolute magnitudes depends on accurate measurements of the distances of a few stars, but since these stars are distant, direct parallaxes are not available and the distances must be estimated in more indirect ways. Each recalibration of luminosities for these stars might change the spectral classifications (since numerical values were assigned to absolute luminosities), and different groups—those at Mt. Wilson versus those at the Dominion Astrophysical Observatory in Victoria, British Columbia, for instance—used different calibrations. It was beginning to seem that the whole field of spectral classification might forever remain in a state of confusion and flux.

The classification of O and B stars was especially problematic. They have weak spectral lines, and because of the great distances of these stars the Harvard astronomers often had difficult seeing the lines at all. Thus, Miss Cannon had classified some heavily-reddened O and B stars as A or even F. In 1936, when Morgan became interested in spectral classification, he shifted his interest to high-luminosity stars because they were precisely those where, as Donald Osterbrock notes,

> ... the Harvard classification was so bad. Before Morgan, people were using spectral types out of the Henry Draper Catalog that were not very good. If you take the spectral types as published in the HD and try to use them today, they're terrible.[11]

Morgan wanted to find away around this, and in the end, he decided to take what Allan Sandage calls "... the drastic step of abandoning the assignment of numerical values to absolute luminosities."[12] With his colleagues Philip Keenan and Edith Kellman (the latter providing clerical assistance), he began work on a new classification scheme. It was a huge project, and was finally published in 1943 as the *Yerkes Atlas of Stellar Spectra, with an Outline of Stellar Classification* (known as the MKK and, more recently, as the MK Atlas).

Instead of the continuous absolute magnitude numbers of the Mt. Wilson continuum, Morgan and Keenan sorted the Mt. Wilson main sequence into discrete bins forming five luminosity classes: Ia and Ib supergiants, II bright giants, III giants, IV subgiants, and V dwarfs (later a category VI, for subdwarfs, was added). Thus the Sun, a yellow dwarf, was classed as G V. Morgan—with some of the psychological craving for permanence and security of one who had endured a childhood of constant change and uncertainty—wanted a classification system that would remain secure, true for all time. He achieved this by choosing a set of standard stars—"specimens," he called them—main sequence stars whose spectra were obtained using the same dispersion, depth of exposure on the photographic plate and method of development. These stars defined a "box" or reference frame. All other normal stars could then be classified by comparing them to these standards.

Since only temperature (or color equivalent) and luminosity were needed to uniquely locate a star's spectrum, Morgan insisted that by simple visual inspection of a spectrogram these parameters could be determined and the star appropriately binned or boxed. There was no need to measure anything. If the standard spectrum of the star looked like a 08Ia, then it was intrinsically like his standard 08Ia star. It was up to the astrophysicists to turn these spectral types into temperatures and gravities—after their luminosities had been calibrated using his method.

Morgan's essentially qualitative approach was unusual in a field that was dominated by quantitative methods, and he was once savaged by a colleague who referred to what he was doing as "celestial botany." But Morgan himself always considered himself as much artist as scientist, and his method (like Hubble's tuning-fork diagram for galaxies) withstood the test of time. Stellar astronomer James Kaler writes:

> The standards become embedded in memory, and the typing of stars can proceed with impressive speed. There is a very important place for

quantitative methods.... Visual classification, however, is at present still useful in surveying in a reasonable amount of time the vast numbers of stars readily accessible to us.[13]

A Passion for Pattern-recognition

One finds countless examples in Morgan's notebooks of his keen passion for visual pattern-recognition tasks (among other things, he used to like to put together jigsaw puzzles, and was famous for making them more challenging by turning the colored sides of the pieces face down and assembling them from their shapes alone). As soon as he could afford it, in the 1930s, he began acquiring art books, and made deep philosophical comments—all completely confidential and private—which often convey a sense of loneliness, on the artists or art-works which were most meaningful to him:

> Sunset. May 19 [1942]. I want to be the man of the Rembrandt self portraits no 40, 41, and 58. I want to look at women the way he looked at Hendrickje Stoffels and at the woman in no. 366. [The numbers refer to drawings in John C. Van Dyke, *Rembrandt Drawings and Etchings,* New York: Scribner, 1927.]
>
> I want to look at the earth as Ma Yuan ... and to feel like the sculptors of the Tang Bodisattuas and to feel as does the head of Buddha.
>
> Sunday afternoon. June 7 [1942]. Elementary images can be created only by sacrificing the individual phenomena, the individual value of the human figure, the tree, the still-life subject. There is one characteristic of Cezanne's mode of representation which one may describe as aloofness from life, or better, as aloofness from mankind. In Cezanne's pictures the human figure often has an almost puppet-like rigidity, while the countenances show an expression bordering almost on the mask.
>
> July 25 1943. Everything—objects and people—are shadows enduring for an instant. No one can come inside where I am. Where I live nothing can touch me. [14]

Throughout his long career he remained deeply responsive to the patterns and forms he saw in the breathtaking environs around Yerkes Observatory, which he attempted to document in numerous photographs and paintings. The following passage, written in later years while awaiting his daughter's arrival at Walworth train station, is typical of countless passages written by Morgan, which really are unlike the writing of any other astronomer we know:

> June 16 [1963]. Ah, another enchanted, cool-brilliant day; another communion; another sharpening of the senses, the vision, the physi-

cal response. How like delicate flower stems are these distant telephone posts. A progressive entrance into the world of reality—the World of the Self—during the past hour. Deeper and deeper, more and more removed from the ordinary. How far will it go?—how far can it go? There seems to be no limit in the Possibility—and no limit set by Time.[15]

Another Problem in Calibration

In 1939, while using the 40-inch Yerkes refractor with a low-dispersion spectrograph that had been at Yerkes since the 1920s and had largely been abandon after failing in its first intended application, the measurement of the radial velocities of stars, Morgan realized that he could identify the different luminosity classes of B-type stars even from low-dispersion spectrograms. This was a major breakthrough since, as noted above, B-type supergiants, together with their even brighter but rarer cousins the O stars, are true stellar beacons visible, in principle, from great distances. Morgan and other astronomers grasped that these stars might be used to map galactic structure, provided the calibration problem could be solved, i.e., their absolute brightnesses needed to be matched to their spectral type.

Unfortunately, this is not so easy to do in practice. Because there are so few of these stars–and only a few within a few hundred light years–they are all dimmed and reddened by interstellar dust. The current belief is that the dust component of the interstellar medium is actually very small—only about 0.1% of the total mass density. Its low abundance means that the dust distribution is essentially determined by the distribution of gas in which it is embedded, and the gas is best described as being concentrated to the plane of the Milky Way. Since the O and B stars, being young hot massive stars which form in discrete molecular dust clouds, also hug the galactic plane, even these luminous stars cannot be seen much beyond the nearest spiral arms—farther out, they are completely obscured—while even the relatively near ones we do see are more or less reddened by dust, and appear dimmer than they really are. Attempts to adopt them as "standard candles" to estimate distances across the Galaxy lead, therefore, to overestimates of their distances.

This dimming (or extinction) by dust must somehow be corrected for if we are to use these stars (even the relatively nearer ones) as distance indicators, and fortunately there are reasonably good ways of doing so. It turns out that extinction doesn't occur uniformly across the spectrum—it occurs about twice as efficiently at the blue end as at the red—by measuring the stellar magnitude in two different wavelength regions and taking the difference (i.e., the Color Index) one can, at least in principle, work out the effect of the dust and estimate the brightnesses of the stars were

they not reddened, and Morgan applied this correction to all his O and B stars.

Baade's Breakthrough

In 1944, as Morgan finished his Atlas and was mulling the problem of calibrating the absolute luminosities of his B stars, a crucial breakthrough took place at Mt. Wilson. The war in Europe was raging on. Hubble had left for the Aberdeen Proving Ground in Maryland. In his absence, the 100-inch telescope was left in the hands of Walter Baade, a gifted observer who had done solid work with a 1-meter reflector at the Hamburg Observatory but had yearned to work with the large American instruments.[16] He had come to the United States in 1931 with the intention of applying for citizenship but lost his paperwork and never followed up. In a way, that was fortunate, since as a German national Baade was classified as an enemy alien, which precluded him from taking part in war work as so many astronomers including Hubble did. By using the remarkably fastidious observing techniques for which he was famous, during nights when the lights of Los Angeles and Hollywood were turned out because of wartime blackout conditions, Baade used the 100-inch to obtain deep plates which resolved the faint red stars in the nucleus of the Andromeda Nebula (M31) and its elliptical companions M32 and NGC 205.

He described this work in a paper submitted to the *Astrophysical Journal* in 1944.[17] Morgan at the time was assisting Otto Struve as editor, and realizing that Baade's plates would not reproduce well, he did as Barnard had done with his *Atlas of Selected Regions of the Milky Way,* personally producing and inspecting direct photographic prints of Baade's plates which he then bound into every individual copy of the journal (its circulation at the time was between 600 and 800). These plates showed what Baade described as two stellar populations. In the spiral arms were large complexes of nebulae, later identified as HII regions (diffuse nebulae of the Orion type consisting of regions of hot, ionized, interstellar hydrogen) and hot massive bright young O and B stars. These hot young stars made up what Baade called Population I. By contrast, the faint red old stars of the galactic nucleus and globular clusters made up his Population II.

As we briefly discussed in chapter 10, Baade's discovery of the two stellar populations would lead to his realization that Shapley's calibration of the period-luminosity relation of the Cepheids, universally adopted since 1918 and used by Hubble in setting up the extragalactic distance scale, was wrong. Admittedly, there were already growing reasons for being suspicious about it. Its adoption had led to the conclusion that the Milky Way was much larger than any other known galaxy, and that the globular clusters of M31 were systematically fainter than those in the Milky Way; also, in a conundrum disturbingly similar to that posed by Lord Kelvin about the age of the Earth before the discovery of radioactivity, the expansion age of the universe derived from Hubble's constant yielded an age much smaller than the age of the Earth determined from radioactive isotopes. But as Baade showed in 1952, each population had its own pulsating variables, with different period-luminosity relations. The Population I variables were the classical Cepheids, like delta Cephei and those studied by Henrietta Leavitt in the Magellanic Clouds. The Population II Cepheids, also known as W Virginis stars, after their prototype the variable star W Virginis, are the ones found in globular clusters. Both follow a period-luminosity relation, but for a given period, the classical Cepheids are about two magnitudes brighter than the other type. Unwittingly, Shapley had combined data from the two. By recalibrating the zero point of the classical Cepheids, Baade showed that the distance to M31 found by Hubble had been underestimated by a factor of 2, which meant that the great spiral was larger than had been believed—in fact, it would prove to be somewhat larger than the Milky Way—and also its globular clusters were just as bright as those of the Milky Way. The time scale of the universe was also doubled and the expansion age of the universe was now consistent with the age of the Earth determined from radioactive isotopes. Scientists could breathe easier again.

These developments still lay in the future. For now, let's go back to 1944 and to Morgan. A crucial point, revealed clearly in Baade's plates, was that concentrations of the O and B stars—the very same bright young host stars that Morgan had been studying in the Milky Way for several years—were the tell-tale markers that defined the spiral arms. The reason they were concentrated in the spiral arms was that they were very young. As extremely bright and massive stars, they would burn out before they had time to migrate very far from their place of formation (as less massive, older stars, of the kind Kapteyn and Bok had been statistically crunching, had done). Thus they remained close to the swaddling clothes of gas and dust in which they had begun. This connection between stellar evolution and galactic structure was the essential clue that would ultimately produce the breakthrough leading to the recognition of the Milky Way's spiral-arm structure.

In the fall of 1946, Oort was a visiting professor at the Yerkes Observatory, and gave a series of lectures that Morgan attended. The latter's pencil notes still exist, so it is possible to follow Oort's reasoning at the time Morgan was rapidly beginning to formulate his own ideas. Oort was interested in identifying spiral arms in the Milky Way. He was just beginning to study the possibility of using the methods of radio astronomy for the purpose (by mapping neutral hydrogen gas by means of van de Hulst's

21-centimeter line) but continued to explore the possibility of doing so by optical observations. He focused on the high-luminosity B stars, and had been corresponding with Baade about the problem. Baade had already suggested to Oort (in a letter dated September 23, 1946) that an extraordinary aggregation of B stars in Scorpius and Centaurus (now known as the Scorpius-Centaurus OB association) might be "in reality a short section of a spiral arm, the more so because in its orientation and motion it would fit perfectly into the expected picture (the arms trailing)."[18] Oort responded in early 1947:

> I quite agree that a study of the early B-type stars would be one of the most important steps for finding the spiral structure of the Galactic System. I have been discussing this subject with Van Rhijn [Kapteyn's successor at the Kapteyn Astronomical Laboratory at Groningen for some time], and when [Gale Bruno] Van Albada left Holland in order to pass a year at [Warner and Swasey Observatory in] Cleveland we suggested to him that he should try to star a program with the [24-inch] Schmidt camera for finding faint [i.e., dust-reddened] B-type stars in the Milky Way.... This is a large programme, however, and I don't think the Warner and Swasey people are sufficiently interested yet to start it on a sufficiently big scale.[19]

Unbeknownst to Oort, Morgan had already been teaming up with Jason Nassau from the Warner and Swasey Observatory on that very project. Morgan began to spend part of each year as a Visiting Professor of Astronomy in Cleveland, where he and Nassau identified the stars on plates with the Schmidt camera. Back at Yerkes, Morgan used the 40-inch refractor to obtain spectrograms and to classify the stars rigorously by spectral type and luminosity. (This work was later extended to more southerly regions of the Milky Way by astronomers at Mexico's Tonantzintla Observatory.)

Morgan and Nassau had hardly gotten underway with this ambitious survey when, in December 1947, Baade spoke on the two stellar populations at an American Astronomical Society meeting at the Perkins Observatory in Ohio. By then it seemed increasingly likely that the spiral-arm structure of the Milky Way—if it existed—could best be mapped using the B stars. Baade later confided to Michigan astronomer Leo Goldberg that star-counts and statistical analysis had not led astronomers "... much beyond old William Herschel." Nassau and Morgan were of the same mind as everyone else at the time, and fully expected that once they had mapped the B stars they would have a good chance of working out the spiral arm structure.

Nassau and Morgan were finishing their "galactic survey for high-luminosity stars" in the spring of 1949 when Morgan visited Pasadena and discussed his progress with Baade. Shortly after he returned to Yerkes, he wrote a long and detailed letter to Baade which summarized how far his thinking had progressed by then. The letter, not published until 2008, shows that Morgan already had the solution almost within his grasp:

> … After thinking the matter over, it appears that the high luminosity stars which are observed within a fairly narrow range of true distance modulus [a formula relating absolute and apparent magnitudes directly to the distance] may well define a spiral arm located at a distance around 2-2.5 kpc. outside of the Sun. I have always been puzzled at the extent of the super giants surrounding the double cluster in Perseus; the concentration is probably explicable in terms of a spiral arm rather than as a physical cluster. In this respect, the region of Cepheus appears to be different in that high luminosity objects are observed over a greater range in the distance; this might be explained as a foreshortened effect for the outer spiral arm…
>
> Could the nearby extended dark nebulosity in Ophiuchus and diametrical[ly] opposite in Perseus and Taurus be considered the tattered outer remnants of the general extinction stratum of the spiral arm immediately within the position of the Sun?
>
> It seems to me that within the next year it should be possible to reach a definite answer…[20]

Baade, absorbed in trying to arrange a great sky survey using the new 48-inch Schmidt camera at Palomar, did not write back for several months, but when he did he made it clear he thought Morgan was on the right track:

> … Your interpretation of the large number of supergiants surrounding the double cluster in Perseus would be in line with the findings in the Andromeda nebula. There supergiants of very high luminosity are always bundled up in large groups which stand out as prominent condensations in the spiral arms.
>
> The nearby extended dark nebulosities in Scorpius-Ophiuchus and Perseus-Taurus seem to be indeed manifestations of a single dark cloud ("streamer") which is tilted against the plane of the Milky Way and partly engulfs the solar neighborhood (both the Ophiuchus and Taurus dark cloud are at a distance of only 100 parsecs).
>
> The distribution of the B stars which you first pointed out to me … leaves no doubt that the Sun is either *in* or close to the *inner* edge of the nearest spiral arm… I still think that the B-star program will be the first to lead to definite information about the spiral structure in our neighborhood and that you will push it as far as you can.[21]

In July 1950, a symposium on galactic structure, led by Baade, was held at the University of Michigan Observatory. Morgan and Nassau were both there, and reported on the progress of their survey. Within a galactic belt 16° wide, they had identified 900 O and B stars. For most of these stars, the distances had not been determined, but for 49 OB stars and 3 OB groups they had been able to estimate distances. When Morgan and Nassau plotted these stars, however, they were disappointed—expecting to detect definite signs of the Milky Way's spiral-arm structure, they found only the well-known distribution of "Gould Belt" stars, a ring of brighter hot stars tilted at about 16° to the galactic plane that had been first described by John Herschel and then by Benjamin Apthorp Gould in the 19th century. As we now know, these are hot young massive stars that have formed

Fig. 12.2. One of Don Osterbrock and Stewart Sharpless's 120-degree wide images with the Greenstein-Henyey camera, that helped Morgan work out the distribution of HII regions and OB associations in the Perseus Arm. *Courtesy: Yerkes Observatory.*

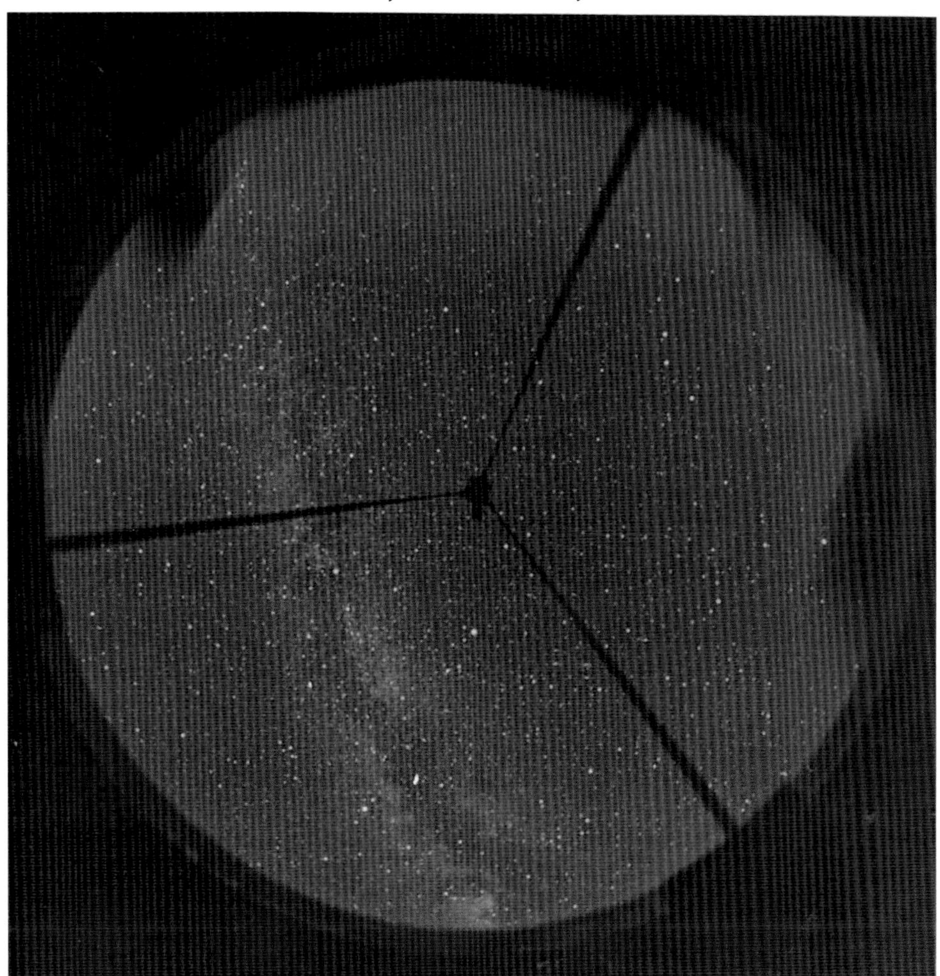

recently in a "spur" of the Orion Arm of the Galaxy (the reason for the sudden efflorescence of these hot young massive stars about 40 million years ago isn't entirely clear, and continues to be debated).

Eureka!

It was hardly a breakthrough to have replicated a result that was over a century old. The frontal assault on the spiral arms by using reddening-corrected B stars had failed. Morgan quickly regrouped, however, and formulated a grander strategy that he hinted at in another paper presented at the same meeting. From its title, "Application of the principle of natural groups to the classification of stellar spectra," its importance was hardly likely to have been apparent to the other participants.[22] Here Morgan coined the expression "OB stars" to designate a category—a "natural group," he called it—consisting of both O supergiant and early (young) B stars. What he realized was that these stars occupy a very small narrowly-defined area of the HR diagram, and he could tell even from low-dispersion spectra, "by just a glance, [by looking] just a few seconds at each spectrum … to tell if a star was located…" there.[23] Remember, the HR diagram relates color or temperature to luminosity. Crucially, stars in this narrowly-defined part of the HR diagram varied by only 1.5 or 2 magnitudes on either side of the means, which were around absolute magnitudes -5 or -6. Thus he didn't have to use color-corrected spectra for the distant ones at all.

Morgan was groping toward the concept of "OB star associations" (the term itself was introduced later by the Armenian astronomer, Victor Ambartsumian). The O and early B stars are found in loose aggregations, typically consisting of a few dozen stars, the majority of type B, which might be spread over a volume as small as an ordinary cluster or as much as a few hundred parsecs across. With a fair-sized group of moderately discordant values of the luminosities Morgan could pick the mean (-5 or -6) and end up with a fairly reliable distance for the group as a whole. In this manner he was able to plot their positions along the Galactic Plane out to much greater distances than he and Nassau had reached with their earlier survey. At the same time, Morgan added another component to his quest. He was solving a puzzle, like the jigsaw puzzles he put together with the faces down. He needed another clue. He now remembered the complexes of nebulae like those imaged in the spiral arms of Andromeda. These complexes of nebulae had been identified as regions of hot ionized hydrogen glass, their ruby-red glow being that of the emission line of ionized hydrogen (H-alpha) of the Balmer series; among notable examples in the Milky Way were the California Nebula close to ξ Persei, the so-called Barnard Loop in Orion, and the Rosette Nebula in Monoceros. The HII regions

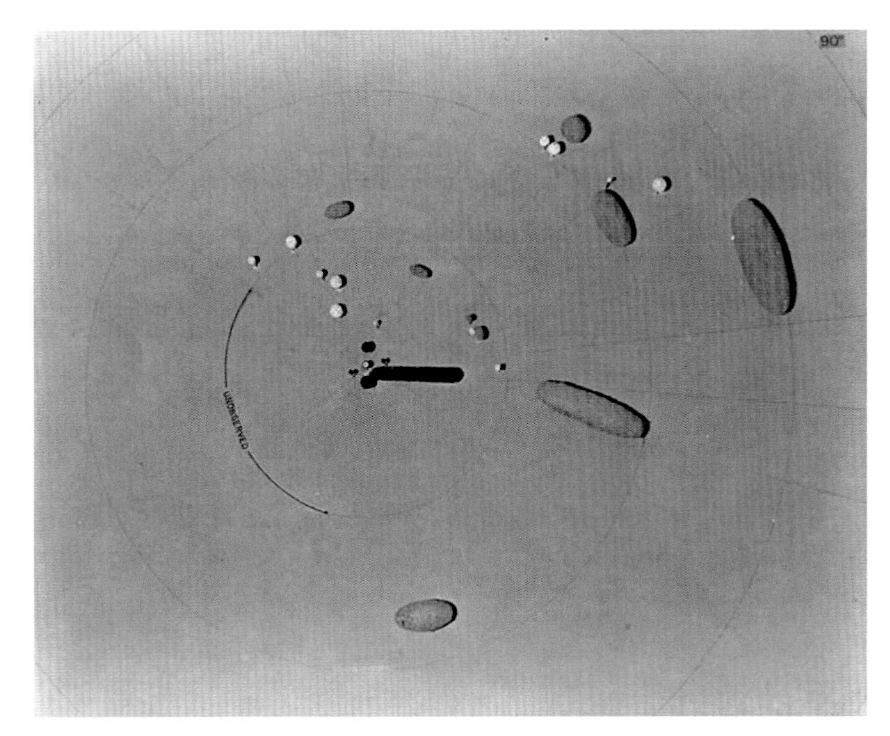

Fig. 12.3a. Morgan's original model of the spiral arms, in which sponge rubber was used to depict OB groups.

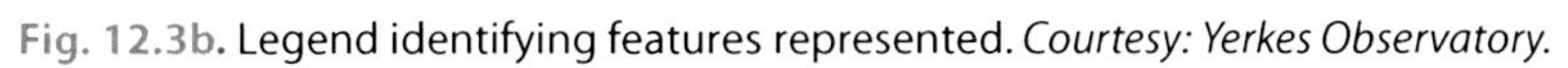
Fig. 12.3b. Legend identifying features represented. *Courtesy: Yerkes Observatory.*

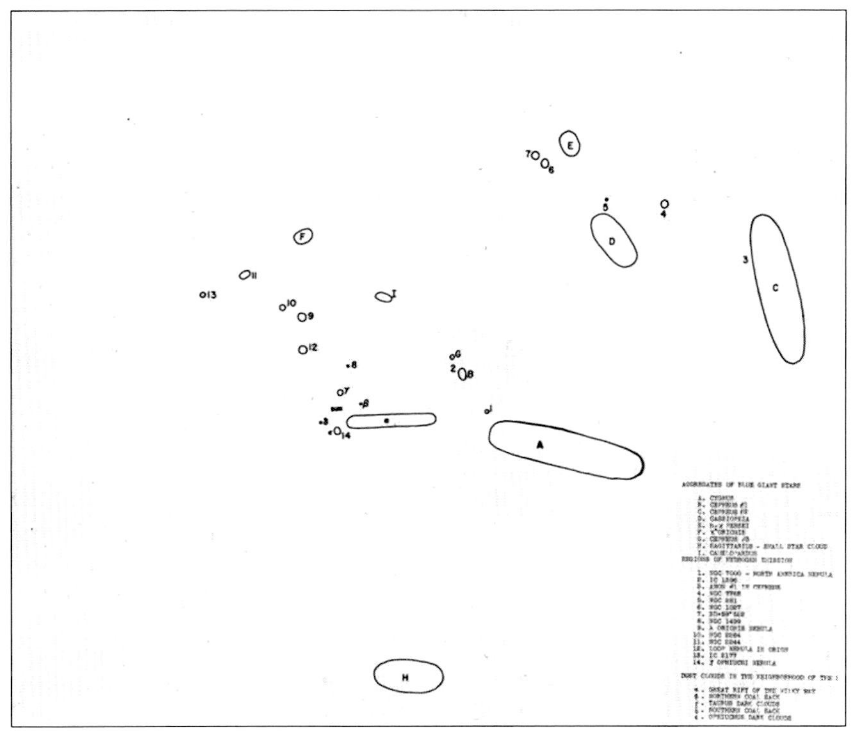

and the OB associations would, Morgan hoped, reinforce one another, and finally allow him to put the puzzle together by tracing out the Milky Way's spiral arms.

At Yerkes at the time there was a wide-angle camera with a field of 140°, which had been developed during the war by Jesse L. Greenstein and Louis G. Henyey for use as a projection system to train aerial gunners. It could equally well be used the other way around, as a camera, and under Morgan's direction two graduate students, Donald E. Osterbrock and Stewart Sharpless, began using it to photograph the Milky Way with narrow-band (hydrogen alpha) filters (which had become available only after the war) in the search for HII regions.[24] Many of the HII regions were already well known, but some were new; moreover, because of the very wide-field of the photographs, they showed up as the important, extended objects they are.

By the fall of 1951, Morgan had been immersed in the problem of trying to find the spiral arms of the Milky Way for at least four years. He had laid out what was, essentially, the correct approach to be taken to the problem, and since then had pursued it in a diligent and systematic way. Already at the time he wrote about the B stars to Baade, he was getting close to the solution. According to his recollection, however, the pieces of the puzzle finally fell into place rather suddenly, in a sudden flash of recognition, as he walked from his office to his house on an October night.

Figs. 12.4. The moods of a genius: a. (left) Morgan concentrating at playing cards, b. (right) Morgan ecstatic. Informal snapshots. *Courtesy: Judith Bausch, Yerkes Observatory.*

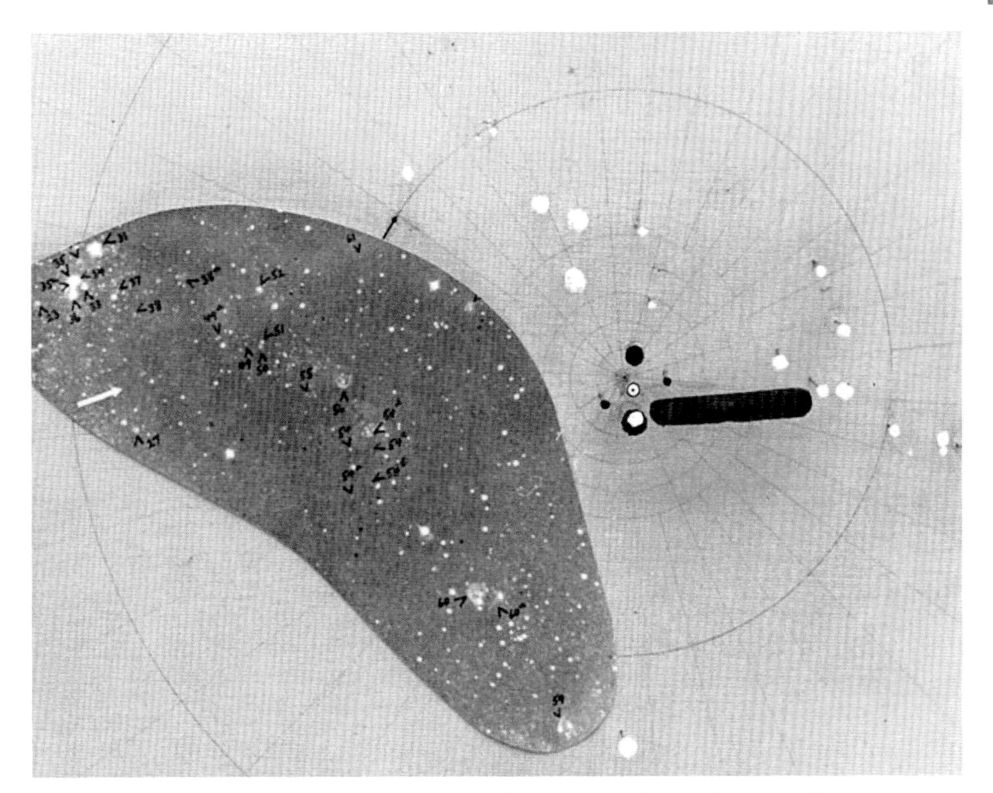

Fig. 12.5. Morgan's second model. juxtaposed with a section of one of Baade's 100-inch plates of OB associations in a spiral arm in M31. This composite image appeared on the cover of Sky & Telescope in 1952. *Courtesy: Yerkes Observatory.*

His most complete account of what happened that fall evening is given in an August 1978 American Institute for Physics interview with David De Vorkin:

> This was in the fall of 1951, and I was walking between the observatory and home, which is only 100 yards away. I was looking up in the sky ... just looking up in the region of the Double Cluster [in Perseus], and I realized I had been getting distance moduli corrected the best way I could with the colors that were available, for numbers of stars in the general region.... Anyway, I was walking. I was looking up at the sky, and it suddenly occurred to me that the double cluster in Perseus, and then a number of stars in Cassiopeia, these are not the bright stars but the distant stars, and even Cepheus, that along there I was getting distance moduli, of between 11 and 12, corrected distance moduli. Well, 11.5 is two kiloparsecs ... and so, I couldn't wait to get over here and really plot them up. It looked like they were at the same distance.... It looked like a concentration.... And so, as soon as I began plotting this out, the first thing that showed up was that there was a concentra-

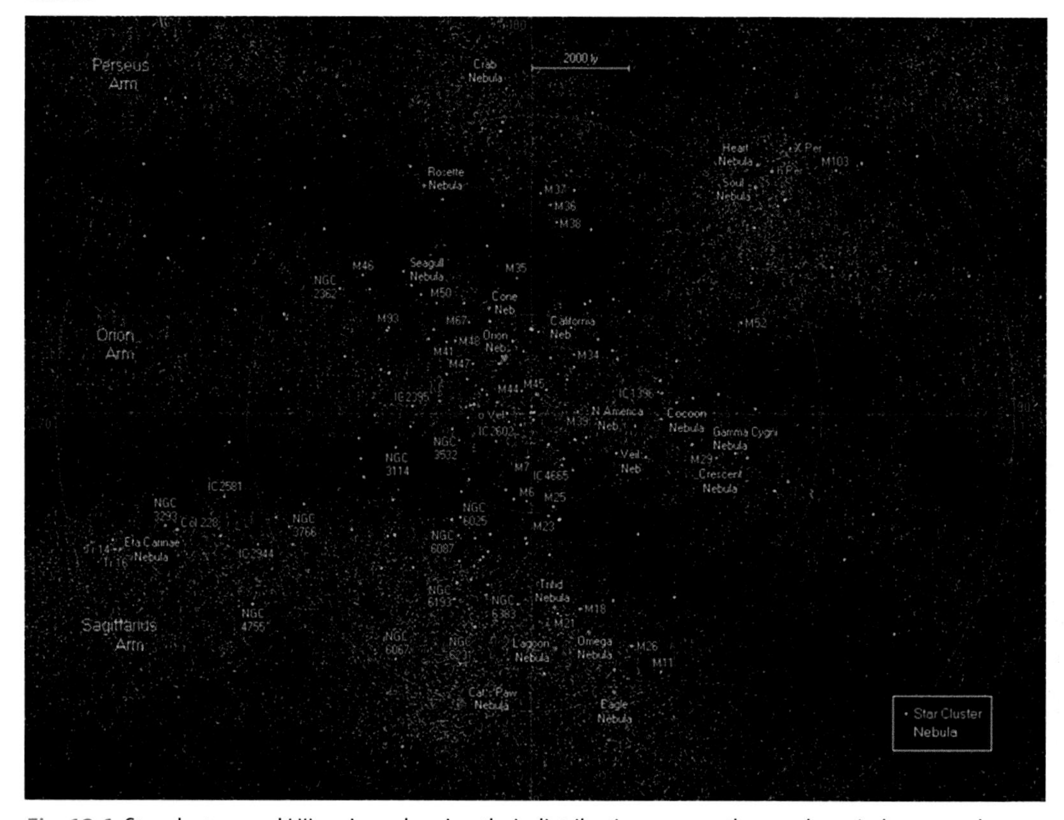

Fig. 12.6. Star clusters and HII regions showing their distribution among the nearby spiral arms to the Sun. *Image: Wikipedia.*

> tion, a long narrow concentration of young stars.... There are HII regions along there too ... And that was the thing that broke [the problem] down.[25]

This first spiral arm – the Perseus arm – was traced between galactic longitudes about 100 and 17° (according to the modern convention of galactic longitudes). As he plotted the OB stars out, Morgan found that in addition to this arm there was another, the Orion arm, extending from Cygnus through Cepheus and Cassiopeia's chair past Perseus and Orion to Monoceros, i.e., between galactic longitudes 50° and 210° or 220°. The so-called Great Rift of the Milky Way marked a part of the inner dark lane of this arm; the Sun lay not quite at the inner edge but 100 or 200 light years inside it. It was the Sun's proximity to – almost immersion in – this arm that had made it so difficult to identify. Indeed, Morgan pointed out: "The hardest thing is to know what's going on if you're in the middle of something, or if it's going right through you."

There is no reason to doubt Morgan's account of that October night at Yerkes. As he walked home from the observatory under the autumn stars, he experienced a "revelation-flash," a moment of sudden pattern-recognition. As so often happens with those who have experienced a "Eureka!" of "aha!" experience (an insight-based solution to a seemingly impossible problem), Morgan came to see it as something impossible to define in words, an inspiration breaking through in a flash from the subconscious mind.[26]

Morgan's discovery was incarnated at first in a model in which old sponge rubber was used to depict the OB groups he had identified. Later he added some concentrations of early B stars from the southern hemisphere (stars Annie Jump Cannon had classified as BO stars, those with hydrogen lines weak in their spectra that turned out to be a close approximation to Morgan's OB stars). This more detailed scale model, constructed using balls of cotton, he presented in a slide shown at the American Astronomical Society in Cleveland—the meeting at which he received the standing ovation. Morgan, who was a profoundly insecure individual (understandably, given the circumstances of his childhood), had finally received the recognition of his astronomical peers—and yet, paradoxically, within a few months of the most ecstatic moment of his life he suffered a mental collapse, a "complete personal crisis."

That spring he was so depressed he was unable to work, and his condition deteriorated to the point where he was admitted to Billings Hospital that summer. He never published a complete account of his discovery of the spiral arms (brief notices appeared in early 1952 in the *Astronomical Journal* and in the popular magazine *Sky & Telescope;* the latter including on the cover an image juxtaposing Morgan's cotton-ball model with a section of one of Baade's plates of OB associations in a spiral arm of M31). By the time he could return to work again, the radio astronomers had rushed in and stolen much of his thunder.

Fig. 12.7. Donald Osterbrock in older age, as his many friends and associates remember him. *Photograph by William Sheehan, 2005.*

Oort, financially-strapped after the war in which Holland had been occupied by the Germans, was delayed in getting hold of the proper equipment, but the Americans, beneficiaries of a crash wartime radar research program, were under no such limitations. The first detection of van de Hulst's 21-cm line of neutral hydrogen was thus accomplished by Americans, Edward M. Purcell and H.I. McEwen

Fig. 12.8. The North America nebula (NGC 7000), discovered by William Herschel in 1786 and named by German photographic pioneer Max Wolf, and nearby Pelican nebula (IC 5070), in Cygnus near the 1st-magnitude star Deneb (outside the field of this image). These are part of the same emission nebula (HII region). Deneb itself, at a distance of 1800 light years, is thought to be the star responsible for ionizing the hydrogen so that it emits light, and if so, its diameter is about 100 light years across. Imaged by Klaus Brasch from Flagstaff, Arizona, on September 29, 2013, with a TMB-92 at f/5.5 apo refractor, an IDAS LPS-V3 filter, and Hutech modified Canon 6D. *Courtesy: Klaus Brasch.*

at Harvard, in the early spring of 1951; six weeks later Oort, in collaboration with the radio engineer C.A. Muller, emulated the feat with an antenna at Kootwijk. Now that the 21-cm line had been detected, the mapping of galactic structure using radio was no longer theoretical, and Oort, together with Dutch and Australian colleagues, began working in earnest. They systematically mapped clouds of neutral hydrogen gas, including those in hitherto inaccessible regions on the far side of the Galaxy. Morgan's accomplishment was overshadowed, at least until 1970 when it was finally realized that the radio maps were not as definitive as they had seemed in the early 1950s.

Morgan returned to Yerkes—now keeping up, as therapy, the fascinating personal notebooks (which he personalized as his "Dear Book"), in which he undertook a kind of Freudian self-psychoanalysis, set down his thoughts about philosophy (Nietzsche and Wittgenstein were particular influences) and recorded his responses to the great works of art that moved him (especially those of the *Trecento,* the period from Cimabue to Giotto, that he believed had gone furthest in probing "deepest reality"). In astronomy, he grappled heroically with the classification of the galaxies through

Fig. 12.9. OB associations and HII regions in spiral arms in the barred spiral galaxy M83 in Hydra. The outer reaches of this galaxy are undergoing vigorous star-formation. *Courtesy: Hubble Heritage Team (STScI/AURA).*

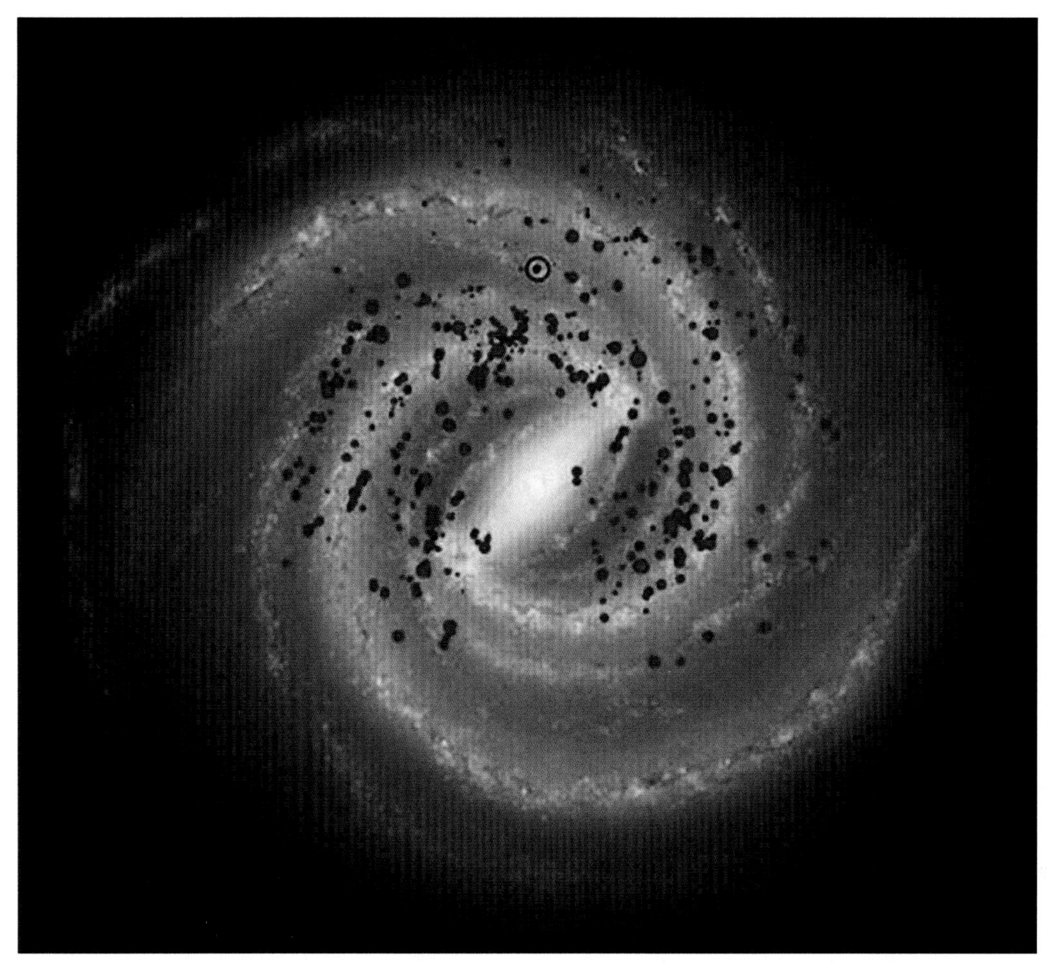

12.10. The background reconstruction of the structure of the Milky Way, based on infrared observations by the Spitzer Space Telescope, clearly showed two main arms, Scutum-Centaurus and Perseus, emerging from the ends of the central bar. However, more recent work, involving a 12-year radio survey of 1,650 young massive stars (OB stars, like those included in Morgan's optical survey), indicated by red dots, clearly show in addition the Sagittarius and Norma arms. The reason the OB stars give such good delineation of the spiral arms is that, as Morgan realized, in contrast to cooler low-mass stars like the Sun the young massive stars haven't had time to move from their birthplaces in the spiral arms. This makes them better tracers. *Courtesy: OB star data J.S. Urquhart et al., Monthly Notices of the Royal Astronomical Society/ background R. Hurt, Spitzer Science Center.*

altering cycles of creative elation and let-down, struggled, not entirely successfully, to be a good husband to his wife Helen, who suffered from chronic ill-health and died in 1963, and father to his daughter and son. Though he was happier after he remarried in 1966 (his second wife, Jean Doyle Eliot, was a teacher),[27] he continued to struggle with self-doubt and low self-esteem, the legacies of his traumatic childhood, all his life. In the last decade of his life (he died in 1994) he finally gave up writing his personal notebooks as he succumbed to Alzheimer's disease. Before the darkness closed in, he recorded a conversation he had with Donald Osterbrock, who had collaborated on the discovery of the spiral arms so long before:

> He [said] he would like to write my life... In the following conversation, I said I was not a genius; he said he was not sure I was right—that I had made "Conceptual Breakthroughs." The implication seemed to be that I might be... I told him that he had just given me the highest honor of my entire life.[28]

Edmond Halley said on discovering the comet bearing his name, "if it comes to be discovered, be it known that it was first seen by an Englishman." Let it be remembered likewise that the spiral arm structure of the Milky Way was first recognized by an optical astronomer.

The association of molecular clouds, HII regions, and star forming regions (and brilliant young OB associations) is a leitmotif repeated in many other spiral galaxies, not just in the Milky Way. As for the Perseus arm, the quintessential spiral arm Morgan pieced together, it is (according to the most recent measures) 6400 light years farther from the center of the Galaxy than the Local Spur inhabited by the Sun and all the bright naked-eye stars that form the familiar constellations—those are all foreground objects. The Perseus Arm appears in projection far beyond these neighboring suns, and stretches – through condensations and star-knots among its dusty lanes— from the Rosette Nebula and its OB association in Monoceros, which includes the sparkling jewel-like cluster NGC 2244, through the splendid open clusters of Auriga (including Messier clusters M36, M37, and M38), across the relatively dust-free regions of Cassiopeia and Perseus itself, with lovely HII regions like NGC 281, the open cluster M52, and finally the unrivalled double cluster h-χ Persei in the sword handle of Perseus, containing a host of OB stars, each a rival to Rigel in the foot of the Giant. It is also, according to data from the Spitzer Galactic Legacy Infrared Mid-Plane Survey Extraordinaire (GLIMPSE), one of two main arms that emerge from the ends of the central bar of the Milky Way; the other, the Scutum-Centaurus arm, is partly visible from the Southern Hemisphere, but mostly hidden from view as it swings behind the Galaxy's central hub.

Morgan, walking that night under the stars, measured the distance to them with his mind. He received an inspiration worthy of the great artists he so much admired, which took place, in his words, in "... the hours of stillness—with supple brain—deep in the vistas of space, time, and form—that Heavenly World of Form."[29]

* * *

1. But see: Anonymous, "Spiral arms of the Galaxy," *Sky and Telescope,* April 1952, 138-139; and W.W. Morgan, S. Sharpless and D.E. Osterbrock, "Some features of galactic structure in the neighborhood of the Sun," *Astrophysical Journal,* 57 (1952), 3.
2. W.B. Burton, "The morphology of hydrogen and other tracers in the Galaxy," *Annual reviews of Astronomy and Astrophysics,* 1976, 275-306.
3. Donald E. Osterbrock. *Walter Baade: a life in astrophysics* (Princeton, New Jersey: Princeton University Press, 2001), 147
4. Owen Gingerich, "The discovery of the spiral arms of the Milky Way," in H. van Woerden, R. J. Allen, and W. Butler Burton, eds., *The Milky Way Galaxy: proceedings of the 106th symposium of the International Astronomical Union held in Groningen, the Netherlands, 30 May-3 June 1983* (Dordrecht: D. Reidel, 1985), 61.
5. Ken Croswell, *The Alchemy of the Heavens: searching for meaning in the Milky Way* (New York: Anchor Doubleday, 1995), 75.
6. W. W. Morgan, Personal Notebook No. 242, 1987; contains a short autobiography; Yerkes Observatory Archives.
7. Ibid.
8. W. W. Morgan, 1978; Interview with David DeVorkin on August 8. Niels Bohr Library, American Institute of Physics.
9. W.W. Morgan, "A Descriptive Study of the Spectra of the A-type Stars," *Publications of Yerkes Observatory, 7* (1935), 133-250.
10. Croswell, *Alchemy of the Heavens,* 74.
11. Ibid., 78.
12. Sandage, *Mount Wilson Observatory,* 254.
13. James Kaler, *Stars and their Spectra* (Cambridge: Cambridge University Press, 2002), 112.
14. W. W. Morgan, Journal, 1942; Yerkes Observatory Archives.
15. W. W. Morgan, Personal Notebook No. 104; Yerkes Observatory Archives.
16. The authoritative biography of Baade is: Donald E. Osterbrock, *Walter Baade: a life in astrophysics.* Princeton, New Jersey: Princeton University Press, 2001.
17. Walter Baade, "The Resolution of M32, NGC 205, and the Central Region of the Andromeda Galaxy," *Astrophysical Journal, 100* (1944), 137-146.
18. W. Baade, letter to Jan Oort, September 23, 1946; Jan Oort file, Niels Bohr Library, American Institute of Physics.
19. J. Oort to W. Baade, January 11, 1947; Jan Oort file, Niels Bohr Library, American Institute of Physics.
20. W.W. Morgan to W. Baade, July 12, 1949; Yerkes Observatory Archives.
21. W. Baade to W.W. Morgan, August 20, 1949; Yerkes Observatory Archives.
22. The paper was published as: W.W. Morgan, "Application of the principle of natural groups to the classification of stellar spectra," *Publications of the Observatory of the University of Michigan, 10* (1951), 43-50.
23. DeVorkin interview with Morgan, 1978.
24. The camera and another of its early applications, photography of the *gegenschein,* are well described in: Otto Struve, "Photography of the Counterglow," *Sky and Telescope,* July 1951, 215-218.
25. DeVorkin interview with Morgan, 1978.
26. Ibid.
27. Jean's first husband had been Henry Ware Eliot, brother of T.S.Eliot the poet and cousin of Charles Eliot the president of Harvard University.
28. Morgan, personal notebook no. 230; Yerkes Observatory Archives. Osterbrock was true to his word. He did write Morgan's biography. See: Donald E. Osterbrock, "William Wilson Morgan, 1906-1994," *Biographical Memoirs of the National Academy of Sciences, 72* (1997), 289-313.
29. W.W. Morgan, personal notebook no. 15, 1956; Yerkes Observatory Archives.

13. To Forge a Galaxy

The eternal silence of these infinite spaces frightens me.

—*Blaise Pascal, Pensées*

Edith Sitwell recalled [an] afternoon in Hubble's study, when he showed her plates of "universes in the heavens" millions of light years away. "How terrifying!" she had remarked. "Only at first," he replied. "When you are not used to them. Afterwards, they give one comfort. For then you know that there is nothing to worry about—nothing at all."

—*Gale E. Christianson, Mariner of the Nebulae, p. 359*

As noted in previous chapters, the discovery of the universe of galaxies was accomplished through slow but steady progress. However, since 1945 the pace of change in understanding galaxies has increased at a rapid rate. In the earlier days, as we have been discussing so far, certain individuals—for example, Messier, the Herschels, Rosse, Huggins, Barnard, Shapley, Hubble—made towering contributions, and therefore command the attention of the chronicler of the nebulae.

In more recent years—the subject of this chapter—progress has depended less on individual astronomers and more on the development of a large and very specialized group of expert individuals in the rather technical areas that need to be mastered in order to put together a complete pictures of galaxies, and how they form throughout cosmic time. Since the end of World War II, astronomers and cosmologists have been trying in a quantitative way not possible before, to address some of the oldest and most profound questions humans have ever asked, such as: where do we come from? And how did the universe we see around us originate?

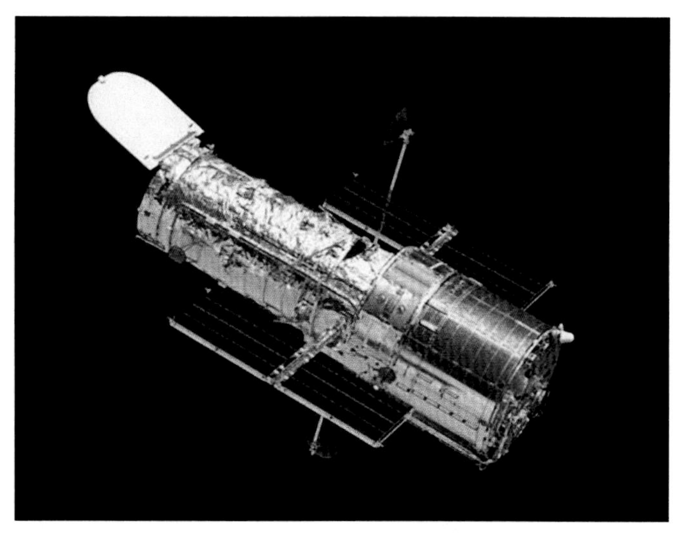

Fig. 13.1. Hubble Space Telescope, as photographed during the fourth servicing mission by the Space Shuttle. *Courtesy: NASA.*

The idea of what defines the universe has constantly changed through time. For centuries, up until the end of the 18th and early 19th centuries, it was mostly the Solar System that concerned astronomers, as the stars were so far away that their distances were immeasurable. Even at the beginning of the 20th century, the universe was still generally thought to consist only of what we now call the Milky Way Galaxy (admittedly, there were always a few who believed that the spiral nebulae were "island universes" in their own right, but this was probably never the majority view until the late 1910s or even the 1920s).

Beginning with Hubble's great investigations with the 100-inch reflector, it was established that the universe is, in fact, a universe of galaxies, and that galaxies in a practical and real sense define our universe, just as cells define the structure and function of living tissues, or neurons the structure and function of the brain. As with the discovery of stellar evolution and the properties of our own Galaxy, the discovery of the Universe of galaxies in the 20th and 21st century has led —and still is leading—to a new revolution in our understanding of our place in the universe.

When galaxies were first recognized as distinct objects they were mostly used as cosmological probes by those with access to large optical telescopes, namely at Mt. Wilson and Palomar, as described in detail in previous chapters. The way that astronomers used galaxies as test particles—like the marks on the balloon—to gauge the expansion of the universe was by using either the total brightness of galaxies, or of the brightest stars (like the classical Cepheids) within them. The role that galaxies played in this phase of cosmological thinking was similar to that of point masses in celestial mechanics, or ideal gases in gas theory.

In these early days of galaxy studies, little thought was given to the details of galaxies' structure, or to the way galaxies formed or evolved through time. The exception was the Hubble-Jeans formulation of "early-," "middle-," and "late-type" galaxies, which was one possible interpretation of the sequence presented on the Hubble tuning fork; but it was not the only one, and it was criticized as premature even at the time—and it was. As we now know, things are much more complicated than the tuning fork. In any case, in the face of criticism, even Hubble abandoned his original evolutionary model. Not much progress or work on this topic occurred for the next few decades.

The very fact that we can refer to galaxies as probes or point-masses shows just how far away they are, and how faint and small they appear to Earth-based observers. Though one or two—M31 and M33—are visible with the naked eye from dark sites (hard to come by these days!) it takes a fairly large telescope before one begins to appreciate their sheer numbers. However, a deep photographic image like those of the "Hubble Deep Fields", discussed later in this chapter, reveals them swarming the field like the single-celled organisms in a drop of pond water. By talking about distances as astronomers do, referring to Megaparsec-this and Megaparsec-that, it is possible to become inured to the sheer vastness of the cosmos. Another way to think about it is that the Milky Way is itself but one galaxy out of the hundred—or two hundred—thousand million that exist in the observable universe. A typical galaxy like the Milky Way is on the order of 10^{21} meters across, which makes the large scales of even just our own galaxy vastly more remote from human scales of reference than even atoms, which are about 10^{-10} m. Or, to put it another way, if the Milky Way were reduced to the size of the North American continent, the Earth would not only be much smaller than the period at the end of this sentence, it would be invisible without an electron microscope. It's hard to get too big a head when one spends one's time thinking about such things—which is more or less the gist of Hubble's remark to Edith Sitwell cited in the headnote to this chapter.

Our Local Universe – Realm of the Spirals

The Milky Way is a typical example of a "disk galaxy." These are common—about 70 percent of all bright galaxies appear to be of this type. Elliptical galaxies make up about 25 percent. The irregular galaxies of the Magellanic Cloud or Barnard's Galaxy type come far to the rear, making up about 5 percent. (Needless to say, as with stars, dwarf galaxies are by far the most common.)

In disk galaxies, the central part usually consists of a central bulge of cool red stars (Population II stars, mostly M- and K-type dwarfs). The central bulge is surrounded in turn by a flattened disk of material made up of a variety of different types of stars, among which are bright young hot stars of Population I, which for all their brilliance and mass as individuals, make up only about 2% of all stars, together with gas and dust. The proportions of the bulge to the disk are roughly those suggested by the analogy of the yolk and white of two fried eggs slapped together back-to-back; the homely visual savored by many an astronomy lecturer proves, at least on this occasion, to be remarkably apt. Though many disk galaxies resemble the Milky Way and show a spiral arm pattern among their disk stars, not all of them do. In some disk galaxies, the central bulge is very prominent, and surrounded by tightly wound spiral arms (these are Hubble's Sa-type spirals); in others, the bulge is inconspicuous, and the galaxy appears to be all disk, with long loping spiral arms flung far and loosely wound (Hubble's Sc-type). In many of these, a "grand design" is displayed, sometimes "extending over the whole galaxy, from the nucleus to its outermost part and consisting of two arms starting from diametrically opposite points."[1]

Then there are the disk galaxies known as barred spirals, in which the spiral arms seem to twist outward from a thickish bar crossing the center (Hubble's SB-type galaxies). But many galaxies appear barless, with the spiral arms emerging directly from the center without the intervention of any bar. According to present thinking, the Milky Way—whose thickish bar has been discovered only in the last few years—is best classed as an SBb galaxy.

As had been evident as far back as when Lord Rosse obtained the first really good view of M51, the Whirlpool Galaxy, the spiral-form suggests rotation. The disks are in fact rotating; in the case of the Milky Way, the orbits of disk stars, like the Sun that are located far from the center, and so beyond most of the Galaxy's mass, are rather straightforwardly Keplerian (though at very great distances, because of dark matter, things become more complicated, as we shall see in the next chapter). The Sun orbits at a distance of some 27,000 light years from the center of the Galaxy (give or take a thousand light years or so). The center of the Galaxy, located in the direction of the constellation Sagittarius, is anchored by the presence of a monster Black Hole, whose position lies close to the "bright" radio source, Sagittarius "A," which may be either an accretion disk or a relativistic jet associated with the acceleration of gas into the Black Hole; the Black Hole itself is referred to as Sagittarius A* (Sgr A*). The huge gravitational field of the Black Hole sucks in gas, dust and stars, and within only 23 arc-seconds (3 light years) of the Black Hole itself, there is a rich cluster of hot, massive, young stars attesting to a recent burst of star formation. By tracking the motion of stars within only 1 arc second of Sgr A*, the Black Hole has been weighed—its mass is 4 million solar masses, all compacted into a small pocket of space only 15 times the width of the Sun. Clearly, this is a rough and violent neighborhood, and we are lucky to be well out from it. As we

shall see in the following chapter, most galaxies have one of these massive black holes lurking in their centers.

Our knowledge of these dense inner regions of the Galaxy depend entirely on observations made in infrared, radio, and X-ray, i.e., entirely outside of the optical window, since a curtain of dust completely hides the inner sanctum from our view. Indeed, among the greatest factors contributing to our recent understanding of the galaxies has been our ability to scrutinize galaxies (including the Milky Way) in regions of the electromagnetic spectrum outside the virtual sliver of the optical range. It is, after all, largely chance that galaxies were discovered by humans at optical wavelengths. If for some reason our eyes were more sensitive to ultraviolet light or to infrared (as is, in fact, the case with some insects and birds' eyes), it is likely that the first telescopes would have been most sensitive at these wavelengths, and our classification of galaxies would have been much different.

Starting with radio in the 1930s and '40s, detectors have been devised that are sensitive to infrared, microwave, ultraviolet, X-ray, and gamma-ray regions of the spectrum. These detectors have often needed to be set up on the peaks of high mountains in very dry conditions in order to avoid absorption in the Earth's atmosphere (especially by water vapor for infrared). Now they are routinely lofted above the atmosphere altogether, and make their observations from orbiting spacecraft. The 1990s and the 2000s saw the era of great space telescopes, including the Hubble Space Telescope (launched in 1990), the Compton gamma-ray observatory (1991), Chandra X-ray (1999), GALEX ultraviolet, and Spitzer infrared (both 2003).

The Dirty, Dusty Universe

We emphasize that what we see of galaxies, including our own, is critically dependent on the part of the spectrum used. The Milky Way, viewed in the optical range, is dominated by the dark obscuring bands of dust that cut through the central bulge and extend out into the surrounding disk, and spiral arms. Similar dust bands are prominent in other galaxies—such as the edgewise spiral NGC 891 in Andromeda. In the case of the Milky Way, the obscuring dust we see is located not far from the Sun, which lies on a spur of the Orion Arm, and hides much of the northern part of the bulge of the Galaxy from our view. We may get the impression when we see the Galaxy with the naked eye, and notice these towering, thick and massive dark markings that Barnard photographed, that we are witnessing dense clouds of something like coal-dust (as suggested by names such as the Great Rift or Coalsack). In fact the dust that does such an effective job of obscuring light beyond it is in actuality little more than a fog—dust makes

Fig. 13.2. M51 and its companion in infrared. This Spitzer Space Telescope image is a four-color composite showing emissions from wavelengths of 3.6 microns (blue), 4.5 microns (green), 5.8 microns (orange) and 8.0 microns (red). At shorter wavelengths (3.6 to 4.5 microns), the light comes mainly from stars, while in longer wavelengths (5.8 to 8.0 microns), we see the glow from molecular clouds. Of special interest is the contrast in the distribution of dust and stars between the spiral and its companion, NGC 5195. Whereas the whirlpool is rich in dust, bright in infrared wavelengths, and actively forming new stars, its blue companion shows little infrared emission and hosts an older stellar population. *Courtesy: Education and Public Outreach team at the Spitzer Science Center, California Institute of Technology.*

up only 0.1% of the total mass density of the interstellar medium, the rest being gas.[2]

The dust, being concentrated in the galactic disk, was always a major obstacle to the investigation of the structure of the Galaxy, as well as other galaxies. It blocked Herschel's view, which was accordingly confined to the 1% of the stars in the Galaxy represented in the grindstone. As we saw in chapter 12, it required nothing less than a *tour de force* effort by a genius like Morgan to use even the visually luminous spiral-arm tracers such as the O and B supergiant stars to make out the neighboring arms. The stars of the bulge and the inner disk, and the galactic center, are virtually completely hidden from us by dust. Admittedly, there are a few "windows," where the dust is sparse enough for some bulge stars to be visible. The largest, Baade's window, discovered by Walter Baade with the 100-inch reflector in the mid-1940s, is about a degree wide, and centered on the globular cluster NGC 6522. But in order to see really well what the Galaxy looks like beyond the dusty "mask," in order to see the old and much more numerous stars in the disk that form the structure's backbone, we need to view it in near-infrared light, at a wavelength of about 2 microns.

In the near-infrared, as if by magic, a kind of *open sesame* to the beyond occurs! The dust becomes almost transparent. In the Milky Way, we see the bulge and inner disk clearly, and discover that the bulge has a boxy shape, something that would never have been suspected from views in the optical window.

Other spirals—some quite messy looking, with spurs and extra octopus-like arms—come clean, and show backbones that look like neat "grand-design" spirals. A particularly dramatic case, but a fairly typical one, is that of the 12th magnitude galaxy NGC 309, in Cetus. Its distance is 300 million light years, and its diameter, at 260,000 light years, makes it one of the largest spirals known. In optical wavelengths, this galaxy has multiple narrow, well-defined, highly branched, and seemingly interwoven arms, but in the infrared, all of this is revealed to be little more than froth, overlying a normal two-arm grand-design spiral, with a small central bar barely hinted at in the optical images.[3]

Clearly, our views of the Milky Way and other spiral galaxies in optical images are heavily weighted in favor of regions of star formation and spiral formation. The HII regions, dust, and the bright young stars of the spiral arms dominate our view of other spirals no less than of the Milky Way. But these bright young stars make up less than 2% of the stellar population. They are, in Walter Baade's phrase, like the candles and frosting on the birthday cake. When we look at galaxies in the near-infrared, the dust becomes transparent, and our view is no longer skewed to the exceptional stars but instead reveals the "backbone" of older stars. Thus near-infrared images, though not in general as pretty as those that emphasize the star-forming regions, provides us with a view of the distribution of the stars that constitute most of the galaxy's mass. That this kind of dramatic transfiguration wreaks complete havoc with the optical-biased Hubble classification is obvious. We'll say more about that in a moment.

From whence the Spirals?

The beauty and majesty of galaxies lies in the beautiful spirals exhibited by so many of them. The form is not only ubiquitous in the realms beyond, it is ubiquitous in the manifestations of nature in the plants and animals of Earth—to give but one example, consider the Nautilus, whose shell is formed on the logarithmic spiral. A whole book—and not a short one—has been written on these spirals in nature, and its author, in introducing the vast subject, notes that "with very few exceptions the spiral formation is intimately connected with the phenomena of life and growth."[4] By analogy, it is natural to ask what significance the spiral pattern may have in the phenomena of life and growth of galaxies.

At some level, the spiral form seems inevitable. Any linear feature in a galaxy's disk will wind up into an ever-tighter spiral if the disk undergoes differential rotation—in which the inner portions of the disk orbit the center of the galaxy more frequently than the outer ones. Thus, spiral arms might seem to be a natural consequence of differential rotation of the disk. However, as was noted in the 1950s by the Swedish astronomer Bertil Lindblad, the problem with this explanation was that the galaxy should simply wind up on itself after a few turns. Unless, then, they were extremely evanescent, astronomers had to explain not only what formed, but also what sustained, them.

These fundamental problems of galactic structure proved to be surprisingly intractable. In the 1950s and 60s, it was thought that spiral arms were tubes of ionized gas magnetically bound by the interstellar magnetic field. However, when the interstellar magnetic field was finally measured, it was found to be too weak to accomplish this. Since the mid-1960s, the most promising idea, based on earlier (not very well articulated) insights by Lindblad, has been that spiral arms in galaxies are gravitational "spiral density waves." This theory, first developed in detail by two Chinese-American physicists at MIT, Chia-Chiao Lin and his then undergraduate student Frank Shu, was that the spiral arms aren't material entities at all; instead the arms trace gravitational "density waves," traveling zones of compression—the analogy that is usually given is to the cars in a traffic jam. Though there is an increased density of cars in the area of the traffic jam, the cars are different at any given time. Instead of cars moving along the freeway, in the disk of the galaxy stars we are looking at waves of stars moving in and out of the disk. The increased density along the wave gravitationally attracts passing stars, gas and dust, and causes some of the gas and dust to collapse into dense molecular clouds that may give birth to the bright, young hot stars highlighting the spiral arms—the "candles and the frosting" on the birthday cake.

Lin and Shu predicted that the spiral patterns should be relatively long-lived, and that basic characteristics such as the number of arms should last over most of a galaxy's lifetime. This is because they assumed that the waves were standing waves. Standing waves arise wherever waves reflect off a barrier. A wave reflected from the barrier reinforces one approaching, and wherever there is a steady supply of mutually reinforcing waves going both waves, there will be a seemingly frozen pattern of crests and troughs. In the case of a galaxy, it was presumed that the waves were being reflected off the inner bulge of stars at one end, producing constructive and destructive interference patterns.

In contemplating the beautiful spiral patterns of galaxies, one thinks back to Pythagoras of Samos, the Greek mathematician and mystic of the 6th century B.C. and his wonderful discovery of the octave—the relation between simple ratios in the lengths of vibrating strings and musical tones. As Pythagoras learned, a string can be used to produce a set of harmonics. If you touch it at the nodal point, dividing the string in half, you produce A, the fundamental harmonic. If you subdivide the string again, you produce the fifth, and so on. Of course, the same principle applies to other instruments—a piano, or a violin. When the musician, the pianist or violinist, hammers or plucks his instrument, he or she generates waves—pulses of energy.

In a field perhaps too long dominated by the analogy of the tuning fork, we hesitate to make the following observation, that as in music, so in spiral galaxies, we seem to encounter fundamental modes or harmonics in

Fig. 13.3. The Local Group. *Image: Wikipedia.*

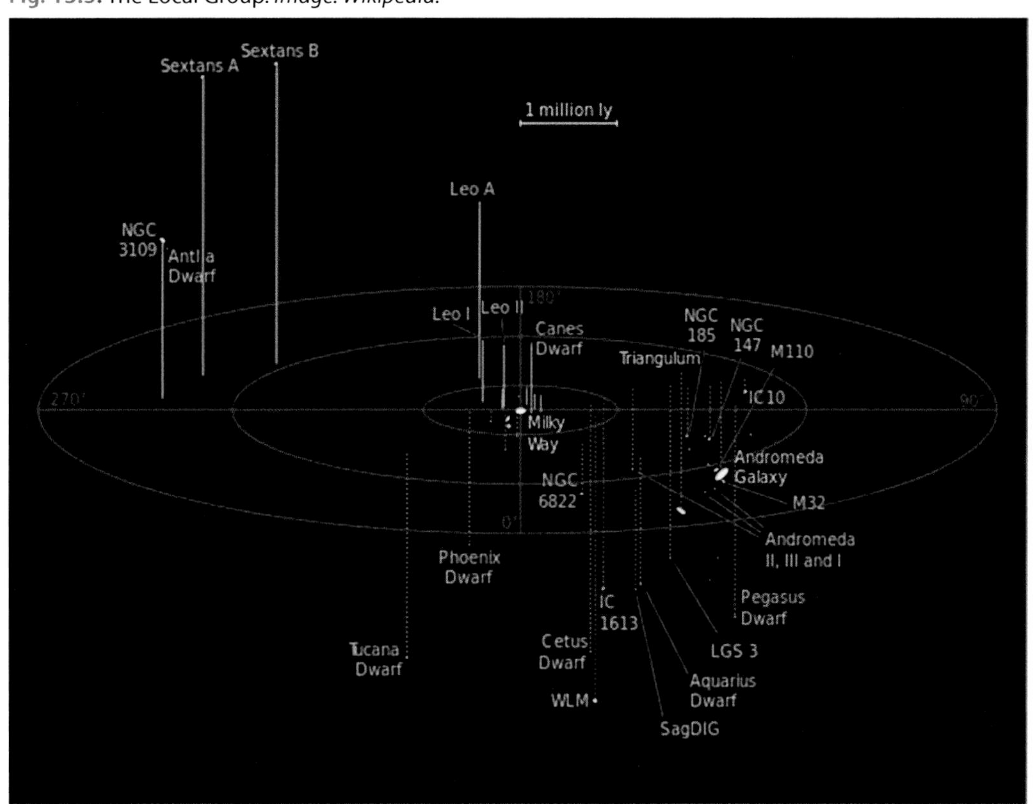

the spiral patterns. Is it possible to imagine the Milky Way as a Stradivarius, 100,000 light years across, and the disk of stars as the vibrating string that is plucked, giving rise to the wavelets that, wherever they reinforce, give rise to the standing spiral density waves?

Perhaps, but if so who is the Maestro, who—or what—does the plucking? Lin and Shu were always a bit vague about what caused the spiral density waves in the first place. Basically, they left it to other investigators to sort that out. It has become clear that there isn't any one thing that produces the spiral density waves in galaxies. Just about any odd disturbance, if it is large enough, will do, which explains why the majority of the large galaxies we find in the local universe are spirals. For a while it seemed that the bars near the centers of many galaxies including the Milky Way might be doing the plucking, since galaxies with bars are more likely to have bold and beautifully symmetric spiral arms like the Milky Way's. Bars are tube-like structures within which the stars are marching in lockstep. They look rather like rotating lawn sprinklers, with streams of stars seeming to shoot out of the ends into the spiral arms, though since the spiral arms are winding up, the stars are presumably moving inward—toward the center of the Galaxy—rather than shooting out.

It is not yet clear how bars form, or what role, if any, they may play in spiral-arm formation. Moreover, they are probably not the major cause of spiral density waves. Recently, astronomers have realized that interactions with smaller galaxies—including outright collisions—are likely the most important plucker of the galactic disk and generator of the splendid harmonics we see as spiral arms, as well as other phenomenon associated with these galaxies. We will return to this topic shortly.

Our Galaxy: Love it or Leave it

Naturally, since the Milky Way is our home Galaxy, we're rather partial to it. It simply wouldn't do to print a message on a t-shirt, "Our Galaxy—love it or leave it"—as the slogan would be utterly unactionable. We must accept the Galaxy as our one and only galactic home, the only one we'll ever know from the inside-out—and may well say, as the bilious Scottish writer Thomas Carlyle did when he heard that the New England transcendentalist Margaret Fuller's favorite phrase was, "I accept the universe!"—"Gad, she'd better!"

Of course, we know many other galaxies much better than we know our own Milky Way. We will probably never see it from the outside looking in, only from the inside looking out, so that our predicament is rather like that evoked by the Scots poet Robert Burns,

O wad some Pow'r the giftie gie us
To see oursels as ithers see us.

As far as we can tell, though, the Milky Way is a rather nice typical galaxy. It's not only an excellent example of a barred spiral, but it's also an unusually elegant and attractive one—as radio astronomers have recently determined by mapping clouds of neutral hydrogen gas on the other side of the galactic center from us, the Milky Way has rare symmetry, with one half being a mirror image of the other. The major star-forming arm our side of the Galaxy, the Perseus arm, has an exact counterpart on the other side of the Galaxy, the Scutum-Centaurus arm. Surprisingly, the latter was identified as a complete arm only in 2011. The arm segment leading to its identification was so long delayed because it turns out that the Milky Way is warped in its outer reaches, so that this outer arm segment did not line up exactly with the midline of the Galaxy.

At present, the Milky Way's structure is believed to look like what is shown in the map, though admittedly, given the difficulty of piecing things together, it's probably not quite as symmetric as appears there. The reality is likely to be much messier. Nevertheless, as viewed from somewhere far out in extragalactic space, with its imperfections smoothed out with distance, it would doubtless appear as a lovely object indeed—a spiral of the classic grand-design type.

Fig. 13.4. One of the dwarf galaxies belonging to the spiral M101. *Courtesy: Hubble Heritage Team (STScI/AURA).*

To complete the picture, the disk of the Milky Way is embedded in a spherical halo of ancient Population II stars that include the globular clusters that, as Shapley realized, form a kind of scaffolding to it. It is generally thought that the halo was the first part of the Galaxy to form, and so contains many of the oldest stars of the Galaxy—a typical age for stars in globular clusters is 12.7 billion years, which makes them almost as old as the universe itself. Naturally, these stars have been relatively little enriched by the contributions of puffing and huffing dying dwarf stars like the Sun and supernovae, and so are lower in heavier elements (metals) than the younger disk and the Sun. In addition to the visible stellar halo, the galaxies are immersed in a far more substantial, but until recently unsuspected because invisible, dark-matter halo (see below).

The Milky Way is not solitary as it travels through the universe. It travels in the company of a small association of other galaxies, which Hubble named the "Local Group."

The Milky Way, together with the Great Spiral in Andromeda, are by far the most massive and dominant members. There is one other spiral, the smaller (but more average) M33. There are also numerous low-luminosity (dwarf) galaxies, raising the total number of members of the Local Group to about 54 at present count, although this number seems to rise every year or so.

The galaxies in the Local Group are gravitationally bound; in other words, the gravity between members is sufficient to resist the general expansion of the universe. Thus, rather than dispersing into the general field, the galaxies stay together like a flock of geese or a school of fish. Some of the dwarf members are satellites of the Milky Way, including the Sagittarius Dwarf, Canis Major Dwarf, Ursa Minor Dwarf, Draco Dwarf, Carina Dwarf, Sextans Dwarf, Sculptor Dwarf, Leo I, Leo II, Ursa Major I Dwarf and Ursa Major II Dwarf; the Magellanic Clouds are also usually mentioned as satellites of the Milky Way, though the situation is actually complicated, and the orbits of the Clouds relative to the Milky Way are not yet known with certainty. They also have many of the earmarks of systems that are remnants of past collisions. Andromeda's satellite system is even more numerous, and includes several galaxies visible in amateur-sized telescopes—M32, M110, NGC 147, and NGC 185. A figure of the Local Group clearly shows the smaller systems tending to cluster around the dominant Milky Way and Andromeda systems.

As numerous as the members of the Local Group are, in terms of its mass, it is a small and sparse association. One might describe it as rather rural. There are much larger and richer associations—true metropolises of galaxies, which may contain thousands of disk galaxies (in contrast to the

Local Group's three) and giant ellipticals. These are clusters of galaxies of which the nearest is in Virgo—the home of several of Messier's objects, and of the rich fields of nebulae discovered by William Herschel in his sweeps. The Virgo cluster contains some 2000 bright member galaxies, centered on the giant elliptical M87. It marks, in turn, the heart of an even larger association, the so-called Virgo supercluster, a cosmic structure so massive that its gravitational effect slows the rate with which nearby galaxies (including the Local Group, which appears to be an outlying member) recede with the general expansion of the universe.

Fig. 13.5. M87, the giant elliptical galaxy marking the center of the Virgo cluster and located 50 million light years from the Earth. A jet of high-energy electrons and sub-atomic particles streams out of the active galactic nucleus at nearly the speed of light, originating in a disk of superheated gas swirling around a supermassive black hole with a mass equivalent to 2 billion solar masses. *Courtesy: J.A. Biretta, W.B. Sparks, F.D. Macchetto, and E.S. Perlman, NASA and the Hubble Heritage Team (STScI/AURA).*

It was for a member of the Virgo cluster, M104, that V.M. Slipher first measured a large redshift in 1913, thus supplying the first evidence of the expanding universe. It lies at a distance of 29,350,000 light years—a fantastic distance. (It is sobering to think that light from M104 just reaching us now left it when the Great Rift Valley of Eastern Africa was beginning to open up.) Almost half of V.M.'s redshifts plotted by Hubble in his first velocity-distance diagram in 1929 were to Virgo cluster objects, while the detection of Cepheids in galaxies as far out as the Virgo cluster so as to better calibrate the velocity-distance relation was a chief motivation of the Hubble Space Telescope.

Hubble was lucky, by the way, that he lived in a sparse association like the Local Group, and not in a large cluster like that in Virgo. Otherwise he might never have recognized his law. As Iowa State University astronomer Curtis Struck points out, "If the Milky Way resided within a giant cluster of galaxies, as about half of all galaxies do, the observations might have realized the simple expectation of just as many galaxies coming toward us

Fig. 13.6. Optical image of M51 and its companion NGC 5195. The two curving arms of M51 are the hallmark of "grand-design" spiral galaxies. The prominence of the Whirlpool's spiral arms is thought to be due to the effects of its close encounter with the smaller galaxy, which at first glance appears to be tugging on the arm; however, the Hubble Space Telescope's clear view shows it is actually passing behind the Whirlpool, and has been gliding past it for hundreds of millions of years. As it passes, it stirs up spiral-density waves in the Whirlpool, squeezing the molecular clouds along each arm's inner edge (which look like gathering storm clouds); the clouds collapse, and give rise to a wake of star birth, which is taking place in the bright reddish HII regions. Eventually the large massive blue stars sweep away the dusty cocoons within which they form in a torrent of hurricane-like stellar winds, and emerge as bright blue star clusters along the Whirlpool's arms. *Courtesy: S. Beckwith, NASA, ESA and the Hubble Heritage Team (STScI/AURA).*

as moving away. This is because the enormous self-gravity of such a cluster overcomes the universal expansion, and most member galaxies are moving randomly (and at high speed) relative to one another."[5] With all this data about galaxies we can start to ask how and when these systems formed in the universe.

Currently, there are two major theories for explaining how galaxies formed. The first theory is simply that galaxies form like stars, only on a much larger scale, through the collapse of gas. This theory predicts that the gas is converted into stars in a very short time, and also—and this is the critical feature—that very little or no addition mass is added to a galaxy once it is formed in the very early universe. In this model there is some star formation occurring in galaxy disks, such as we see within the Milky Way in HII regions and molecular clouds, but the basic size and parameters of galaxies in this model are set very early in the universe. The other idea is that galaxies form through the accretion and merging with other galaxies throughout time.

This competition is a bit like the classic nature/nurture controversy: does a person's characteristics, such as intelligence, personality, etc., depend merely on the genetic inheritance they receive at birth, or is it a product of one's experiences, education, etc? Viewed in terms of this analogy, the gas-collapse theory stands on the side of nature. The collapse of gas occurs, and the galaxy is set on its course for the rest of its evolutionary lifetime. This was the dominant idea about how galaxies formed up until the 1980s.

In the 1950s, it seemed reasonable to believe that galaxies including our own Milky Way formed in a rapid collapse. This model—which was top-down—was given classic form in an influential research paper written in 1962 by Olin Eggen, Donald Lynden-Bell and Allan Sandage who com-

Fig. 13.7. An early computer model of the tidal production of spiral arms in the M51 system by Alar and Juri Toomre. The large shaded dots represent the centers of the gravitational potentials of the two galaxies, and representative stars, initially in circular orbits around the center of the primary galaxy, are shown by open circles. *Courtesy: Professor Alar Toomre, MIT.*

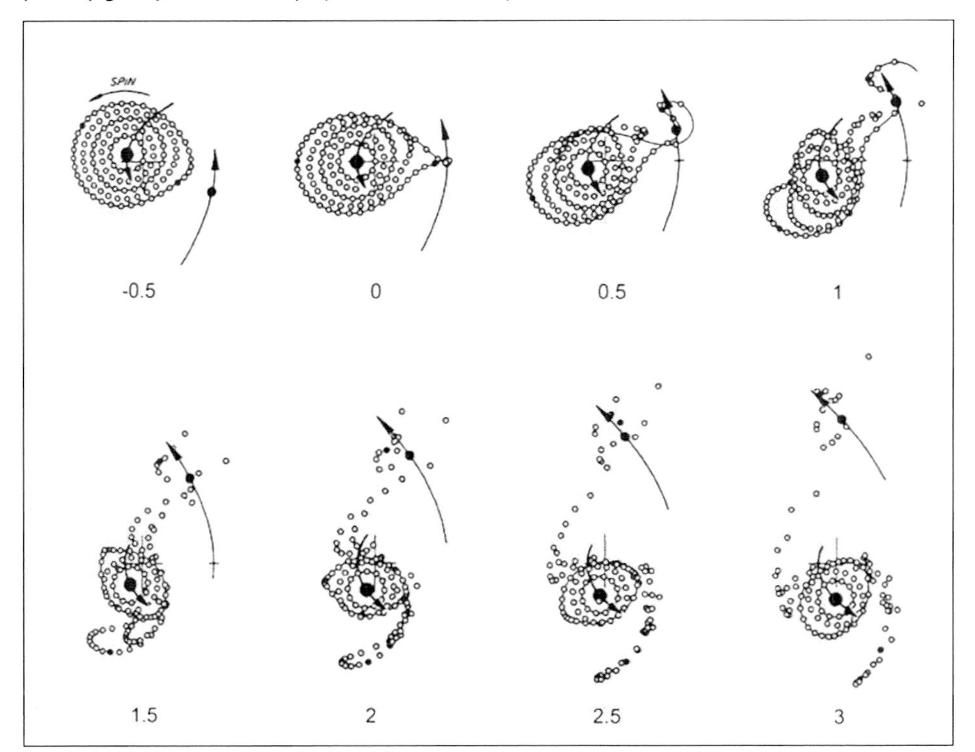

bined theoretical ideas and data to deduce that the Galaxy formed in one of these "monolithic" rapid collapses. While this mode of thinking about galaxy formation is now passé, there is evidence that this process is potentially partially correct for some galaxies, especially the first to form in the universe.

The second major theory offered to explain the formation of galaxies is what might be described as the nurture theory. Here, the galaxy's experience, and "personal history," do matter, because galaxies don't exist in splendid isolation from one another; they interact gravitationally, and two galaxies may even merge to form a new more massive system. These kinds of interactions are still going on in the nearby universe—and even in the Local Group itself. Some of these might be described as "minor interactions," an interaction of a smaller galaxy with a larger one, where others appear rather cataclysmic. The classic case of an interaction is our old friend M51, the "Whirlpool," and its satellite NGC 5195. We can see at a glance what's going on: NGC is clearly moving past, and giving a torque to M51, twisting it. Material has been flung out and scattered around during their extended interaction, triggering starbursts both in M51 and in the satellite (which appears like a wreck in optical wavelengths, but as a small and surprisingly regular barred spiral in the infrared). The interaction between M51, the classic spiral first recognized by Lord Rosse, and its satellite have often been studied by researchers using computer simulations.[7] In fact, it proves quite easy to set up parameters that accurately model the observed features, such as the greatly intensified and boldfaced arms of M51, which make it such a beautiful system.

The Milky Way's own lovely arms, in one of which the Sun travels around the center of the Galaxy, may owe something to a satellite galaxy whose existence was not even identified until Rodrigo Ibata, Mike Irwin and Gerry Gilmore discovered it in 1994. The satellite galaxy is Sagittarius Dwarf Elliptical Galaxy (SagDEG) (not to be confused with the Sagittarius Dwarf Irregular Galaxy (SagDIG), a Local Group member a safe 3.4 million light years away). The reason it was so difficult to discover is that most of the constituent stars, though covering a large area of the sky, are on the far side of the galactic core, and therefore very faint. SagDEG has four known globular clusters, of which one, M54, resides in its core. It moves in a highly elliptical orbit around the Milky Way, and appears to have made several brushes through the Milky Way, the first 1.9 billion years ago and the second 0.9 billion years ago. Though the Milky Way's effect on SagDEG are obvious—tidal forces from the larger Galaxy have wrenched it apart, leaving in the wake of its destruction a faint trail of old red giant stars—SagDEG's effect on the Milky Way are far from insignificant. Simulations have shown that its plunges through the Milky Way have likely stirred powerful spiral density waves that have enhanced the strength of the Milky Way's spiral arms. A number of other stellar streams have been identified in the Milky Way that may be remnants of other small galaxies now merged with it—thus the Milky Way really is like an archaeological site. It is even possible that some globular clusters may be remnants of early mergers, although most do not have the dark matter content associated with true galaxies.

The Milky Way's counterweight in the Local Group, M31, has undergone perhaps even more violence than our galaxy. Infrared astronomers have found evidence of structures that could have been produced in a head-on collision with M32, an anomalous, and perhaps disturbed, dwarf elliptical companion. There is also a faint gas bridge between M31 and its large companion M33, so that, quite possibly, M31, M32, and M33 are all actively involved in collisions at present.

Collisions between a large galaxy and one of its satellites, or a smaller dwarf system, typically leaves the larger galaxy intact. In fact, the stars of the one galaxy pass through those of the other system, and most of the pile-up involves gas. The exact nature of the interaction, however, depends on the relative motion of the galaxies. If two galaxies collide but do not have enough momentum to keep traveling after the collision, they will fall back into each other and eventually merge, forming one galaxy. This, in fact, is expected to occur to the Milky Way and M31. As V.M. Slipher found in 1913, M31 is approaching the Milky Way; this is a result of the mutual gravitational pull between them, and eventually—about four billion years from now—they will embrace and form a single massive galaxy, perhaps even transforming into an elliptical. The Sun will still be on the Main Sequence by then (though nearing the depletion of its hydrogen-burning phase), and so will experience this major merger event from the inside, assuming it isn't ejected from the system in the process. Fortunately, it's far enough in the future that we don't need to worry about it much.

Interacting and "colliding" galaxies, if the galaxies involved are both large ones, may appear to be enmeshed with one another in a deadly embrace, or distorted into bizarre forms. Hubble didn't know where to put them on the tuning fork, and so tossed them aside into a bin of "peculiars." At the end of his classic 1943 book *Galaxies,* Harlow Shapley summed up the thinking at the time:

> There are ... frankly "pathological" types (as Baade calls such freaks) like NGC 5128 [in Centaurus] ... and the ring-tail system, NGC 4038[-39].

> The theories that sufficiently explain the relatively simple looking Sc spiral Messier 33, and the most common galaxies in Virgo, must have sufficient flexibility to take care of these aberrant types. The interpreter may need to resort to the assuming of collisions to find satisfactory causes. He will find some justification, because the individual galaxies are not so far separated but that encounters may have been fairly numerous, if the time scale has been long enough.[8]

The First Objects in the Universe

So, what do we know about the formation of the first structures in the universe? The first stars are thought to have formed about 100 or 200 million years before the first galaxies. As explained in chapter 9, theories of star formation in the present-day universe assume that stars form from the collapse of Molecular Clouds. In the very early universe, however, there was no dust; there were giant blobs of dark matter (see next chapter) and hydrogen and helium, initially mixed with the dark matter, that were created in the Big Bang, 13.8 billion years ago. All heavier elements, or metals (again, in the astronomers' sense of elements heavier than hydrogen and helium) and dust formed in later generations of stars.

Fig. 13.8. M101 in Ursa Major. This is a close cropping of the image shown in Fig. 11.10. Individual dust lanes are clearly visible, and bright, hot HII regions where star formation is taken place dot the spiral arms. *K.D. Kuntz, NASA, ESA, and the Hubble Heritage Team (STScI/AURA).*

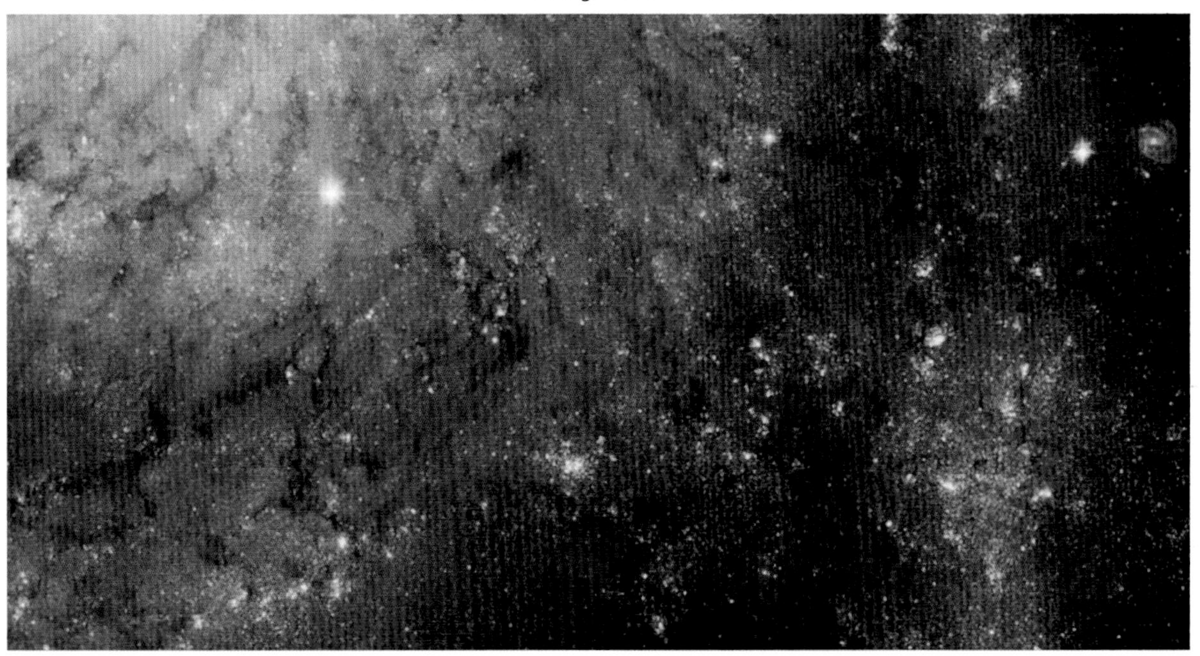

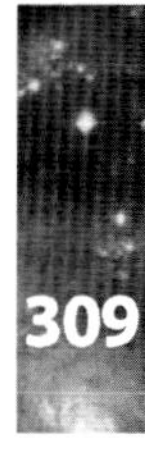

As the universe cooled by expanding rapidly, the ordinary baryonic matter contracted, whereas the dark matter remained dispersed. Denser regions of gas contracted into star-forming clumps hundreds of times as massive as the Sun, and began to collapse into the first generation of stars. They were very massive, 100 times as massive as the Sun and far more massive than any stars found today. The very large, massive, brilliant stars of this first generation, many of which may have originated as binary systems, evolved rapidly, as such stars are bound to do, and ended up, after only perhaps 2 million years or so, exploding as supernovae and strewing their guts into space. The most massive among them collapsed into black holes (see next chapter). A mix of heavier elements, with silicate- and carbon-dust, was scattered in a wispy dust-veil across these galaxies—dust which, using infrared imaging devices, has actually been detected in very-early galaxies. Some of the earliest black holes perhaps merged to form supermassive black holes, and gravitational attraction pulled clouds gas and dust toward one another, and as they collided, presumably they began to trigger the formation of new generations of stars just as galactic mergers still do. In this way, the first generation of massive stars may have laid the groundwork for the first galaxies.

This scenario is probably right in its broader features although the details still have to be worked out. For the earliest galaxies, at least, rapid collapse must have been involved, broadly along the lines deduced in the Eggen-Lynden-Bell-Sandage paper for the initial formation of a galaxy's halo.

However, for a long time this rapid-collapse concept dominated our thinking about the way all galaxies evolved. It was thought that, once formed, a galaxy was set for life. If they started out an Sa or Sb or Sc, or an SBa, SBb, or SBc, or an elliptical or an irregular galaxy, that is what they remained (barring the occasional collision or merger event). Their characteristics and further evolution were completely determined by their initial mass, angular momentum, and the like. They were, like Victorian children, born little adults. However, astronomers now think the situation is considerably more complicated, and that biography is destiny for galaxies as for people; a great deal of what has gone into the making of the galaxies we find populating today's universe occurred over the entire course of the universe's history, not just in the quick initial formation stage

In our own Galaxy, we can indeed observe stellar population and kinematic features fitting the 'monolithic' collapse model, and any idea about galaxy formation must account for the fact that there are old stars in every galaxy that has been studied in any detail thus far. However, there are many features of galaxies, including our own, which are not explainable by a quick formation, but require a continuous one of star formation and assembly that we can see in nearly all galaxies that we can study in detail. It is in fact clear that galaxy formation is not a simple process, but one that started early and continues even today.

Part of the reasoning behind the idea that galaxies form in a rapid and early collapse is due to the fact that in the nearby local universe nearly all galaxies have a shape or morphology which is either that of a disk galaxy or an elliptical. Both of these are relatively stable structures and therefore could have existed in their present form for more than 10 billion of years or so.

Fig. 13.9. (left) M66, the largest member of the Leo "triplet," and notable for its unusual anatomy, consisting of asymmetric spiral arms and an apparently displaced core. The peculiar anatomy is most likely caused by the gravitational pull of the other two members of the trio, M65 and NGC 3628. (For an amateur view of the triplet, see Fig. 2.10.) *Courtesy: S. Van Dyk and R. Chandar, NASA, ESA and the Hubble Heritage Team (STScI/AURA)*

Fig. 13.10. (right) M64, a Virgo cluster member known as the "Black-eye" galaxy and notable for its unusually prominent dust lanes. Detailed studies in the 1990s showed that, while all the stars in M64 are rotating in the same direction (clockwise as seen in this Hubble image), interstellar gas in the outer regions of M64 is rotating in the opposite direction from gas and stars in the inner regions, and active formation of new stars is occurring in the shear region where the oppositely rotating gases collide. In this image, hot blue young stars and HII regions of hydrogen gas fluorescing when exposed to ultraviolet light from these newly formed stars can be clearly seen. Astronomers believe the oppositely rotating gas is left over from a satellite galaxy that collided with M64 perhaps a billion years ago. Though the small galaxy has now been completely destroyed, signs of collision persist in the backward motion of gas at M64's outer edge. *Credit: S Smartt and D. Richstone and the Hubble Heritage Team (STScI/AURA).*

Fig. 13.11. NGC 2841, located at a distance of 46 million light years away in Ursa Major. An example of the remarkable variety of galaxies. In contrast to some of the galaxies featured here, blazing with HII regions (emission nebulae), NGC 2841 currently has a low rate of star formation, probably because radiation and supersonic winds from hot young blue stars cleared out the remaining gas and hence shut down further star formation in the regions where they were born. *Courtesy: M. Crockett and S. Kaviraj, R. O'Connell, B. Whitmore, and the WFC3 Scientific Oversight Committee, NASA, ESA and the Hubble Heritage Team (STScI/AURA).*

Galaxy Evolution becomes a Science

The first really convincing evidence that galaxies may not be static systems after all was provided by the PhD thesis work of the New Zealand/American astronomer Beatrice Tinsley. As a PhD student at the University of Texas in the 1960s, Tinsley showed that over time the brightness of galaxies should change, and significantly so, due to the evolution of their constituent stars. Tinsley showed that, by making simple assumptions about the star formation history of a galaxy, the initial light would be dominated by massive blue and bright O and B stars.

Since, however, as we noted in chapter 8, these massive and bright OB stars do not live for very long, and because they are massive enough, they end cataclysmically in supernova explosions. As they self-destruct, their light essentially is subtracted out. As the blue stars are removed, the galaxy becomes fainter and redder than before, dominated eventually by the Population II stars—the long-lived M and K dwarfs.

The luminosity of a star is in fact proportional to its mass to the power of 3 or 4. In other words, for the most massive stars, if you double the mass, the luminosity of the star increases by $2^3=8$ times or even $2^4=16$ times. Also, since the brighter a star is, the more energy it uses up, the most massive stars also have the shortest lifespans. (They are like the Greek hero Achilles, who chose glory and a short life over obscurity and a long one.) As time progresses, the next lower-mass stars will die and subtract out their light, and the next, and the next, and so on. Tinsley used this process to try to explain why elliptical galaxies, which have very few young stellar populations, are so red, while spiral galaxies and especially galaxies with active star formation such as irregulars and peculiar, are often very blue.

Fig. 13.12. Interacting galaxies Arp 273, in Andromeda, at a distance of about 300 million light years. The large spiral galaxy, UGC 1810, has a disk distorted into a roseate shape by gravitational tidal interactions with the companion galaxy below it, UGC 1813. Detailed analysis indicates that the smaller companion actually dived deep, but off-center, through UGC 1810. The inner set of spiral arms is highly warped out of the plane, with one of the arms going behind the bulge and coming back out the other side. *Courtesy: Hubble Heritage Team (STScI/AURA).*

Tinsley first carried out calculations along these lines in the late 1960s and established that a galaxy over a relatively short time span, cosmically speaking, may come to have a very different brightness. This meant that none of these systems could be used as a "standard candles." (We will say more about that particular problem and how it was eventually surmounted in chapter 15.) This fact was not fully appreciated by Sandage and his collaborators, who continued to do as Hubble had done, assuming that galaxies or features within galaxies have a constant luminosity, which they clearly do not. With Tinsley's important work it became clear that before galaxies could be used as probes of cosmic distances, a detailed understanding was needed of the evolution and the properties of their stars. This is still an active area of research even today.

Fig. 13.13. Interacting galaxies NGC 2207 and IC 2163, in Canis Major. Here, tidal forces from the larger and more massive galaxy, NGC 2207 (left) has distorted the smaller one, IC 2163, on the right, flinging out stars and gas into long streamers stretching out a hundred thousand light years toward the right edge of the image. Computer simulations by Bruce and Debra Elmegreen have shown how leisurely these collisions are; IC 2163 is swinging past NGC 2207 in a counterclockwise direction, having made its closest approach 40 million years ago. It does not have sufficient energy to escape from the gravitational pull of NGC 2207, and is destined to be pulled back and swing past the larger galaxy in the future. Trapped in their mutual orbit around each other, these galaxies will continue to distort and disrupt each other until, billions of years ago, they will merge into a single more massive galaxy. It is believed that many present-day galaxies, including the Milky Way, were assembled through such a process of coalescence of smaller galaxies occurring over billions of years. *Courtesy: D.M. Elmegreen and B.G. Elmegreen, NASA and the Hubble Heritage Team (STScI/AURA).*

Enter Hubble – The Telescope

The Hubble Space Telescope, which was launched in 1990, also challenged and effectively destroyed the view that galaxies formed quickly, in a rapid collapse, when it began to take photographs of the distant universe in the mid-1990s (after an embarrassing focusing problem caused by the primary mirror being polished to the wrong curve by contractor Perkin-Elmer was corrected by a Space Shuttle servicing mission in 1993). Until Hubble, we were only able to study in detail galaxies in the nearby universe, and for the more remote ones—those that appeared as little more than dots on Hubble's deep plates—we could learn no more than how much light they were emitting. To get at the kinematic motions of their stars, which is needed to study their dark matter haloes, as well as their morphologies and internal structures, such as bars, star forming regions, and even individual stars, we needed much higher resolution than was available from ground-based instruments. What we have learned from Hubble, and other large telescopes, has given us our modern view of how galaxy formation has occurred over cosmic time.

The Hubble Space Telescope discovered in the years 1994-1998 that distant galaxies were most certainly very different from nearby galaxies. Indeed, as ought already to have been clear by the dramatic shift of Hubble type when a galaxy like NGC 309 (described above) was viewed in infrared instead of optical wavelengths, the Hubble classification is heavily beholden to the froth, the bright young stars, HII regions, and spiral arms, which are decoupled from what's going on in the backbone of older stars representing most of a galaxy's mass.

The Hubble classification, as convenient as it is, is thus merely descriptive and does not provide useful information about a galaxy's evolution. Even Hubble's terms "early," "middle," and "late" are merely conven-

Fig. 13.14. A hit, a palpable hit. This galaxy, AM 0644-741, lies at a distance of 300 million light years away in the southern constellation Volans. It is an example of a ring galaxy, a striking example of how collisions can dramatically change their structure while triggering formation of new stars. In ring galaxies, one galaxy (the "intruder") plunges directly through the disk of another (the "target"). In the case of AM 0644-741, the "intruder" is outside the field of the image. The resulting gravitational shock imparted due to the collision drastically changes the orbits of stars and gas in the target's disk, causing them to rush outward rather like ripples in a pond after a large rock has been thrown in. As the ring plows outwards, gas clouds collide and are compressed, leading to the formation of new star (which explains why the ring is so blue). *Courtesy: J. Higdon and I. Jordan, NASA, ESA and the Hubble Heritage Team (STScI/AURA).*

tions in the descriptions, and nothing more. Moreover, even as description, the Hubble classification fails completely when it comes to galaxies in the distant universe—out past a relative wavelength redshift of z=1, when the universe was roughly half its current age, and even more so when we get out past a relative wavelength redshift of z=6, where we are observing the universe's early days indeed.

The most important survey for showing this was the Hubble Deep Field, a single deep pointing of the sky (taken in optical wavelengths) taken by Hubble over ten days in December 1995. The Hubble Deep Field is a very deep look at a tiny, apparently empty, area of the sky, 2 ½ minutes of arc across, in Ursa Major, which is a composite of 342 separate images taken with the Wide Field and Planetary Camera-2 – probably Hubble's most scientifically productive camera to date. It is one of the iconic images of modern astronomy, and has been cited, as of this writing (August 2013), in nearly a thousand publications. Despite the smallness of the field—it covers only 1/24-millionth of the whole sky, and it contains several thousand galaxies. What is seen after a quick examination is that the fainter, presumably more distant and younger galaxies, look stunningly different from the galaxies in the modern universe, as we describe below.

There were several other successor "Deep Fields" with Hubble over the next 15 years, each requiring several weeks of telescope time and largely due to the initiatives of the Space Telescope directors who used their personal director's 'discretionary' time to carry out these large public programs.

The galaxies imaged by the Hubble Space Telescope in these deep fields (whose spectra were eventually measured, confirming their great distances back to when the universe was only a few hundred million years old) appear to be very distorted and irregular in appearance. They look like slivers, twisted shapes, crescent moons, fuzz-balls among other hard to describe shapes and forms. We might almost imagine that we have the optics reversed, and that we are looking through a microscope rather than a telescope, at the protozoa, foraminifers and infusoria in a drop of pond water. It is hard to believe that these tiny little beasties are actually galaxies, not animalcules locked in one another's grips. They look to be engaged in a feeding frenzy, a fierce and ferocious fight to the death. These are the denizens of the very early universe; these—the pathological ones, the freaks—and not the stately and majestic and well-behaved systems we see in the modern universe, ruled the era of cosmic dawn.

Understanding what these observations of peculiar galaxies in the distant universe mean, and how we can understand the physics of galaxy formation from these observations, has been a major goal in astronomy ever since. One idea is that these galaxies are actively forming either

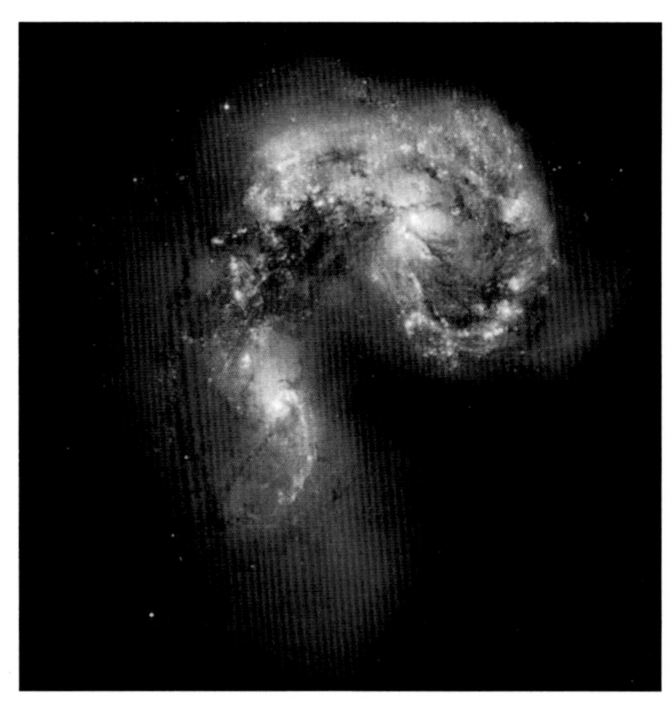

Fig. 13.15a. (left) The Interacting galaxies, the "Antennae," NGC 4308/4309, are among the nearest and youngest examples of a pair of colliding galaxies. This image, by John Drummond using a 25-cm Schmidt-Cassegrain, whose the antenna-like "arms," which extend far out form the nuclei of the two galaxies. They are tidal tails formed during the initial encounter of the galaxies some 200 to 300 million years ago, and give a preview of what may happen when the Milky Way and Andromeda collide in a few billion years from now. *Courtesy: John Drummond.*

Fig. 13.15b. (right) Detail of the above, from the Hubble Space Telescope. Nearly half of the faint objects in the Antennae image are young clusters containing tens of thousands of stars; the orange blobs to the left and right of the image are the two cores of the original galaxies and consist mainly of old stars criss-crossed by filaments of dust. Only a fraction of the newly formed star clusters will survive the first 10 million years; the vast majority will disperse, with the individual stars becoming part of the smooth background of the galaxy. It is believed that about a hundred of the most massive clusters will survive to form regular globular clusters, similar to those that form the framework of globular clusters found in our own Milky Way. *Courtesy: B. Whitmore, NSA, ESA, and the Hubble Heritage Team (STScI/AURA).*

through intense star formation and/or interactions with other galaxies. The most obvious hypothesis is that these are galaxies smashing into other galaxies by their mutual gravity.

Astronomers refer to these events as galactic collisions or mergers—rather placid terms for what actually seems to be going on out there. We might more aptly refer to grasping, clawing, disemboweling, chewing, grinding. It appears to be natural selection—survival of the fittest—writ large. The strong prevail, the weak perish. The early universe was (or should we say is?—after all, in the telescope we are reading messages in the light from an eternal present, where it is always now) is dominated by a few

Fig. 13.16. A collision involving the spiral galaxy pair NGC 3314, in Hydra, at a distance of 140 million light years. By a chance alignment, a face-on spiral galaxy lies precisely in front of another larger spiral. This allows an unusually clear view of the dust belonging to the foreground galaxy as it absorbs light from the more distant system. The bright blue stars that form a pinwheel shape near the center of the foreground balaxy have recently formed from interstellar gas and dust, but in addition, one sees numerous additional dark dust lanes not associated with bright young stars. *William Keel, NASA, and the Hubble Heritage Team (STScI/AURA).*

(relatively) gargantuan specimens. Those behemoths, the leviathans of deep space, seem to be trying to suck the rest into their insatiable gravitational maws.

We are allowing ourselves a fair amount of poetic license here, but a large fraction of what we know about the formation of the galaxies we see today originates from these precious observations of the most distant galaxies. This understanding comes from interpreting the distribution of the stars in them, and examining their ages and star formation histories.

Finding the Most Distant Objects

Unfortunately, this is not as easy as it sounds, as in modern astronomy—in all of science for that matter—understanding the biases and errors in our calculations are critical for making progress. This was clearly relevant to studies throughout the history of astronomy, including Herschel's efforts to map the Galaxy due to the "bias" of dust. It is even more relevant to studying distant galaxies, as there are many effects that can distort our view of the early universe. Much of what we now know about the early formation of galaxies has involved trying to remove this bias from our observations, just as much of what we know about galactic structure has come from viewing the galaxies in wavelengths that "remove" the dust.

Fig. 13.17. Hubble Deep Field. Hubble Heritage Team (STScI/AURA). This color image, one of the deepest images ever taken by the Hubble Space Telescope, combines images in four different optical wavelengths taken by the WFC2 camera. Dating from 1995, it was the first true "deep" image taken by the Hubble Space Telescope, one where it took exposures of the same part of a blank portion of the sky for ten days reaching a depth into the universe that had never been achieved. The size of this image is 2.5 arcmin on a side, giving it an area about 0.1% of that of the full moon. Nearly every source in this field is a distant galaxy, with some seen at just a billion years after the big bang, with others in the nearby universe, and everything in between. Astronomers have used this image, and other similar follow up ones, such as the Hubble Ultra Deep Field (UDF), to redefine our view of how galaxy formation and evolution has occurred.

It is, however, easy to be fooled. This is particularly so when we are trying to study the first galaxies in the universe. Even the first step in studying these systems is difficult—which is to locate them reliably.

When we examine distant galaxies what we seeing are not necessarily representative ancestral forms of the galaxies we find in the local

universe. We are in fact biased to the brightest systems, the cosmic celebrities, as those are the easiest to find. The situation is not that different from going out on a dark night in a city and seeing only the brightest stars. These stars are not the typical or average ones, and so the naked-eye sky—with all the bright Gould Belt stars—might make us imagine that A-type and B and O-type giants and supergiants are the commonest stars in the Galaxy, when they are actually very rare.

What we need is a way to connect galaxies we see in the distant universe with those we see today. One way to do this is to examine the luminosities of these galaxies. This method can be applied by simply finding galaxies that emit the same amount of light at different distances, and then assuming that these are statistically the same galaxies. Ideally this would work, but there are a few things that hinder this approach. One is that until recently—we will explain the caveat presently—we had no choice but to view distant galaxies using optical light instruments, the traditional wavelength in astronomy.

As a result, measuring galaxy stellar masses has become the standard way to trace galaxy evolution through time. The reason for this is that a galaxy cannot lose its stellar mass—it can gain stellar mass through various processes, but it cannot lower its stellar mass over time. This makes the stellar mass a very robust quantity in which to find statistically the same galaxies over many cosmic epochs.

As an aside, we know that galaxies have two other types of masses besides that represented in its current complement of stars: the amount of mass in gas which can form into stars which can increase or decrease with time, and the dark matter mass. These two types of mass are however difficult or impossible to measure in many distant galaxies, unlike stellar mass, which can be measured for even the most distant galaxies.

Because distant galaxies are extremely faint it is hard to study them in any detail. Typically distant galaxies have magnitudes of 25 in the blue, which is about 100 million times fainter than the faintest stars that can be seen by the human eye (by comparison, the limiting magnitude of the galaxies Hubble and Humason were studying with the 100-inch reflector was 17.5 or so). We are, however, able to study in detail the most massive galaxies—again, this is a bias effect, and due to the fact that these systems are typically much brighter than lower mass systems at the same distances.

Indeed, and rather incredibly, with the Hubble Space Telescope, and with 8-10 meter ground-based telescopes, we are able to study these distant systems with a level of detail needed to determine a great deal about how galaxy evolution in the early universe occurred.

What the Hubble imaging surveys described above showed is that structurally, faint galaxies look "peculiar," that is, they are not spirals or

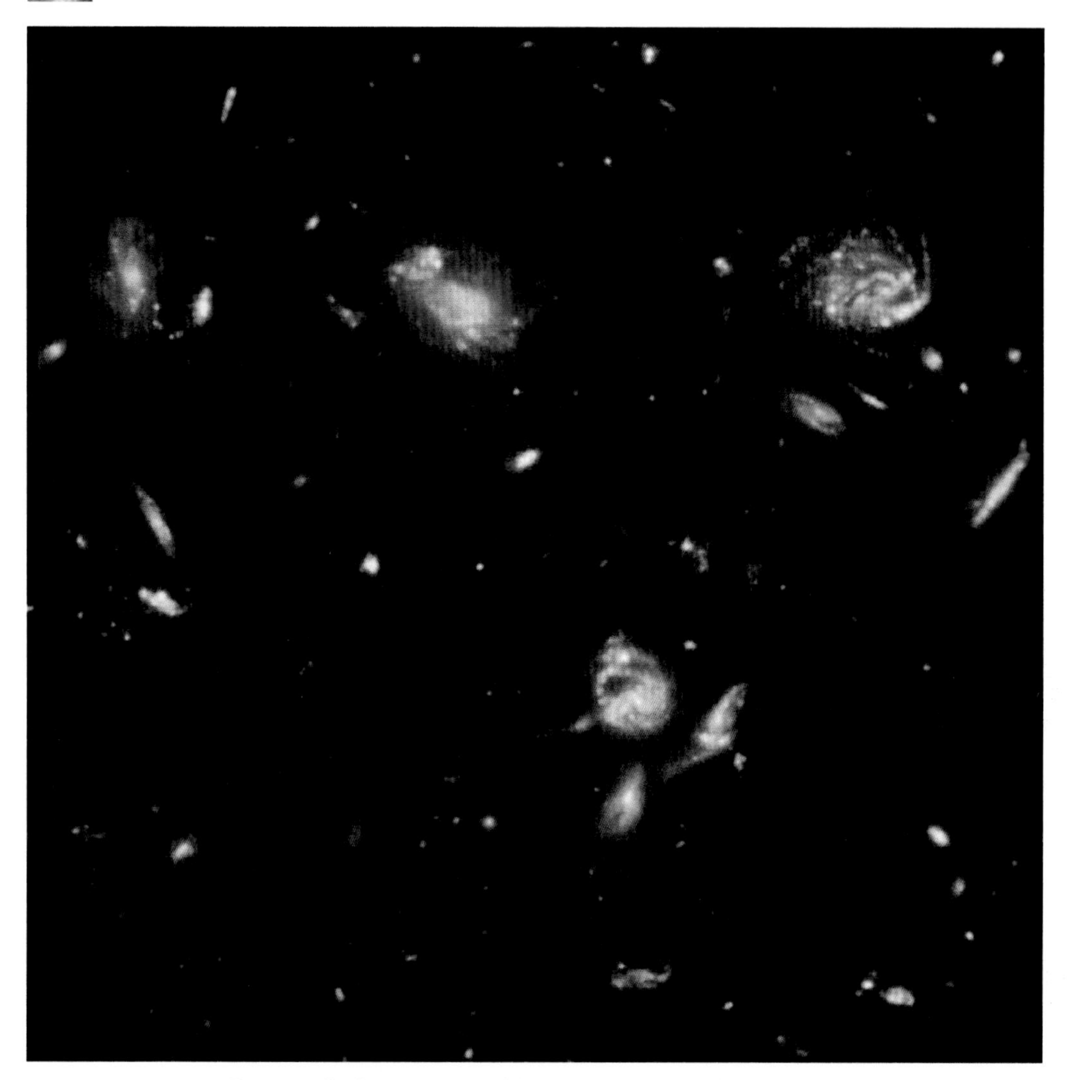

Fig. 13.18. Close up of galaxies in the Hubble Ultra Deep Field from a color image of this field. Shown are some of the closer galaxies in this field, which are at $z = 0.5$ or thereabout, to when the universe was several billion years younger than it is today. Most of these brightest galaxies are spiral type systems but differ from nearby systems in that they have intense star formation, and show the location of massive star formation knots in their structure. Some tidal interactions can also be seen between some galaxies. *Courtesy: Hubble Heritage Team (STScI/AURA).*

elliptical, as in the Hubble Sequence, but look as if they are "train-wrecks" or galaxies in some obvious formation mode.

These galaxies have structures and shapes which show they are fundamentally different from galaxies we find in the nearby universe. The formation process clearly produces the distorted structures we see, structures which gravitationally are not in equilibrium. When this was first discovered, it was one of the biggest clues for trying to answer how these galaxies formed.

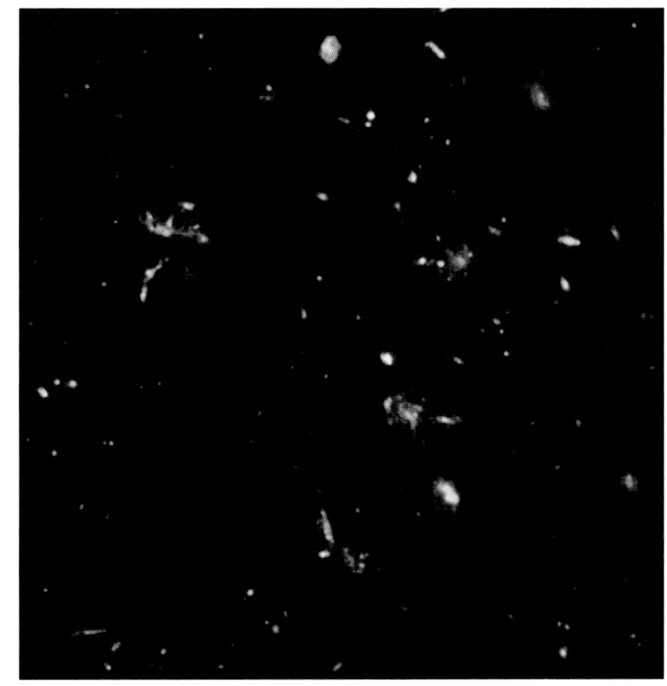

Fig. 13.19. (left) A closer view of some of the spiral galaxies seen in the Hubble Ultra Deep Field. This field shows one of the few stars found in this Ultra Deep Field. Nearly every other point of light in this field, even the faintest barely seen ones, are distant galaxies, some of which are only a few hundred million years old. *Courtesy: Hubble Heritage Team (STScI/AURA).*

Fig. 13.20. (right) Another close up view of the Hubble Ultra Deep Field, showing in this case some examples of galaxies involved in interactions and mergers, which are much more common in the distant past than they are today. The bright white areas in these galaxies show the location of intense star formation. Note the diversity in color of these galaxies, showing how the star formation histories for these systems can differ by significant amounts. *Courtesy: Hubble Heritage Team (STScI/AURA).*

When we measure the number of stars in these distant galaxies we find that they contain a similar amount of mass (to within a factor of a several-fold) to the most massive galaxies in today's universe, and therefore these systems must somehow manage to evolve over time into the normal galaxies we see in our immediate cosmic surroundings. Furthermore, kinematic studies of these galaxies show that many of these systems do not have an ordered internal motion similar to nearby disks and ellipticals—strong evidence for a violent origin.

The role of mergers and acquisitions in galaxy formation

Detailed studies of these galaxies reveal that a significant fraction of these peculiar systems in the distant universe are undergoing some type of mass assembly, producing their distorted structures. It was hypothesized early on by the first observations that these structures were produced through a merger process. What this implies is that the galaxies that we see in the distant universe become the larger massive galaxies in the nearby uni-verse simply through smashing together to create larger and more massive galaxies.

One primary way that this hypothesis has been tested is through examining the quantitative structures of these galaxies using new imaging analysis tools to quantify how these structures evolve through time. As mentioned, Hubble observations, beginning with the Hubble Deep Field, showed that galaxies were very distorted compared with today. Distorted structures can however arise from several causes.

Clumpy star formation can produce structures like the ones seen in distant galaxies, as can the smashing together of already existing galaxies. There is, however, one key difference between these two modes, and that is that within mergers, the older stellar populations are in a non-equilibrium distribution, as well as the younger stars that may be forming. For mature galaxies with ongoing star formation the older stellar populations are not distorted, but symmetrical, as equilibrium in a stellar system is reached rather quickly.

Starting in the 1990s astronomers were able to develop imaging tools using data from the new CCD imagers to determine whether a galaxy is undergoing a major merger based on its structure. The primary way this is done is through using the distribution of light within these distant galaxies resolved by Hubble, and by analyzing them in a quantitative way. That is, with this method we can quantify the structures of galaxies rather than use visual estimates of morphology as has been done for decades previously.

By using these imaging methods we can measure which galaxies are undergoing mergers with other galaxies. From this we can measure directly the merger history of galaxies. The main parameter used to determine this is the 'asymmetry' parameter, which quantifies how much a galaxy deviates from perfect symmetry. It turns out that the higher this parameter is, the more likely a galaxy will be in a merger state.

However, using these structural parameters requires that we have the right kind of data to analyze distant galaxies with. As mentioned earlier, galaxies can look very different, particularly distant ones, when examining their structures in 'rest-frame' optical and ultraviolet light. What is needed for examining distant galaxies is therefore an instrument that allows us to probe the rest-frame optical light originating from these young galaxies.

This was possible due to advanced near-infrared instruments on the Hubble Space Telescope – the first of which was the Near-Infrared Camera and Multi-Object Spectrograph (NICMOS) installed on a space shuttle mission in 1998, which allowed astronomers for the first time to examine galaxies in the first half of the universe in terms of the bulk of their stellar mass, as opposed to simply examining the location of their star formation, as revealed in the ultraviolet. Using this camera to take deep exposures of the distant universe, astronomers were able to examine forming galaxies in terms of their bulk structure for the first time.

What astronomers found was rather astounding. A large fraction of the massive galaxies— similar in mass to the Milky Way or Andromeda— seem to be undergoing mergers a few billion years after the Big Bang. For the most massive galaxies, the fraction is as high as 40%. In comparison, the fraction of similar mass galaxies undergoing a merger in the nearby universe is less than 1%. Between the peak merger rate which occurred a few billion years after the Big Bang until today, we can clearly see the merger history decline. That is, most of the assembly history for galaxies occurs early in the universe's history in the first few billion years.

How else do galaxies in the early universe differ from those we see today in great abundance in the local universe? Another important difference is that these distant galaxies are undergoing vigorous star formation dwarfing what we see occurring in galaxies today.

Galaxies need to produce and contain stars for them to be seen, and this process does not happen quickly in the early universe. Observations show that this star formation spans the history of the universe, and is in fact still occurring in many galaxies even today. The star formation history is such that it ramps up from the start of the universe, and peaks at around 2-3 billion years after the Big Bang, after which it gradually declines. Today the star formation rate is roughly a tenth of what it was at its peak. Understanding why the star formation rate is so high in the early universe is an important question. So is why it later declines.

For example, our own Milky Way produces on average about a single solar mass unit of stellar mass on average per year, while these distant galaxies are often star forming at over 100 to one thousand times this rate. Integrated over a few billion years, this star formation adds a significant amount to the masses of these systems.

Another major clue, still not totally understood, is that galaxies in the distant universe are much smaller, by up to an order of a magnitude, and have stellar mass densities around 50 times higher, than galaxies of similar mass we see today. Furthermore on average, the masses of these early galaxies are also lower than galaxies of similar luminosity in today's universe. These are two observables that show how galaxies are evolving significantly. What is producing this evolution is debated still, but we have some good ideas.

It is easy to show that the observed merging between galaxies will allow the small, lower mass systems we see in the early universe to become in due course the more massive and larger systems that we see today, i.e., the bulk of galaxy formation for some galaxies is occurring through the merger process. This merger process is also potentially the trigger for the intense star formation seen in the early universe, as well as the way gas is driven into the centers of galaxies to drive the formation of the central black holes universally found in these systems (as discussed in more detail in the next chapter).

While mergers can account for the bulk of galaxy formation for some massive galaxies late in their history (as will be the case with the Milky Way and Andromeda, which as we noted above may merge to form a massive elliptical galaxy several billion years from now), there are galaxy formation modes that are just starting to be investigated. These include the accretion of gas from the intergalactic medium onto galaxies, which is later converted into stars, as well as minor mergers (involving dwarf systems like those drawn like moths to a lamp around the Milky Way and Andromeda in the Local Group), a process which can also build up the masses of galaxies. Understanding the relative role of these processes, as well as observing the very first galaxies, will be the focus of galaxy studies over the next ten years or so, using new telescopes and instrumentation, many of which are just now in the planning stages.

Fig. 13.21. Close up view of one of the highest redshift galaxies known as of late 2013. This image shows a field from the Hubble Space Telescope CANDELS survey, the largest Hubble program ever carried out. The close up view shows the location and zoomed in structure of this system, which is at redshift $z = 7.51$ or 700 million years after the Big-Bang. *Credit: V. Tilvi, S.L. Finkelstein, C. Papovich, CANDELS Team and Hubble Space Telescope/NASA*

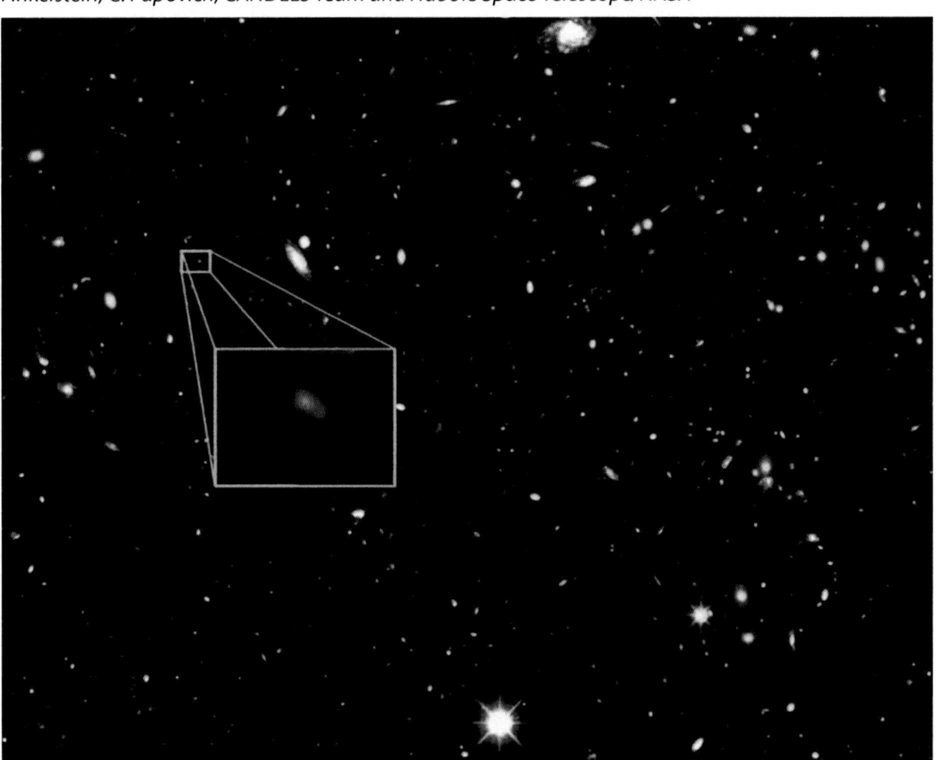

One particularly important project that is addressing detailed issues in galaxy formation, gas accretion and galaxy merging is the CANDELS survey, among others. This is a survey being carried out with the new near-infrared camera on the Hubble Space Telescope, known as the Wide-Field Camera-3 (WFC3) and installed by Space Shuttle astronauts in 2009. This camera is exquisitely sensitive in the infrared, and has been used to obtain two Ultra Deep Fields—the first in 2009, the second in 2012—which show galaxies so remote, and so redshifted by cosmic expansion, that visible and ultraviolet light from them has ended up in the infrared. These galaxies appear as they were only 560 million to 780 million years after the Big Bang, and yet (according to the effect Tinsley described) they are redder than expected and so are already somewhat evolved, showing that galaxy formation must have occurred extremely early. They consist of stars that are already at least 100 million to 200 million years old, which means that the galaxies imaged by Hubble had already been in existence for at least that long. We are getting close to looking so deep in the universe that we are seeing the very first galaxies—approaching the horizon that Milton Humason looked for in vain.

We will not reach them with the Hubble Space Telescope. However, Hubble is due to be superseded by the James Webb Space Telescope. With a 6.5-meter (21 foot) diameter mirror, it will represent a huge step up from Hubble. It is scheduled for launch in 2018. Though the WFC3 results of Hubble have given us a few whiffs of insight into what the first galaxies may have been like, the James Webb Space Telescope will allow us to push back into the period "when Beldam Nature in her cradle was," the period from 200 million years to 500 million years, to see how the earliest ancestral forms of galaxies—including our own Milky Way—formed stars and assembled into the prototypes of the majestic systems we see around us in the universe today. That will allow us to be practically present at, if not the first day of the Creation, the second or third.

* * *

1. J.H. Oort, quoted in Giuseppe Bertin, *Dynamics of Galaxies* (Cambridge: Cambridge University Press, 2002), 26.
2. For an excellent text on dust, at a high technical level, see: D.C.B. Whittet, *Dust in the Galactic Environment*. Bristol and Philadelphia: Institute of Physics Publishing, 2nd ed., 2003.
3. D.L.Block and R.J. Wainscoat, "Morphological differences between optical and infrared images of the spiral galaxy NGC 309," *Nature, 353* (Sept. 5, 1991), 48-50.
4. Theodore Andrea Cook, *The Curves of Life* (London: Constable and Co., 1914), viii.
5. Curtis Struck, *Galaxy Collisions: forging new worlds from cosmic crashes* (New York: Springer, 2011), 9.
6. O.J. Eggen, D. Lynden-Bell, A.R. Sandage, "Evidence from the motions of old stars that the Galaxy collapsed," *Astrophysical Journal,* 136 (1962), 748-766.
7. A classic paper on the subject is: A. and J. Toomre, "Galactic bridges and tails," *Astrophysical Journal,* 178 (1972), 623-666.
8. Shapley, *Galaxies,* 172-173.

14. Over to the Dark Side: Dark Matter, Black Holes, and the Origin of the Universe

Now is the time, the walrus said, to talk of many things...

—Lewis Carroll, "The Walrus and the Carpenter"

In the previous chapter we outlined the story of how galaxies have likely assembled into the forms we see in the modern universe. This formation is largely based on observational properties of these systems that allow us to construct how these galaxies were put together.

However, astronomers are interested in much more than the way galaxies assemble and form. The problem of galaxy assembly ultimately ties in with some very deep physics, including the existence and type of dark matter contained in these systems, as well as to the existence of black holes. The last few decades have seen an explosion in our understanding of the internal physics of galaxies as well as learning how they form.

Galaxies it turns out are not simple systems in the way that, say, stars are. The formation histories of galaxies covers over 13 billion years, and involves around at least a dozen different processes that happen at various times over the course of the history of a galaxy, and many of these are still going on today. In some ways it makes as much sense to ask when and how a galaxy forms as it is to ask when and how large cities such as London or New York assembled.

However, there are major clues to how the process has occurred, and we can identify some of the major aspects, detailed in the last chapter. However it is possible, if not likely, that there are still galaxy formation processes we have not yet identified.

One of these is that galaxies do not form in a random way, but follow the structure of the universe, and their assembly is largely determined in a bulk way on the properties of the universe. One of the major ways this is done is through the dark matter, which provides the structure or scaffolding in which galaxy assembly takes place.

In fact, it has become clear in the last few decades that the vast majority, up to 90% or more, of the mass in a galaxy is in a dark form, whose nature is unknown, but whose gravitational force can profoundly influence the way that galaxy formation has occurred.

Dark Matter Takes the Stage

Hints for the existence of dark matter in galaxies have been around almost as long as we have known about galaxies. Historically, the first clue that there was something else within the universe beyond the stars, dust, and light that can be detected directly originated from measurements of the motions of, and within, galaxies.

Shots in the Dark

The first time astronomers found something amiss in the observations of galaxies that suggested there was a dark component to galaxies was in 1933. In that year, the astronomer Fritz Zwicky, in one of his first observational astronomy papers, took deep spectra of galaxies in the Coma cluster to measure the motions of individual galaxies.

Zwicky came relatively late to astronomy and had started his research career as a physicist. Born in Bulgaria in 1898 to a Swiss father, he attended the Eidgenossiche Technische Hochschule (ETH) in Zurich, the same institution where Einstein was a student two decades earlier, completing his diploma thesis (first degree) under mathematician Herman Weyl and his Ph.D. dissertation in the theory of crystals under future Nobel laureate Peter Debye. In 1925 he moved to Caltech at the invitation of its president Robert Millikan (earlier, Millikan had been one of the young Edwin Hubble's teachers at the University of Chicago, and had won the Nobel prize in physics two years earlier for his measurement of the charge on the electron by studying the behavior of oil droplets in electrical fields). Though Zwicky spent nearly his entire career in the United States, he remained a Swiss citizen, once remarking that "a naturalized citizen is always a second-class citizen."

Though Millikan had expected Zwicky to work on the quantum theory of solids and liquids, which was closer to his dissertation topic, Zwicky already showed his independent streak, and though he did some work along the lines expected, his interests gradually turned to astrophysics. At first he was interested in trying to understand the origin of cosmic rays—high-energy protons and atomic nuclei coming from outside the Solar System. Then, after Hubble published his 1929 paper on the velocity-distance relation, he proposed the first of the many unorthodox ideas for which he would become famous—he suggested that the redshifts of distant galaxies might be explained by "tired light," that is, the photons simply became less energetic after covering great distances. This was an idea that could not finally be ruled out for sixty years, when it was finally shown that the light curves of Type Ia supernovae are all broadened (time-dilated) as expected according to the predictions of the expanding universe.[1] Though Zwicky himself remained suspicious of large redshifts throughout his career, much of his own work tacitly accepted that the standard redshift-distance relation was cosmological.

In 1933—the fateful year in which Hitler came to power in Germany—Zwicky married Dorothy Vernon Gates, the daughter of a California state senator and successful businessman and railroad man. Though the marriage later ended in divorce, Gates's money helped fund the first telescope at Palomar, the 18-inch Schmidt, in 1936. Meanwhile, Zwicky was now becoming active in observational astronomy. He joined forces with Walter Baade, the German national who had joined the Mt. Wilson staff in 1931. Baade was the meticulous observer, Zwicky the ideas man. In 1933-34, they studied the subset of fourteen or so exceptionally bright novae that had been observed up to that time (going back to S Andromedae in 1885) and concluded, as had Knut Lundmark a few years earlier, that these were no ordinary novae; they were far more energetic events, involving the annihilation of an appreciable fraction of a star's mass, and Zwicky even realized—correctly, as it turned out—that they might be a fruitful source of at least some of the cosmic rays. In their paper summarizing their results, Zwicky and Baade introduced the term "supernova" for these extraordinary celestial conflagrations, and as an afterthought suggested that their energy source might be the collapse of massive stars into neutron stars, cores consisting entirely of neutrons.[2] (Discovered only in 1932, neutrons are neutrally charged particles of about the same mass as the proton found in the nuclei of atoms.) Like many of Zwicky's ideas, neutron stars long remained a theoretical construct; their actual existence was confirmed only with the discovery of pulsars in 1968. The most celebrated case is the pulsar in the Crab Nebula, a neutron star remnant left over from the supernova event observed by Chinese astronomers in 1054. Strange as neutron stars are, even stranger are black holes, produced in the collapse of more massive stars still. We shall discuss these objects a bit later.

More important for our story, Zwicky also tackled the clusters of galaxies. With the Mt. Wilson 100-inch reflector, he measured Doppler redshifts for galaxies in the Coma Cluster, and found that there was a large spread (dispersion) in their velocities. The average velocity dispersion was 1000 km/sec (very close to the modern value), and Zwicky used this result to effectively "weigh" the Coma cluster, giving a kinematic measure of the

Fig. 14.1. Gravitational lensing: light from galaxies in the background is being bent by dark matter associated with the giant elliptical galaxy in the center of the field. *Hubble Heritage Team (STScI/AURA).*

cluster's mass. The calculation is quite straightforward and involves basic Newtonian physics of the same kind used to determine the mass of the Sun by observing how fast the Earth and other planets revolve around it. However, when Zwicky performed this calculation, it didn't come out as expected: the amount of mass needed to hold the Coma Cluster galaxies together, he found, was 400 times more than expected from the stellar mass estimated from the amount of light visible from galaxies in the cluster. It seemed a strikingly anomalous result for the early 1930s, but it has stood the test of time. Revised estimates of the distance to the Coma cluster obtained since then have reduced this to a factor of 50, but the problem Zwicky identified has not gone away– only 2% of the Coma Cluster is made of stars we can see.

In a paper published in a Swiss journal (and hence not as widely read as it might have been), Zwicky tried to explain the discrepancy by invoking what he called dunkle materie (German for "dark matter").[3] He did not know what this dark matter might be, but suggested it might consist of some combination of small faint galaxies and diffuse gas. In any case it meant that the average mass of a galaxy must be much greater than was generally believed at the time. Zwicky followed this insight up in 1937 by proposing that galaxies might even be massive enough to give rise to gravitational lensing—an effect Einstein had predicted, though he knew that it would be very difficult to actually observe. Indeed, a case of gravitational lensing was not finally observed until 1975.

As productive as Zwicky was, he was not as widely recognized or honored during his lifetime (he died in 1974) as he perhaps deserved. In part, this was because he was always so far ahead of his time—it often took decades for his contemporaries to catch up with his ideas. But also, like many men of his type, he had a violent temper—he once called Baade a "Nazi" to his face, and routinely referred to colleagues who crossed him as "spherical bastards" (in other words, they remained bastards from whatever angle they were viewed). His colleagues, for their part, were tempted to dismiss him as something of a crackpot. In later years, after he had moved on to Palomar, he routinely groused that he was being denied observing time proportional to his merits, and once, notoriously, proposed a scheme that literally involved firing shots in the dark at Palomar (somehow Zwicky had worked out that firing artillery shells into the air in front of the shutter of the 200-inch dome would improve the seeing!).

In retrospect, Zwicky seems one of the great astronomers of the 20th century, who made many contributions to the field but whose greatest achievement remains his 1930s era recognition of the existence of dark matter.

More evidence

Zwicky's anomalous result did not stand quite alone. As early as 1936, Sinclair Smith, also at Mt. Wilson—an instrument-man whom Virginia Trimble has called a "widgeteer par excellence," but is little remembered today partly because he died so young, of cancer, at age 39—used a fast spectrograph of his own design on the 60-inch reflector, which allowed him to record spectra of galaxies even fainter than those being obtained at the time by Milton Humason with the 100-inch. With this set-up, he analyzed the spectra of a number of Virgo members relative to the cluster's center and found the same large mass-to-light ratio that Zwicky had found for the Coma cluster. In order to account for the enormous amount of missing mass, Smith concluded that there must be either large quantities of "inter-nebular" material or enormous faint extensions of the visible galaxies.[4] Around the same time Horace Babcock, then a graduate student at Berkeley, whose thesis study at Lick Observatory involved obtaining large-slit spectra to investigate the trailing nature of the spiral-arm pattern of M31, found a similar result for the HII regions in a single galaxy. His thesis was published as a *Lick Observatory Bulletin,* and met with distrust—so much so that Babcock never worked again in extragalactic astronomy. He turned instead to solar astronomy, and succeeded in discovering the general magnetic field of the Sun (Hale's long-sought holy grail). His contribution to the history of dark matter is only a footnote to a long and distinguished career.

Shedding Light on Dark Matter

The observations of Zwicky, Smith, and Babcock were all decades ahead of their time. Most astronomers ignored them, and continued to act as if the only matter in the universe was the ordinary, baryonic (atomic) type that can be directly observed from electromagnetic radiation they emit and which makes up the gas in the diffuse nebulae and the substance of solid bodies such as stars, planets, and comets. As a case in point: one of us remembers that the advanced astrophysics curriculum at the university he attended in the early 1970s had only two offerings, "Astrophysics of diffuse matter" and "Astrophysics of condensed matter." At the time, that seemed to cover the waterfront. How far away that era seems now! We've become inured to the existence of dark matter now, but in the 1970s, when evidence for it finally became too overwhelming to ignore, it seemed a shock—as with the 1930s discovery in cosmic rays of particles (muons) with masses between electrons and protons and neutrons that seemed to have no role at all in nuclear reactions, many an astronomer and astrophysicist must have seconded the spirit of I.I. Rabi's famous remark, "Who ordered that?"

When it finally came, the acceptance of another kind of matter that, unlike ordinary baryonic matter, was not directly visible, was largely owing to measures of rotation curves of galaxies similar to that Babcock had found for M31. The astronomer most closely associated with this breakthrough was Vera Rubin. After receiving a BA degree at Vassar, an all-women's college in upstate New York, Rubin attempted to enroll at Princeton, but never even received the graduate catalog. Women were simply not allowed in Princeton's graduate program in astronomy until 1975. Instead Rubin went to Cornell, where she received a Master's degree in physics—among the outstanding teachers she had there were Hans Bethe and Richard Feynman. She obtained her doctorate at Georgetown, where her thesis adviser was George Gamow, a professor at George Washington University nearby. It had been Gamow who, among other things, had helped put Georges Lemaître's speculative 1920s "primordial atom" theory of the origin of the universe—now referred to as the Big Bang theory—on a quantitative basis by showing that it could explain the observed cosmic abundances of hydrogen and helium. In Rubin's thesis, completed in 1954, she proposed that galaxies clumped together rather than being distributed randomly through the universe, an idea that was generally pooh-poohed by others in the field. Being a woman astronomer at the time was difficult—she later reflected that getting ahead in astronomy "required more luck and perseverance for a woman than a man."[5] She bided her time. For a while she taught at Montgomery County Junior College, then worked at Georgetown as a research assistant. By 1962, she was an assistant professor there, teaching a course in statistical astronomy to six students, and already deeply interested in understanding the motions of stars in galaxies and the motions of galaxies in the universe. She proposed to her students that she and they try to obtain a rotation curve for stars distant from the center of the Milky Way. The paper—submitted to the *Astronomical Journal*—was accepted, then withdrawn when the editor refused to publish the names of the students that had worked on it with her, at which point the editor relented and accepted it again. It contained the result that for a radius from the center of the Milky Way greater than 8.5 kiloparsecs, "the stellar curve is flat, and does not decrease as is expected for Keplerian orbits." This was, in fact, Rubin's first flat rotation curve. However, the result was not widely accepted. "Following its publication," Rubin recalled, "the many comments I received were negative and some very unpleasant; it couldn't be correct, or the data were not good enough."[6]

She turned to observational astronomy, and tried to study the rotation properties of the Milky Way by obtaining radial velocities of stars in the anticenter of the Galaxy. Her early work was done at Kitt Peak National Observatory in 1963, and two years later, at the beginning of 1965, realizing that observing was her priority, she gave up teaching and became an astronomer in the Department of Terrestrial Magnetism at the Carnegie Institution of Washington. There she began to collaborate with another staff member at Carnegie, W. Kent Ford, Jr., who had developed an image-tube spectrograph that decreased observing time by a factor of ten compared to the conventional photographic plate. That summer Rubin and Ford tested the device on galaxies with the Lowell Observatory's 69-inch (soon to be 72-inch) Perkins reflector at its remote site Anderson Mesa, and by the end of the year—through the intervention of Allan Sandage—Rubin was invited to observe with the telescopes at Palomar. (Until then, continuing a tradition dating back to Hale and the "monastery" at Mt. Wilson, women had never been "allowed" to use the telescopes there. She was the first.).

When Rubin, Ford and their colleagues set out to determine the rotation curves for galaxies, the expectation was that they would find very simple dynamics—basically, material just going around in circles, as with the planets going around the Sun. In the case of the Solar System, since 99.99% of the mass is in the Sun, the velocities of bodies in orbit around it

drop off according to a very simple relation (from Kepler's third law, the velocities fall off proportional to $\sqrt{(1/r)}$; thus Mercury's velocity in its orbit is 47.88 km/sec, and Pluto's 4.74 km/sec). The material in the disk of a galaxy like the Milky Way is also undergoing differential rotation—in other words, the speed is a function of distance. In the case of a star like the Sun, whose orbit is far from the galactic center and which lies outside the bulk of the mass of the Milky Way, it will even be roughly Keplerian (similar to the case of that of the planets circling the Sun). The upshot is that as one gets farther and farther out in the disk, the velocities should fall off, and very far out the rate of drop off should be proportional to $\sqrt{(1/r)}$.

That, at any rate, was the expectation, and it could hardly have been any clearer. But as Rubin had shown in 1962 for the Milky Way and now, using Ford's image-tube spectrograph to determine rotation curves for nearby galaxies, she and her colleagues showed that this was not the case at all. The rotation curves did not decline at large distances from the centers of galaxies, but either remained at the same maximum velocity or continued to rise out to larger and larger radii.[7] This result, first established by Babcock for M31, was now found to be a general result, holding for all spirals. In other words, far from the center, the galaxies were rotating much too fast.

Moreover, this rise in the rotation curves occurred well beyond the light-emitting part of these galaxies. The only explanation for this finding—unless one wanted to throw out the Newtonian law of gravity altogether, which was unthinkable—was that the gravitational field of the stars was too small by a factor of 10 or so to explain how fast the galaxies were rotating. But that meant that there must be a lot of material out there (what kind of material one couldn't say). This "dark matter" not be detected directly, but was considerably more preponderant than ordinary dark matter, and furthermore, if there were some kind of telescope that allowed imaging of mass without imaging light, the galaxies would look completely different than the pictures of them we were used to. There was the disk of stars, which is on the order of 20 parsecs across, and then far outside that would be this enormous thing extending out to maybe 150 parsecs.

The preponderance of dark matter in galaxies—and indeed, in the universe—has been confirmed in measure after measure ever since. According to the most recent measures of the Cosmic Microwave Background radiation by the Planck satellite, ordinary baryonic (atomic) matter accounts for only 4.9% of the mass-energy of the universe. Dark matter accounts for 26.8%, with the rest, 68.3%, being in the form of the even more mysterious dark energy (described later). We can now speak of the everyday, ordinary (baryonic, i.e., atomic) universe as the 4—or 5% universe.[8] Clearly, in the jargon of accountants, early astronomers and astrophysicists had been guilty, in assuming that what they saw was all that was, of what might be called a "significant rounding error."

The Preponderant Form of Matter in the Universe

Thanks to Rubin and her collaborators, by the 1980s, there was no longer any doubt as to the existence of dark matter, even though astronomers could only speculate about what it might be. The observers had done their part, splendidly. Now it was time for the theorists set to work.

They were soon able to show that large dark matter "haloes" – big blobs of dark matter in which galaxies were embedded—were necessary to keep the structures of many spiral galaxies stable. Moreover, the recognition of the existence of dark matter had many implications beyond the simple idea that the light we see from galaxies is just a small part of the mass in these systems. Indeed, not only is it now clear that it is the dark matter that determines when and how galaxies are formed as well as their distribution across the universe, it even seems to be the case that without dark matter, the earliest galaxies, whatever they were, could never have formed at all, or would have formed in a much different way

Even without knowing the identity of the dark matter particle, astrophysicists can ascertain some of its features. It does not emit or absorb electromagnetic radiation to any significant degree (hence it is "dark"). It is also "collisionless," in the physicist's jargonish sense; the interaction cross-section of particles is so low that collisions between dark-matter particles have no significant effect on the system as a whole (in this sense, stars in a galaxy can also be described as collisionless). But the most important feature concerned the temperature of dark matter. Is it "hot," or is it "cold"? This question has a bearing on the distribution of galaxies across the universe, which astronomers refer to as its "large-scale structure."

The Large-Scale Structure of the Universe

In his longest but perhaps least influential paper, Hubble wrote in 1934, on the basis of deep studies with the 100-inch over 1000 small fields thinly distributed over the sky north of declination -30°, that the distribution of galaxies on very large scales is homogeneous, i.e., there was no indication of a super-system of the galaxies.[9] But though there was no such "super-system," it was clear that the galaxies associated in clusters and groups rather than existed as individual systems. This is nothing less than a tribute to the fact that gravitation is the prime actor in astronomy and cosmology, and acts and isolates clumps of matter on all scales. Hubble himself bore witness to this trend, and in the *Realm of the Nebulae* concluded that "the groups (such as the Local Group) are aggregations drawn from the general field, and are not additional colonies superposed on the field."[10] These groups and clusters ranged from small associations like the Local Group, with three galaxies large enough to be spirals, and a couple score dwarf systems, up to gigantic clusters containing thousands of large galaxies, such as the Virgo cluster.

That was as far as anyone could say in the 1930s. But since Hubble's survey—his nebula-gages, the counterpart of William Herschel's star-gages—sampled only small fields that were very widely distributed, he was not able to address the question of the large-scale structure and the distribution of the clusters themselves. The question remained: on the large scale, did they tend to form large associations—Superclusters—or were the clusters randomly distributed in the universe?

Other imaging surveys of distant galaxies, including an important one done by Lick Observatory showed through the next few decades that galaxies were indeed very clustered into the groups and clusters Hubble found. An important result towards answering this question in more detail came through the Center for Astrophysics (CfA) redshift survey, carried out at the Harvard-Smithsonian center for astrophysics led by Margaret Geller and John Huchra in the 1980s.[11] With data mostly coming from the CfA 1.5 meter telescope on Mt. Hopkins (near Amado, Arizona), Geller and Huchra and their collaborators surveyed the redshifts of several thousand galaxies in a thin slice of the sky. When they looked at this distribution of galaxies in redshift space (a parameter kindred to distance) relative to sky-location, they found that, instead of a random distribution of galaxies and galaxy clusters, as they expected to find, an intricate patterned structure emerged consisting of well-defined "walls" and "bridges" of galaxies,, nestled between large empty "voids" where few galaxies were found.

The discovery of this astonishing filamentary structure in turn inspired theorists to try to understand what produced it. It turned out that the structure they got was highly dependent on the type of dark matter they assumed in their models. At first, there was a strong suspicion that dark matter might be made up of massive neutrinos, which were referred to "hot dark matter" because these particles would be characterized by very large (relativistic) velocities. However, by the mid-1980s it was clear that hot dark matter produced a universe that looked very different from what the CfA data suggested. The dark-matter particle could not be a heavy neutrino, even if a neutrino with the desired mass could be found (which proved not to be the case). Whatever it was, the dark-matter particle had to be cold—i.e., a particle more massive than 1 keV. Thus the concept of Cold Dark Matter was born, which remains the dominant idea in cosmology today. Though the specific identity of the dark-matter particle remains unknown, the best candidates were, until recently, the theoretical particles known as WIMPs (an acronym for *weakly interacting massive particles*). Unfortunately, WIMPS are extremely difficult to detect in the laboratory, because they rarely interact with ordinary matter, and in their search for them physicists have built giant underground experiments, in which detectors are shielded by soil and rock from other particles. Such detectors have been located in the Tower Soudan mine in Northern Minnesota, USA and under a mountain located between the towns of L'Aquila and Teramo, Italy. A few potential WIMP detections were reported from those experiments, but they seem to have been false positives; the most sensitive experiment to date—the Large Underground Xenon detector, or LUX, in which the detector is located in a former gold mine 1.5 kilometers beneath Lead, South Dakota—has turned up empty-handed. The nature of dark matter remains a mystery. Fortunately, for our present purposes, we are interested only in how dark matter—whatever it may turn out to be—affects galaxy formation.

Simply put, if the dark matter particle is moving quickly, near the speed of light, as neutrinos do, then the first structures in the universe would have been relatively larger than any of the superclusters found today. These large structures would then condense and fragment to form smaller systems as the universe aged. This would imply that structures such as galaxy clusters would form first, then the most massive galaxies would form next, and finally the smallest low mass galaxies would form last. It's a top-down approach similar to that envisaged in the Eggen-Lynden Bell-Sandage scheme described in the previous chapter.

If, on the other hand, the dark matter particle—WIMP, or whatever it turns out to be—is cold, meaning it is moving slowly with respect to the speed of light, then the first structures to form will be lower mass. The evolution of the universe will then proceed in such a way that large structures are built up from the merging of these smaller ones. That is, the universe's history will be one of building up larger and larger galaxies and galaxy clusters through the assembly of already existing systems. The first galaxies should therefore be mostly small, lower mass systems that merge and assemble into larger galaxies and clusters over time. Cold matter gives us a bottom-up approach.

Already during the early days of looking at galaxies at high redshifts with the Hubble Space Telescope, there was circumstantial evidence favoring this picture of 'bottom-up' formation. Not only were the earliest galaxies lower mass systems on average, but also the most massive structures in the universe—galaxy clusters—do not seem to exist then, although they are common today. The overall scheme has been supported by subsequent and more detailed modeling as well as direct observations in the Hubble Deep, Ultra Deep, and Extremely Deep Fields. But if the massive clusters we see today did not exist in the early universe—say, the universe in the first few billion years after the Big Bang—it follows that these clusters had to grow through time. All this goes to prove that the bottom-up formation of structure was the rule in the universe, and that the dark matter is Cold Dark Matter.

Building on these findings, theoretical astrophysicists have proceeded to create more and more models, based on different initial assumptions about the temperature of dark matter in the early universe and using basic equations of the physics of gases to produce star formation. With only these modest ingredients, a Cold Dark Matter universe can be modeled that will predict correctly or nearly so, properties of nearby galaxies such as their stellar masses and luminosity distributions, how fast these systems are rotating, how these galaxies are distributed, even their star formation rates. The success of these models gives us confidence that we are on the right track. The first of these models was by Martin Rees and Simon White from the University of Cambridge published in a paper in 1978 in the *Monthly Notices of the Royal Astronomical Society.* These types of models have been refined over the years and are now one of the dominant ways that galaxy evolution and formation is studied theoretically.

Dark Matter—to Black Holes

Admittedly, the models are still a bit wobbly when it comes to accurately predicting the observed changes in the galaxy population at high redshifts, probably because their assumptions are a bit too simplistic. Real galaxies consist of more than just dark matter and stars. They also contain gas and black holes, which can profoundly affect their formation. In particular we know that the formation of galaxies is driven in significant ways by black holes—the remnants of the collapse of even more massive stars than those that form neutron stars, and, in supersized form, the mysterious denizens lurking in the centers of large galaxies.

As exotic and even mildly frightening as they are, black holes are often the subject of science fiction, and have always stirred the imagination of the public. These objects are entrees into realms where the laws of physics are completely dominated by gravitation. They are created whenever a large amount of mass is compressed into an extremely small volume, and the only force able to do this is gravity.

Black holes are a consequence of Einstein's General Theory of Relativity, which was completed at the end of 1915. The first calculations related to these strange objects occurred within months of when Einstein's theory began circulating among European physicists, and were carried out by a German astrophysicist, Karl Schwarzschild. Solving Einstein's gravitational equations would be a tremendous achievement under any circumstances, but to do so under the conditions Schwarzschild was faced with—he was then serving with the German Army on the Eastern Front—seems almost incredible. Despite being comfortably ensconced as director of the Potsdam Astrophysical Observatory, where among other things he had put the problem in Ejnar Hertzsprung's way that would lead to the HR-diagram, and being already in his forties, Schwarzschild saw it as his patriotic duty to volunteer for the German army when war broke out in August 1914, and spent the year 1915 as a technical artillery expert, first in Belgium and France (fast sliding into a black-hole like morass of the Western Front) before being transferred to Russia.

Somehow, Schwarzschild managed to remain in touch with scientific developments in Germany, and followed Einstein's work well enough to penetrate the complex mathematical formalism of Einstein's theory to achieve an exact solution to Einstein's equations—admittedly, it was for an idealized case, that is, for the gravitational field surrounding a static, spherical, symmetric mass. By the time he did so, he was already suffering with an incurable skin disease (pemphigus), which led to his being invalided out of the army, hospitalized, and completing his work literally on his deathbed. He sent this paper, with others, to Einstein, who read it at a meeting of the Berlin Academy of Sciences in early 1916, just weeks before he died, at the age of only 43.

Schwarzschild was the first person to calculate the formula for the radius of the "event horizon" around a black hole (though the term "black hole" was coined only in 1967, by Princeton astrophysicist John Archibald Wheeler). The event horizon or Schwarzschild horizon is the distance from the black hole's center from which light cannot escape. This distance—the Schwarzschild radius R—is given by $R = 2GM/c^2$, where G is the gravitational constant, M the mass of the black hole, and c the speed of light. Schwarzschild himself did not expect that any such object could actually exist in nature, and neither did Einstein, since there was no known object in the universe as small as the Schwarzschild radius. However, we now know that they not only exist but are common throughout the universe, and play a critical role in the formation and evolution of galaxies.

For our purposes a black hole is simply an object that is so dense that even light cannot escape from its gravitational hold, and thus is not directly detectable. In this respect, black holes are similar to dark matter in having a non-electromagnetic presence, but there the similarity ends. Unfortunately, we cannot follow the detailed physics of black holes here—given the attractiveness of the subject matter, there is an extensive literature for the reader who wishes to pursue the subject further—but must confine ourselves to the effects on the galaxies harboring them.[12] (We only briefly advert here to the fact that in general, in contrast to the idealized case Schwarzschild solved, real-world black holes are in rotation. The solution to this more general case was published only in 1963 by the New Zealand-born theoretical astrophysicist Roy Kerr. Since then, astronomers define black holes by the two characteristics of mass and spin. Astronomers have long known how to measure black holes' masses, by the gravitational effect on the orbits of nearby star; but measuring spin is more difficult, since no light escapes from the black hole's event horizons.) Though the black hole itself cannot be seen, its effects on the fabric of space-time around it—outside the Schwarzschild radius—are dramatic. Within the central 2-kiloparsecs or so of the black hole, large quantities of gas swirl inward toward the black hole, to form an accretion disk. Through violent collisions of its particles, this material becomes superheated to the point where it can emit high-energy electromagnetic radiation such as X-rays. It is here—in the accretion disk—that the powerful gravitational field of the black hole drives the highly energetic processes in galaxies astronomers have observed.

Black holes can exist in many different forms and have a variety of masses. The traditional theoretical way to form black holes is through the collapse of a very massive star. Just before World War II, J. Robert Oppenheimer and Harland Snyder calculated that if a massive star collapsed into a neutron star of above 3 solar masses, it would continue to collapse still further, forming a bizarre object so dense that it would become effectively a singularity—a point of infinite density—whose space time would remain forever disconnected from the outside world. The collapsed star case is in fact the explanation for very bright X-ray sources in our own galaxy, such as Cygnus X-1, often considered the first black hole to be identified and discovered only in 1964. These isolated black holes resulting from collapsed stars typically have masses a few times that of the Sun.

However, lurking in the centers of nearly all galaxies (including the Milky Way) are supermassive black holes with masses up to 10 million times that of the Sun. They are monsters, like the Cretan Minotaur lying deep within the entrails of the labyrinth of their home galaxies, or like a black widow waiting in the coils of its tangled, sticky web. Clearly these monster black holes could never form from the collapse of a single star. Even the most massive stars fall short by a factor of 100,000 times of having enough mass to produce such oversized systems

In fact, a big unanswered question is how these supermassive black holes form and furthermore why there are no "middle mass" black holes. All the black holes we know of are either a few times the mass of the Sun, like the collapsed stars (including Cygnus X-1), or a few tens of thousands to a few million times the mass of the Sun, like those found in the centers of galaxies. So far there appears to be nothing between these extremes. A possible clue is found in the fact that some of the supermassive black holes, which can be estimated from the X-rays emitted by their accretion disks, seem to be spinning at more than 90% of the speed of light or more. This suggests that they must have gained their mass through major galactic mergers.

Yet another curious fact is that the mass of black holes found at the centers of galaxies are tightly correlated with the properties of the host galaxy to an amazing degree—so much so that we can say that whatever process formed the galaxy also formed the black hole or vice-versa, and that the process is highly regulated. It turns out that the ratio of the central black hole's mass to the total galaxy mass is about 1 to 1000, and this relationship tends to be similar at all galaxy masses and is robust enough to make us suspect that it must hold at earlier times in a galaxy's lifetime as well.

This tight relationship implies that the black holes in the centers of galaxies grow along with the galaxy that hosts them. The relationship is so good that the correlation must be a law of nature, fundamental to how galaxies form and evolve, but it is still largely unexplained. Moreover, the black holes in the centers of galaxies are still growing, even in galaxies in the nearby universe.

Black Holes show their hand: Seyfert Galaxies, AGNs, and Quasars

The reader at this point may be wondering how we can identify these growing the black holes in the centers of galaxies at all. And how can we possibly know whether they are still growing or not? The answer is not immediately obvious. However, for decades there have been important clues right before astronomers' eyes, whose significance we have only recently come to recognize.

Starting in the 1940s, astronomers discovered that the centers of galaxies were very energetic. As in many cases in the history of astronomy—V.M. Slipher's discovery of the large redshifts of galaxies, for instance, or Zwicky's of dark matter—this discovery was serendipitous. The result was completely unexpected, and for a long time remained unexplained.

By taking spectra of nearby galaxies taken with the Mt. Wilson 100 inch, Carl Seyfert, then a National Research Council Fellow, later a professor of astronomy at Vanderbilt, discovered that the cores of many nearby galaxies have intense energy output and unusual emission lines. Seyfert outlined his discovery in 1943. It is the one thing Seyfert did for which he will always be remembered.

Emission lines had, of course, been observed in nebulae within the Milky Way going back to Huggins's pioneering study of the Cat's-eye nebula in Draco in the 1860s. They are produced wherever gas is being excited into higher ionization states. Most typically, they are associated with young, hot, massive stars in star-forming regions, like the O stars of the Trapezium that excite the gas in the Orion Nebula. Some examples of important emission lines are the H-alpha line [HII], which is given off when hydrogen ions recombine with electrons, H-beta [HIII], and [OII] and [NII] (two 'forbidden' transitions from Oxygen and Nitrogen which were interpreted, back in William Huggins's day, in terms of an as-yet unknown element he

Fig. 14.2. The nearest Seyfert galaxy (i.e., galaxy with emission lines superimposed on the stellar spectrum, denoting an active galactic nucleus), M77 in Cetus. *Hubble Heritage Team (STScI/AURA).*

called "nebulium," as described in chapter 6). Emission lines had even, on occasion, been observed in spectra of extragalactic systems. As we saw in chapter 9, Edward A. Fath discovered them in a galaxy, M77 (NGC 1068), in 1908, without knowing what to make of them. Indeed, M77 happens to be the nearest Seyfert galaxy to the Earth. But Seyfert was the first to study such galaxies systematically; he realized that in the very central parts of these emission-line galaxies the lines were not only blooming bright, but also their line ratios appeared to be different from what would be expected if they originated in star-forming regions (in other words, if they were simply HII regions like the Orion Nebula). Something else was going on.

Seyfert also noticed that the light from the very centers of these galaxies was sometimes, though not always, bright in photographic plates, suggesting some kind of violent activity was going on there. This observation led to the recognition that the nuclei of these galaxies are active, and "Active Galactic Nuclei" (AGNs) has been a major area of research in astronomy ever since.

Fig. 14.3. Active galaxy Centaurus A. See also Fig. 4.9. Located in the constellation Centaurus at a distance of 11 million light years, Centaurus A contains the closest active galactic nucleus to the Earth. The warped shape of Centaurus A's disk of gas and dust is evidence for a past collision and merger with another galaxy; the resulting shockwaves have triggered a firestorm of new star formation, in the HII regions (the pinkish patches shown in this close-up). *Courtesy: R. O'Connell and the WFC3 Scientific Oversight Committee. NASA, ESA and the Hubble Heritage Team (STScI/AURA).*

The Quasar connection

Then there were the curious objects known as quasars—"bright" radio sources originally known as quasi-stellar objects, because they appeared star-like on photographic plates. They were shown, by Maarten Schmidt using the 200-inch reflector at Palomar in 1963, to have highly redshifted spectra, so they were also, obviously, extremely energetic objects. Eventually they were also found to be associated with active galactic nuclei, and powered by supermassive central black holes, though it took longer since generally they are much more remote than AGNs. With the Hubble Space Telescope it has at last become possible to visualize their host galaxies, and it has been found that, in addition to being more distant, quasars are generally much brighter, intrinsically, than the nearby AGNs. However, for the most part, they seem to be fundamentally the same kind of objects, with intensity being the only major difference

We now know that the light from the quasars and active galactic nuclei all originate from matter falling into black holes at the centers of galaxies. The way this works is that as gas falls toward the black holes, it becomes excited and emits light at various wavelengths. This produces flux that often dominates the galaxy, and is often variable on short time-scales. When we see this activity in the central parts of galaxies, it is an indication that the system at the center is accreting matter. This implies the black hole is of course growing in mass. At present, the supermassive black hole at the center of the Milky Way, Sagittarius A*, is quiescent, but we know that the Milky Way has been an AGN in the past. A lacy filament of mostly hydrogen gas, called the Magellanic Stream because it trails behind the Large and Small Magellanic Clouds, has recently been identified as a fossil relict of an eruption at the center of the Galaxy—known as a Seyfert flare—that took place some 2 million years ago.

Fig. 14.4. Active galaxy NGC 1275 in Perseus, located at a distance of 235 million light years. The traces of spiral structure, accompanied by dramatic dust lanes and bright blue areas of active star formation, belong to a spiral nearly edge-on in the foreground, which lies in front of a giant elliptical with peculiar faint spiral structure in its nucleus. These galaxies are involved in a collision, in which gas and dust in the central bright galaxy swirls into the center of the object. This galaxy emits a strong signal at both X-ray and radio frequencies which indicate the probable existence of a black hole. *M. Donahue and J. Trauger, NASA and the Hubble Heritage Team (STScI/AURA).*

One of the important questions we want to resolve is how black holes and galaxies can have such tightly coupled masses. What might be going on to produce this correlation?

One major idea involves feedback. Feedback is most familiar to people in the form of feedback in a musical concert or a microphone at a conference/speech, where sound from a speaker goes back into a microphone producing a unique (and usually unappealing) noise effect. (That is known as positive feedback; negative feedback can also occur—and is familiar in awkward social situations.)

Some kind of energy feedback may be occurring in galaxies. The radiation (including light) emitted from the black holes at the center of galaxies is often very intense and energetic. However, even simple calculations show that the amount of energy being emitted from these active nuclei is on the order of several times the so-called binding energy of the galaxy. This means that the energy from just the central black hole in the galaxy is enough to remove all the stars from its potential well several times over.

Possibly what is happening is that the energy from these black holes heats up and removes gas belonging to their respective host galaxies. For a galaxy to continue to grow it must form new stars. For stars to form there must be cold gas in the galaxy. By cold gas we mean gas which is not moving around much and thus can collapse into dense molecular clouds, like those that form the dark silhouettes on Barnard's Milky Way photographs, and these dense molecular clouds can then collapse still further and form into stars. A galaxy which is full of hot gas cannot form stars until that gas cools, and if the gas is kept hot enough, close to 10^6 K, then the cooling time-scale of the gas will sometimes be much longer than the age of the universe. The result: no new stars.

In active nuclei, the black holes can in principle cause the gas in the galaxy to become so heated that it cannot cool and form new stars. This essentially ends new star formation and keeps the galaxy from developing further. The galaxy may, of course, be rejuvenated through interactions or mergers with other galaxies, in which large quantities of gas may be channeled inwards and, under the right circumstances, trigger massive bursts of star formation, known as starbursts. This has recently been observed in the galaxy NGC 253 in Sculptor.[14] Meanwhile, material that is ejected may not be driven all the way from a galaxy's reach. Instead it may remain in the halo, cool, and then rain down at a later time to trigger a fresh starburst, as seems the case in M82, the "starburst galaxy" that is close companion to the majestic "grand spiral" M81 in Ursa Major. How these processes may square with the problem of coupling of the black hole with the galaxy's growth and development remains, however, far from clear.

Though the details of what is going on remain sketchy, there remain many good reasons for believing that black holes and galaxies form together. One is that the activity seen in black holes, through quasars and active galactic nuclei, peak at largely the same time as does the star formation history of galaxies. That is, the amount of star formation occurring in galaxies in the early universe reaches its higher rate of activity at the same time that the activity in these active nuclei does. This may mean that the two processes are related; it may also just be telling us that, whatever it is, the same overall process drives the formation of both the stars and the active nuclei in these galaxies.

Another—and perhaps the most tantalizing clue—is that the peak epoch of star formation and active nuclei occurs at the same time that galaxies are forming rapidly through the mergers of galaxies, as described in chapter 13. This epoch, roughly 2 billion years after the Big Bang, is clearly the critical time in galaxy formation, yet at the present time we do not understand any of the details. It is possible we have not even begun to ask the right questions. It is hard to escape the sneaking suspicion that, as so often in the history of science, when we finally do start asking the right

questions, the answer nature gives us may lead us on to something new and strange and unexpected. In any event, understanding the underlying causes of all this formation activity in the early universe will be enough to occupy astronomers happily for decades to come.

The Cosmic Microwave Background radiation

As the velocity with which galaxies are receding increases with distance, because of the expansion of the universe, it follows that as we rewind the tape back toward the early universe, the galaxies become more and more crowded together (by crowded together, we mean simply the distance between any pair of typical galaxies was correspondingly less than it is now). This, by the way, supplies the obvious explanation for why collisions and mergers were so common then. And if we go back all the way (how far back we are told approximately by the inverse of the Hubble constant, which gives us the expansion age of the universe, now estimated at 13.8 billion years), we come to the very origin of the graph, the point at which all the matter and energy of the whole universe would have been compressed into no space at all.

As with Schwarzschild's solution to Einstein's gravitational equations for a mass point, which led us to the concept of the black hole, we find ourselves confronted here with one of those singularly awkward mathematical things called a singularity—in this case, we refer to it as the initial singularity. It is awkward because here everything—mass-energy density, pressure, temperature, etc.—becomes infinite. In fact, Einstein's gravitational equations lead to the same kind of result for the universe itself as they do for the case of the black hole; they actually *do* say that you can go back all the way to a time when all the mass-energy in the universe emerged from a point of zero size. That's obviously very awkward—in general, mathematicians dislike such things as infinities and divisions by zero. We can only say that physicists now insist that time and space no longer have meaning for times less than 10^{-43} sec from time zero (the Planck time, the smallest unit into which we can subdivide our concept of time), and Einstein's gravitational equations no longer apply.

The pioneering Belgian cosmologist Georges Lemaître, who independently of (and a little later than) Alexander Friedmann introduced the first realistic (non-static) cosmological models based on Einstein's equations in the 1920s, avoided the problem by pushing the solutions of the relativistic equations not to the singularity itself but to a later time when the entire contents of the universe were pictured as squeezed into a sphere some 30 times the diameter of the Sun, which he called the "primeval atom." (Obviously, it couldn't have been a sphere in the usual sense, since

Fig. 14.5. The 15-meter horn antenna at the Bell Telephone Laboratories near Holmdel, New Jersey, used by Robert Wilson and Arno Penzias (looking up at the antenna) to discover the Cosmic Microwave Background Radiation. *From: Wikipedia Commons.*

there would have been no container "outside" it.) Inside the primeval atom, everything was incredibly hot and dense, and for unknown reasons the primeval atom exploded (again, keep in mind that this is only a figure of speech, since there was nothing for it to explode into), scattering the contents forcefully outward. Eventually these *disjecta membrae* formed the constituents of the universe such as stars and galaxies and the general expansion has been going on ever since.

This model was adopted in the late 1940s, by George Gamow and his colleagues Ralph Alpher and George Herman, who realized that it implied the universe would have started out incredibly dense and hot, and would have produced radiation that would still be around and nuclear reactions that might have produced all the elements in the periodic table. Gamow's idea was ridiculed on a B.B.C. radio program by the British astrophysicist Fred (later Sir Fred) Hoyle, who called it the "Big Bang" and regarded it as "about as elegant as a girl jumping out of a birthday cake." At the time Hoyle was advocating for what he thought was a much more elegant and dignified way the universe might have evolved, the "steady-state theory," but by the 1960s—partly as a result of some of his own calculations—he was forced to resign himself to the idea that the birth of the universe really was as inelegant as all that, and the term "Big Bang" has stuck. (There was a contest a few years ago run by one of the popular astronomy magazines in which the public was asked to propose a better term; in the end, there were no winners. No one could improve on "Big Bang.")

Gamow, Alpher, and Hermann could explain from Big Bang conditions the abundances of hydrogen and helium in the universe, but their theory for the creation of heavier elements broke down. It turns out that are no stable nuclei with a total of either five or eight neutrons and protons, so it is not possible to build heavier nuclei simply by adding protons and neutrons to helium nuclei, or combining helium nuclei together. In fact, this is one of the reasons that astronomers concluded that, though hydro-

gen and helium (and a trace of lithium), formed in the Big Bang, heavier nuclei formed in stars. The major breakthrough here was E. E. Salpeter's demonstration in 1952 of the triple-alpha process, which could occur in the dense helium-rich cores of stars. In the triple-alpha process collisions between two nuclei of helium formed an unstable and extremely short-lived nucleus of beryllium (Be^8), which because of the high density was able to survive long enough to collide with another helium nucleus forming a stable nucleus of carbon (C^{12}). A few years later, Geoffrey and Margaret Burbidge, Willy Fowler, and Hoyle showed how the heavy elements of the periodic table (those heavier than iron) could form in supernovae. Nevertheless, despite these successes, the stellar theory of nucleosynthesis also had shortcomings; most notably, it could not explain the large abundance of helium in the present universe (hydrogen, the primordial stuff of the universe, makes up about 73 percent, and helium 24 percent of the nuclear matter in the universe; all the other elements make up less than 2 percent). This much helium could not have been produced in the stars but could be explained if the universe passed through a hot dense Big Bang.

Apart from the helium abundance, the other important prediction of the Big Bang theory—but not the steady state—was that there ought to be background microwave radiation (later known as the Cosmic Microwave Background (CMB), left over from the time when the universe was very hot and dense.

A number of theoretical cosmologists, beginning with Alpher and Herman, considered the possibility, and according to Steven Weinberg, Alpher and Herman even made some inquiries about trying to observe this background radiation with radar experts at Johns Hopkins University, the Naval Research Laboratory, and the National Bureau of Standards, but were told that the 5 to 10K radiation they expected would be undetectable with techniques available at the time. In 1961, a Soviet scientist, Yakov Zel'dovich (coinventor of the Soviet hydrogen bomb with André Sakharov), also concluded that there should be a background radiation left over of a few degrees. However, after Zel'dovich read a Bell System *Technical Journal* report by E. A. Ohm, who with the supersensitive 6-meter horn antenna at the Bell Telephone Laboratories site at Crawford Hill in Holmdel Township, New Jersey, had reported a radiation temperature of less than 1 K, he gave up the idea.

By 1964, the Princeton physicist Robert Dicke and his colleagues David Todd Wilkinson and Peter Roll were finishing the construction of an antenna at Princeton that they planned to use to measure the CMB. Before they could get started, however, the CMB was discovered serendipitously. Two Bell Labs radio astronomers, Arno Penzias and Robert Wilson, using the same antenna Ohm had used, began in 1964 an effort to detect radio waves from high galactic latitudes (well above the plane of the galaxy).

Fig. 14.6. The Belgian priest and cosmologist Georges LeMaître. *Courtesy: Yerkes Observatory.*

Preliminary tests seemed to show more background noise than expected, but it could be due to excess noise in the amplifier circuits or a similar problem. All instrumental sources of noise had to be carefully eliminated. Nevertheless, even after taking all precautions (including using a "cold load," i.e., comparing the power coming from their antenna with that from an artificial source cooled with liquid helium), and removing some pigeons that had nested in the antenna's throat, they continued to register microwave noise at a wavelength of 7.35 centimeters. It was independent of the direction the antenna pointed, and—in contrast to what Jansky had noted in discovering radio waves from the center of the Milky Way—did not change with the time of day or the season. This meant that it could not be coming from the Milky Way; instead, it was coming from a larger volume of the universe, but its nature remained mysterious to the two men until 1965, when Penzias called another radio astronomer, who had recently heard a talk given by a young colleague of Dicke at Princeton, P.J.E. Peebles, in which a 10K microwave radiation from the Big Bang was predicted. (After receiving a call from Crawford Hill, Dicke is said to have quipped, "Boys, we've been scooped.") Soon afterward, there could be no doubt that the microwave radiation detected by Penzias and Wilson was really coming from the distant universe, though the temperature was more like 3K rather than 10K. Also, it was remarkably smooth and isotropic across the sky. It was this discovery that finally convinced even Fred Hoyle that the Big Bang, for all its inelegance, was right after all, and to abandon the steady state theory. Penzias and Wilson were awarded the Nobel Prize for physics in 1978 for their discovery. They were the first scientists to receive this recognition for a strictly observational discovery in astronomy (Hans Bethe's 1967 Nobel had been for his calculations of the thermonuclear reactions by which stars produce energy). The discovery of the cosmic microwave background radiation (sometimes poetically described as the "afterglow of Creation") was a decisive triumph for the Big Bang theory.

Subsequent research, notably that based on data from the brilliantly successful COBE satellite of 1990, whose team leaders George Smoot of Berkeley and John Mather of NASA's Goddard Space Flight Center, were also awarded the Nobel Prize in physics in 2006,[15] showed that, as predicted, the CMB represents an almost perfect blackbody spectrum. Its precise temperature has been determined: 2.7260 ± 0.0013K. The temperature will, of course, continue to decrease as the universe continues to expand. Once the motion of the Earth through the radiation is corrected for, the CMB proved to be completely anisotropic and remarkably smooth across the entire sky—the intensity of the variation is only about one part in 100,000—a finding that at the moment seems best explained by the "inflation" theory, proposed by Alan Guth, now at M.I.T., in 1981.[16] According to inflation, this remarkable smoothness is a result of a phase of exponential expansion the universe underwent very early in its history, causing it to double in size every 10^{-34} secs. The inflationary era was over by 10^{-32} secs after the Big Bang, by which time it had (as rather bravely described by John Gribbin) grown in size from smaller than a proton to the size of a grapefruit.[17] It obviously lies outside the scope of this book to describe inflation any further, but one intriguing idea to emerge from it is that in a modified form of the theory, called chaotic inflation (due to Andrei Linde, now at Stanford), seems to suggest that the universe might have grown out of a quantum fluctuation in some pre-existing region of space-time, in which it would appear that equivalent processes can create other regions of inflation in our own universe—in effect, budding new universes off our universe, and suggesting that our universe may have budded off from another universe, and so on without beginning or end.

The most important thing Penzias and Wilson's discovery in 1965 accomplished was, as Steven Weinberg points out, "to force us all to take seriously that there was an early universe."[18] It also, as Leon Lederman and David Schramm note, changed astronomy "from being a parasite to physics, using the physics in laboratories and applying it to stellar situations, to being able to make predictions about fundamental physics, telling physicists things as yet unmeasured in the laboratory."[19] In particular, it was appreciated that the kinds of collisions involving elementary particles produced in earthly accelerators (produced in small quantities and at great expense) occurred freely in the early universe in the instants after the Big Bang.

The rate of the expansion of the universe is a balance between the gravitational field, whose energy source is supplied by the photons, electrons, positrons, etc., and the outward momentum of the universe. Moreover, the energy source depends essentially on temperature alone. We can say specifically that the time it takes to cool from one temperature to

another is proportional to the difference of the inverse squares of the temperatures, so that "the cosmic temperature can be used as a sort of clock, cooling instead of ticking as the universe expands."[20]

From well-known theories of physics involving the known properties of matter and radiation in thermal equilibrium, we can calculate these temperatures, and work out, with a high degree of confidence, what the state of matter and radiation would have been at each stage. Thus, at 10^{-6} sec. after the Big Bang, the temperature of the universe had cooled to about 10^{13}K, allowing the quark soup of the earlier universe to condense into a hot plasma of photons, electrons, and baryons. The density of the whole universe was about that of the atomic nucleus today. Three minutes, two seconds after the Big Bang, the universe had cooled to 10^9K, and nuclear interactions were able to take place resulting in the synthesis of deuterium, helium 3, helium 4 and lithium. After this, the universe continued to expand, though nothing interesting happened for several hundred thousand years. Moreover, we cannot even see directly into this era, since electrons in the hot matter remained free, and continually scattered photons. The universe remained completely opaque until some 380,000 years after the Big Bang, when the temperature dropped to around 3000K, the temperature at which protons and electrons are able to combine to form neutral hydrogen. Henceforth, photons could travel unimpeded without interacting constantly with charged particles, and the thermal glow of the hot plasma of photons, nuclei, and electrons that existed just before neutral hydrogen formed could radiate throughout the universe without interference. What we observe as the CMB is the relict of the radiation from that time, but redshifted by a factor of 1100 as the universe expanded by a factor of 1100.

The decoupling of matter and radiation occurring with the formation of neutral hydrogen occurred rather suddenly. This was the decisive moment when, in Steven Weinberg's memorable phrase, "the contamination of nuclear particles and electrons [were able] to grow into the stars and rocks and living beings of the present universe."[21] However, much of this happened out of sight. The neutral atoms created a light-absorbing haze, leading the universe into a period, lasting several hundred million years, which are known as the cosmic "dark ages." We know that this cosmic haze was almost perfectly uniform, but not quite; there were small ripples in temperature, fluctuations of about one part in a hundred thousand, which were also mapped by the COBE satellite in 1990. These fluctuations were due to chaotic sound waves in the matter of the early universe. As the universe expanded, the lumps and ripples served as seeds for future structures. In due course gravity caused material—the one-sixth consisting of ordinary baryonic matter and also the fifth-sixths consisting of dark matter which does not emit or absorb light—to clump together. After several hundreds of

Fig. 14.7. Minute fluctuations in the nearly homogeneous Cosmic Microwave Background radiation, which will eventually grow (with the expansion of the universe) into large-scale structures in the universe, galaxies and clusters of galaxies. Image from the European Space Agency's Planck Space Telescope. *Courtesy: ESA.*

millions of years, the densest of the clumps formed galaxies, which without the dark stuff would have rapidly flown apart as the universe expanded. Indeed, as we have realized only recently, the dark matter dominates galaxies, and clusters of galaxies, so that, as Martin Rees puts it, "The beautiful pictures of discs or spirals portray what is essentially just 'luminous sediment' held in the gravitational clutch of vast swarms of invisible objects of quite unknown nature. Galaxies are ten times bigger and heavier than we used to like to think. The same argument applies, on a larger scale, to entire clusters of galaxies, each millions of light years across."[22] The massive stars formed in these early galaxies must have ionized the neutral intergalactic hydrogen gas that filled space in these early epochs, thereby gradually bringing the cosmic dark ages to an end as the gas became transparent again. (Confusingly, this era is referred to as the "re-ionization era"; obviously, the term is a poor one, since the gas had not previously been ionized.) The earliest stars, 100 times as massive as the Sun, end by destroying themselves in supernova explosions after only a couple million years. It might be supposed that these explosions would disrupt the star-forming process in the interstellar gas clouds from which the next generation of stars forms, but Hubble Ultra Deep Field (HUDF) images, which show the largest and brightest galaxies of their era, from 560 to 780 million years after the Big Bang, suggest that there was little or no time-lag between the death of the first generation of stars and the birth of the second. Apparently even galaxies from this early time—when the universe had reached only 5% of its current age—were chemically enriched with dust and metals (elements heavier than hydrogen and helium), which must have been produced by an earlier generation of stars.

There do not appear to be enough of the larger and brighter galaxies in the HUDF images to have completed the "reionization" process. This implies that there must have been many more small galaxies at the time, which completed most of the work. Though untangling what exactly went on will be the work of future large telescopes, it is clear, at any event, that the cosmic "dark ages" were finally over by about 1 billion years after the Big Bang. Thereafter, the universe entered upon the normal era, in which it has been mostly transparent to starlight—the era that continues up to the present-day.

Stars, galaxies, planets, perhaps on some of the planets life, belong to the story of that era. These are the things with which we are familiar, and view with awe—and pleasure—on dark starlit nights. But at their basis is the profoundly unfamiliar and the terrifyingly strange. At the end of his book *The First Three Minutes,* Steven Weinberg reflects:

> It is almost irresistible for humans to believe that we have some special relation to the universe, that human life is not just a more-or-less farcical outcome of a chain of accidents reaching back to the first three minutes.... It is very hard to realize that this all is just a tiny part of an overwhelmingly hostile universe. It is even harder to realize that this present universe has evolved from an unspeakably unfamiliar early condition, and faces a future extinction of endless cold or intolerable heat. The more the universe seems comprehensible, the more it also seems pointless.[23]

That is the hard sentence of modern cosmology.

* * *

1. G. Goldhaber, S. Deustua, S. Gabi, D. Groom, I Hook, A. Kim, M.Kim, J. Lee, R. Pain, C. Pennybacker, S. Perlmutter, et al. (The Supernova Cosmology Project: III), "Observation of Cosmological time dilation using Type IA supernovae as clocks,"; *arXiv:astro-ph/9602124v1* (1996).
2. Walter Baade and Fritz Zwicky, "On Supernovae," *Proceedings of the National Academy of Sciences,* 20 (1934), 254-259.
3. Fritz Zwicky, Die Rotbershiebung von extragalaktischen Nebeln. *Helvetica Physica Acta,* 6 (1933), 110-127.

4. Sinclair Smith, "The Mass of the Virgo Cluster," *Astrophysical Journal,* 83 (1936), 23-30.
5. Vera C. Rubin, "An Interesting Voyage," *Annual Review of Astronomy and Astrophysics,* 49 (2011), 1-28: Abstract
6. Ibid., 7.
7. Among the seminal papers establishing this result are: Vera C. Rubin and W. Kent Ford, Jr., "Rotation of the Andromeda Nebula from a spectroscopic survey of emission regions," *Astrophysical Journal,* 159 (1970), 379-402 and Vera C. Rubin, W. Kent Ford, Jr. and Norbert Thonnard, "Extended Rotation Curves of High-Luminosity Spiral Galaxies, IV; Systematic Dynamical Properties, SA→SC," *Astrophysical Journal,* 225 (1978), L107-L111.
8. See: Richard Panek, *The 4% Universe: dark matter, dark energy, and the race to discover the rest of reality.* Boston and New York: Houghton Mifflin Harcourt, 2011.
9. E.P. Hubble, "The Distribution of Extra-galactic Nebulae," *Astrophysical Journal,* 79 (1934), 8-76.
10. E.P. Hubble, *Realm of the Nebulae,* 82.
11. Margaret J. Geller and John P. Huchra, "Mapping the Universe," *Science, 246* (1989), 897-903.
12. A good non-technical overview is: Kip Thorne, *Black Holes and Time Warps: Einstein's outrageous legacy.* New York: W.W. Norton, 1994.
13. See: Fulvio Melia, *Cracking the Einstein Code: relativity and the birth of black hole physics.* Chicago: University of Chicago Press, 2009.
14. A.D. Bolatto, S.R. Warren, A.K. Leroy, F. Walter, S. Veilleux, E.C. Ostriker, J. Ott, M. Zwaan, D.B. Gisher, A. Weiss, E. Rosolowsky and J. Hodge, "Suppression of star formation in the galaxy NGC 253 by a starburst-driven molecular wind," *Nature, 499* (July 25, 2013), 450-453.
15. Popular accounts include George Smoot and Keay Davidson, *Wrinkles in Time: witness to the birth of the universe.* New York: Harper, 1993 and John C. Mather and John Boslough, *The Very First Light: the true inside story of the scientific journey back to the dawn of the universe.* New York: Perseus, 2nd ed., 2008.
16. A very accessible account is: Alan H. Guth, *The Inflationary Universe: the quest for a new theory of cosmic origins.* Reading, Massachusetts: Helix Books, 1997.
17. John Gribbon, *Companion to the Cosmos* (Boston: Little Brown and Company, 1996), 219.
18. Weinberg, *First Three Minutes,* 123.
19. Leon M. Lederman and David N. Schramm, *From Quarks to the Cosmos: tools of discovery* (New York: Scientific American Library, 1995), 153.
20. Ibid., 79.
21. Weinberg, *First Three Minutes,* 69.
22. Martin Rees, *Just Six Numbers: the deep forces that shape the universe* (New York: Basic Books, 2000), 74-75.
23. Weinberg, *The First Three Minutes,* 144.

15. Dark Energy

Come, an you get it, you shall get it by running.

—King Lear, Act IV, scene 6.

One of the noteworthy features of Hubble's law is that, with the exception of the static universe with which Einstein and de Sitter began in 1917, it is the only redshift distance law that is isotropic and homogeneous (i.e., looks the same) for every galaxy in the universe. It is this feature that gives rise to attempts to explain it by means of a balloon with marks on it being inflated, where, from the perspective of any mark, all the other marks appear to be receding. The marks on the balloon represent galaxies in the expanding universe.

Obviously, the real universe isn't being inflated, like the balloon, into a surrounding space, or being baked in an oven. Instead we are discussing expansion of the metric of space-time. In general terms, metric refers to the way that distance can be measured between two points in space. In Euclidean (plane) geometry, the distance is obtained simply by measuring a straight line between the two points. Since the 1920s, it has been clear that Einstein's field equations of general relativity imply a non-Euclidean geometry, in which space-time has a curvature that is described by the Friedmann-LeMaître-Robertson-Walker (FLRW) metric. This metric gives the distance between galaxies that are far separated from one another—cosmological expansion actually means expansion of over the scale factor and distance over time. (Parenthetically, we have to specify that the galaxies be far separated because on smaller scales, such as those that hold in clusters of galaxies, the gravitational attraction of the galaxies for one another is able to overcome the expansion forces; this is why things that are gravitationally bound together remain bound together and why even in the present universe, where things are much less crowded than they once were, galaxy collisions and mergers continue to take place.)

The universe's current expansion presumably started with the inflationary period described above—indeed, Alan Guth once wrote an article with the title "Was Cosmic Inflation the 'Bang" of the Big Bang."[1] But once the Big Bang gets underway, what happens as the universe evolves over time depends crucially on the density of matter energy in the universe. Several scenarios can be envisaged:

1. If the density of matter-energy in the universe ρ is greater than some critical value ρ_c, which turns out to be only about 5 atoms per cubic meter, the universe's current outward expansion will slow until it begins to contract and collapse back onto itself. This has sometimes been referred to as the "big crunch" prediction. It corresponds to the case where the universe is unable to escape its own gravity. This scenario was regarded as consistent with an oscillating universe scenario where, after each collapse, the density, pressure and temperature of the universe again would again reach the point where another Big Bang would occur, the universe would begin expanding again, and so on, forever and ever.
2. If the density of matter energy ρ is equal to or less than ρ_c, although the expansion of the universe is slowed by gravitation, there isn't enough matter in the universe to overcome the expansion. The universe thus escapes its own gravity, and the universe continues to expand forever.
3. In the case of ρ equal to ρ_c, the expansion continues forever but the speed of expansion slows to zero as time progresses). This is sometimes referred to as the "big chill" prediction, since as the expansion continues the constituents of the universe end in thermodynamic death and the universe, after a trillion years or so (the time that the lowest-mass stars can live before running out of fuel) becomes cold and dark. (Note that this happens if ρ is less than ρ_c too.)

What we are describing is obviously rather similar to the case of, say, a large asteroid or small planet shattered by an impact. If the impact is extremely violent, and the debris are scattered at high speeds, they will essentially go sailing off forever. But if the impact is less violent, and there is enough material, gravity may be able to hold them together, and the fragments may fall back together into a sphere. For the universe as a whole, we know the rate of expansion at present; what we need to determine, in order to know the universe's fate, is whether there is enough gravity to cause it to recollapse, or whether the material in it will just keep flying apart forever. Here the key parameter is Ω, the ratio of actual density to critical density (ratio of ρ to ρ_c). In general relativity, Ω is related to the curvature of space-time. The case where $\Omega > 1$ corresponds to positive curvature (e.g., a closed or spherical universe); the case where $\Omega < 1$ to negative curvature (an open or saddle-shaped universe); and $\Omega = 1$ is the knife-edge between the two, and corresponds to a flat universe.

Clearly, then, density is destiny.

The value of Ω at the present era is now thought to be 1, to within 1 percent. In other words, the difference between omega and 1 is currently less than 0.01. The tiny difference defines what is called the flatness problem, which was first pointed out by the Princeton physicist Robert Dicke—the same person who had been scooped by Penzias and Wilson in the experiment that led to the detection of the Cosmic Microwave Background radiation, and so had missed winning the Nobel Prize—in 1969. In 1978, he gave a series of lectures at Cornell in honor of Hans Bethe. In one of the lectures he discussed what values for mass-energy density at 1 second after the Big Bang would lead to a universe that in any way resembles our own universe. He deduced that whatever its value now, in the early universe it had to be exquisitely close to 1, because unless the energy of expansion and gravitational energy were in exact balance, any departure of Ω from 1 in the early universe, no matter how minuscule, would have been magnified during the billlions of years of expansion. (As we now know, the universe has expanded in volume by an order of 10^{60} from its volume in the Planck era, so the difference between omega and 1—now less than 0.01—would have had to have been less than 10^{-62} during the Planck era.) As discussed above, for $\rho > \rho_c$, the universe would have been so overdense it would have rapidly collapsed into an incredibly compressed state as in the Big Crunch scenario; for $\rho < \rho_c$, the universe would have been so underdense that it would have thinned out so fast that the galaxies would never have been able to form.

Dicke's argument was inescapable at the time, and is still more so today. It was this flatness problem that led Alan Guth, then a post-doc who was present at that lecture, to propose the inflationary theory, in which the universe expands exponentially quickly during a short period in its early history. Guth jotted in his diary after hearing Dicke's lecture:

> Colloquium by Bob Dicke on cosmology—fascinating—most outrageous claims: velocity [of Hubble expansion] at 1 sec tuned to 1 part in 10^{14} to allow for star formation.[2]

Guth's inflationary theory, which was introduced in 1979, accounted not only for the "flatness" problem but also showed how tiny quantum-mechanical fluctuations in the early universe could grow into galaxies and clusters of galaxies in the universe today. Its successes have led many theorists to believe that Ω must be exactly 1 and that the universe was precisely flat.

Unfortunately, it was already clear at the time that there wasn't nearly enough matter energy density to make the universe flat. The matter in the universe that we actually see—galaxies, stars, glowing gas clouds—make up at most 0.05 of the matter-energy density needed to satisfy the flatness criterion.[3] Where was the other 0.95 to come from? This became known as the "missing mass" problem, and its solution became the holy grail of observational cosmologists for 50 years (Once the existence of dark matter was identified, the value of the matter-energy rose to 0.3 of the critical density; still not enough. On the face of it, it still appeared that the universe was a hyperbolic, underdense universe, that would expand forever.)

The geometry of the universe—hyperbolic, underdense for Ω <1), flat for $\Omega = 1$, and spherical for Ω <1—can in principle be probed by measuring standard candles and standard yardsticks throughout the universe. However, it is not so easy to do in practice. To begin with, finding standards that are reliable enough is very difficult. For a long time, beginning with Hubble himself, galaxies themselves were used as standard candles for the distant universe. Hubble had assumed that galaxies had constant surface brightness, so that fainter objects could be considered physically similar to brighter ones but located at larger distances. This was admittedly rough, but it was good enough to show the Hubble distance-velocity relation for the nearer universe. Later, astronomers used some of the brightest galaxies in rich clusters as "standard candles," but because galaxies evolve over time—and the evolution of the brightest galaxies is a subject of interest in itself—there was no reason to believe that galaxies in the distant universe closely resemble those in the nearby universe.

By the 1970s researchers were considering using supernovae as candles for probing the contents of the universe out to very large distances.[4] These staggeringly violent events involving the demise of massive stars (as discussed in chapter 8) and produce a sudden flare across the universe whose brightness can exceed, briefly, the combined light of all its host galaxy's other stars. (The peak blue absolute brightness of a supernova averages about -19.2, which compares well with the absolute magnitude of -20.3 estimated for our own rather larger-than-average Milky Way.) Unfortunately, however, supernovae are very rare. In a typical galaxy, one visible supernova is expected perhaps once every hundred years. Zwicky, again, was the first to systematically search for supernovae in other galaxies. He and a colleague, J.J. Johnson began exposing plates with the 18-inch Schmidt telescope at Palomar and comparing them with a binocular microscope. Between 1936 and 1941, they discovered 14, and later, using the 48-inch Schmidt telescope at Palomar, they discovered many more, some 200 in all. Each time one was discovered, Zwicky's long-time collaborator Baade made careful observations of the light curves showing the rise and fall of the star's brightness, while another associate, Rudolph Minkowski, obtained spectra with the 100-inch telescope. These observations led to the recognition of several subtypes of supernovae. Types Ib, Ic, and II involve core collapse of a massive star. Those of Type Ia, on the other hand, are unique and proved to be particularly important; they appear to occur in binary systems in which one of the stars is a carbon-oxygen white dwarf with a slow rotation. These white dwarfs are limited to below 1.38 solar masses, and if such an object merges with its companion (a very rare event), the Chandrasekhar limit is surpassed in which electron degeneracy pressure is able to prevent catastrophic collapse. At this point nuclear fusion is triggered, and within a few seconds a substantial fraction of the matter in the white dwarf undergoes a runaway reaction leading to a supernova explosion.

Fig. 15.1. Carl Pennypacker, the first leader of the Supernova Cosmology Project. Here he is striking a thoughtful pose as he tests a new device for infrared in vivo imaging of the human brain. *Photograph by William Sheehan, 2008.*

Though the Zwicky-Baade-Minkowski program was highly successful, typically there was a delay between the exposure of a plate and the recognition that a supernova had occurred, meaning the light curves and spectra were seldom as good as the astronomers would have liked. There were not enough observations early in the development of the explosions, as the supernovae were discovered by accident only after peak brightness.

Moreover, the search for supernovae was extremely laborious and time-consuming, and so for a long time astronomers dreamed of using automated telescopes to search for supernovae. The first person to do so was Stirling Colgate, scion of the toothpaste family, then at the New Mexico Institute of Mining and Manufacturing, who in the 1960s and early 1970s set up a 30-inch telescope on a surplus military radar mount at Baldy Mountain. Unfortunately, having to contend with TV tube imagers and inefficient room-sized computers, his telescope never worked. However, his work inspired astrophysicists at the University of California's Lawrence Berkeley Lab to make a new attempt.

The original inspiration came from Berkeley physicist Luis Alvarez. The son of a famous Mayo Clinic physician, Alvarez, in the late 1950s, had used an innovative detector to discover many new elementary particles, work which earned him a Nobel Prize, and was soon to introduce the idea—at first highly controversial but now generally accepted—that an asteroid or comet hit the Earth 65 million years ago, and wiped out the dinosaurs. In 1978, he learned that the Air Force used automated telescopes to monitor missile launches, and with his protégé Richard Muller wanted to use those on the Kwajalein atoll in the Pacific to search for supernovae in galaxies. The plan was simple. Alvarez and Muller would send Kwajalein a list of galaxy coordinates, and receive back pictures of a few thousand galaxies a week. Comparing the new images with reference images, the team would find out if a supernova had gone off. Discussions with Kwajalein were still underway and seemed promising enough for a team leader to be hired. He was Carleton R. Pennypacker, a cosmic-ray physicist who arrived at Lawrence Berkeley lab in October 1979, and was signed to serve as the project's day to day leader.

Though the Kwajalein idea eventually fell through—not least because the Air Force decided it had no interest in switching equipment every few days—a number of promising technological developments had occurred in the meantime, including the increased availability of automated telescopes, cameras built around CCD chips, and inexpensive personal computers that allowed the utilization of new, more powerful image-processing software to help analyze galaxy images. Proposals for funding were sent out, and were promptly rejected. As Ronald N. Kahn, a member of the Berkeley team during its early years, has recalled: "Some reports from referees were full of praise: 'the basic concept is excellent … this kind of search is clearly the necessary way to proceed … the planning has been thoughtfully done.' Others were clearly negative: 'previous efforts have all failed … the authors are basically physicists with virtually no experience in optical astronomy.'"[5]

During an era of tight budgets, there were some lean years for the initiative which eventually evolved into the Supernova Cosmology Project, and the funding to keep it alive was largely provided out of Muller's MacArthur "genius" grant (he had been recognized as one of the top young physicists in the U.S.) In the fall of 1981, there was a rare success, when Pennypacker successfully tested a CCD chip that had cost $6,000—the team was on such a tight shoestring budget it could afford only one. Fortunately, it worked. There followed several discouraging years in which Pennypacker searched for a suitable telescope. For a while they tried to use the Monterey Institute for Research in Astronomy (MIRA) 36-inch reflector, which had been built and run by Ph.D. astronomers who had struggled to find permanent academic positions and wanted to carry on their own research, but logistical problems—including Monterey's distance from Berkeley—intervened. Not until the spring of 1983 did the team obtain its first galaxy picture!Kahn notes, "The team was observing only one galaxy every two years. They needed hundreds per night to succeed. To some the sky seems like a cold and objective place, to others it seems romantic, but to the supernova searchers it was simply their constant source of very personal anguish."[6]

Meanwhile, as the Berkeley team struggled, the Rev. Robert Evans, an Australian amateur astronomer, suddenly burst onto the scene. With a visual memory so remarkable he merited a mention in one of Oliver Sacks's books, An *Anthropologist on Mars,* Evans had committed to memory the patterns of 700 galaxies, and spent 30 or 40 hours a month searching for supernovae with a 10-inch homebuilt reflector in his backyard. He discovered his first, in 1981. From then on he was turning up several new ones every year—a record that demonstrated that amateur astronomers could still contribute important data to astronomy but whose success rather highlighted the Berkeley group's slow progress.

It was only after the Berkeley team gained access to the University of California's 30-inch Leuschner reflector, not as powerful as the MIRA telescope but much more conveniently located in the Berkeley hills, that its fortunes began to improve. After four years without a useful result, they were soon obtaining 50 galaxy images a night, and by January 1986 they had collected some 2000 reference images of galaxies—more even than the Rev. Evans had stored in his memory—and were collecting new images at a rate of about 200 a month. Though they began using software to try to locate a few pixels, out of a hundred million or so they obtained each night, at first they were frustrated by "false alarms"—artifacts—that would need new software to filter out. In the meantime, humans—fifteen or so volunteers—began scanning galaxies on a computer screen; even Luis Alvarez, now 78, spent a few hours a week in front of the screen. At last Frank Crawford, a Berkeley physics professor who joined the team about a year after Pennypacker had come on board, discovered the group's first supernova—in the spiral galaxy M99.

That June, Pennypacker's picture appeared on the cover of *Sky & Telescope,* in association with Kahn's article, "Desperately Seeking Supernovae," with a subtitle added by the *Sky & Telescope* editorial staff: "the gulf between a good idea and a successful project can be very wide."From this first success, they went on to many more. In the next four years, they discovered more than 20 supernovae with their prototype system, and were ready to take their search to the next level. The discoveries so far had been at non-cosmological distances. What they were really interested in were Type Ia supernovae in very distant galaxies—those whose cosmological

redshifts due to the expansion of the universe were a third and more of the speed of light. The reason was that the Type Ia supernovae really are standard candles. The straightforward nature of these explosions is what makes them so useful: they do not depend at all on the evolutionary history of the star, are independent of the age of the universe at the time and on such conditions as the mass of the star from which the white dwarf began or the nature of the companion star. (They may, admittedly, depend somewhat on such parameters as the metallicity of the white dwarf—i.e., the proportion of elements heavier than helium—but these are hopefully details that don't affect their brightness.) Eventually, during the 1990s, Mark Phillips, an astronomer working in Chile, showed that, after correcting for some variations empirically, Type Ia supernovae really are remarkably uniform. In general, all of them have very similar peak magnitudes and light curves. This makes them ideal "standard candles," provided one can discover and follow them up early enough, because then, simply by reading off how bright it is from the light curves, one can work out the distance.

But that required a larger telescope than the one in the Berkeley Hills. Pennypacker managed to achieve something of a coup by gaining access for the 4-meter Anglo-Australian Telescope at Siding Spring, Australia, and he and his colleagues began using it to start the first search for supernovae at cosmological distances (i.e., at expansion velocities greater than about 0.3 times the speed of light). They needed as large a field as they could get. At the time, the largest CCD available had only a 1000 x 1000 pixel CCD. That is minuscule by today's standards; it is hard to believe that it was the biggest CCD on a large telescope at the time. Indeed, it could only be used by inserting custom optics to de-magnify the image onto the tiny CCD.

Bad weather and other issues at AAT hampered that initiative, and led Pennypacker to transfer what was now known as the Supernova Cosmology project to the 2.5-meter Isaac Newton telescope at La Palma in the Canary Islands. This telescope had had a long and rather troubled history, which has been recently told by astronomical historian Lee T. Macdonald.[7] Inspired by Hubble's comment that "the conquest of the Realm of the Nebulae is an achievement of great telescopes," a Royal Society committee on the needs of British astronomy recommended the acquisition of a large telescope, to be set up in Britain (as regards to doubts about the climate, the work that William Herschel had done with the 48-inch telescope at Slough was mentioned). The new telescope was to be at Herstmonceux Castle, where the Royal Observatory was slated to be moved from Greenwich, and in honor of the tercentenary of Isaac Newton's birth—postponed from 1943 to 1946 because of the War—the telescope was to be named the Isaac Newton telescope. Funding for the telescope was quickly approved by the Government, but the telescope took 21 years to build.[8] Originally, it

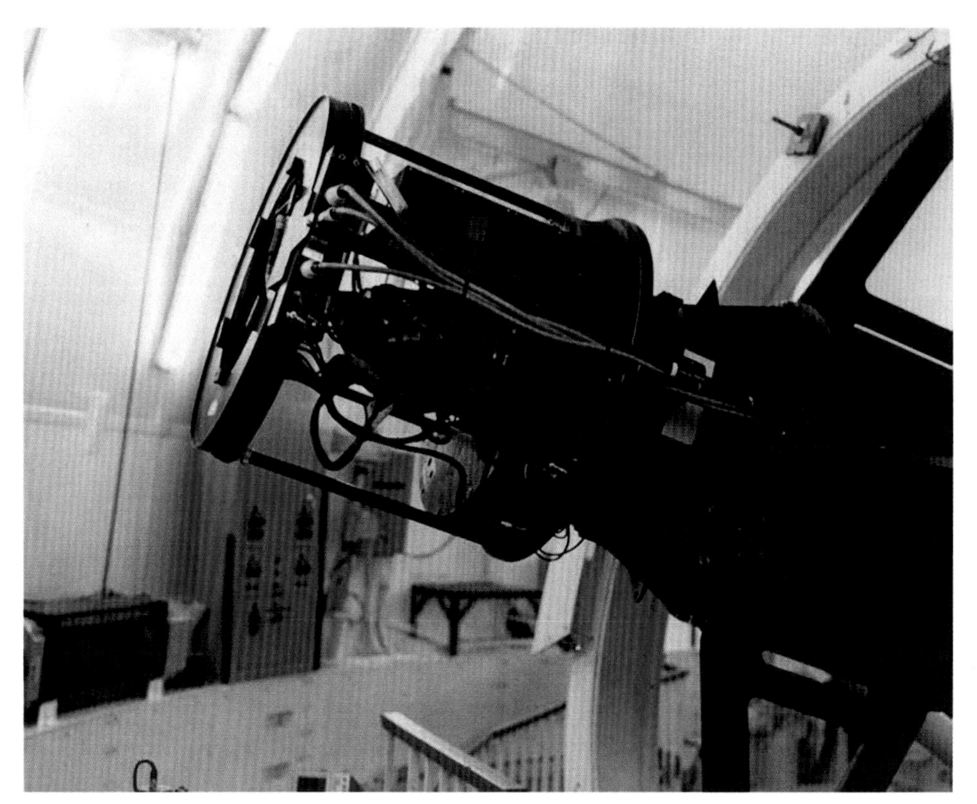

Fig. 15.2. Prime focus camera on the Isaac Newton Telescope, used to observe supernovae at cosmological redshifts for the Supernova Cosmology Project and leading to the discovery of dark energy. *Courtesy: Carl Pennypacker.*

was meant to be an interchangeable Schmidt and Cassegrain instrument, but eventually it was made as a conventional Cassegrain. Because of cracks in the glass, the mirror took a long time to grind; it was figured to an f/3 spherical instrument in 1954, and later reground to a parabolic figure, but because of the cracks, left at f/3. This made it, in the words of Fred Hoyle, "a little runt of an instrument." That was meant as a pejorative statement—like "Big Bang"—but the littleness proved to be a virtue. The telescope was officially dedicated by Her Majesty Queen Elizabeth II at Herstmonceaux on December 1, 1967—a foggy day, so nothing was seen. That, alas, would prove portentous. As soon as the telescope went into regular use, the bad observing conditions became so frequent that it was finally agreed to move it to the Canary Islands. Dismantled at Herstmonceaux in 1979, it reopened on La Palma in 1984.

The telescope—precisely *because* it was a runt—had a very wide field, and proved to be a good fit to the needs of the Supernova Cosmology Project. It was used to make the group's first discovery of a cosmological Type Ia supernova in 1992 (cosmological redshift $z = 0.0457$).[9] At this point Pennypacker turned the reins over to his colleague, another young Berkeley astrophysicist, Saul Perlmutter, and turned his efforts over to something else, the educational project, "Hands on the Universe." Having devoted over a decade and a half of his life to the project, and endured some of the years of great frustration, he had seen it mature with the advent of technological advances (notably the exponential increase in the size of CCD arrays) to the point where it was ready to take off, and within a few years, scores of cosmological supernovae were being registered by the Supernova Cosmology Project led by Perlmutter and the rival High-z Supernova Search Team, organized by Brian Schmidt, a post-doc at Harvard, and Nicholas B. Suntzeff, a staff astronomer at Cerro Tololo Inter-American Observatory in Chile, who began searching for cosmological supernovae with the 4-meter Victor M. Blanco telescope. Preliminary results from the teams began to trickle out from both teams between 1995-97, but they contained large errors. It was not long, however, before definitive results were published. Perlmutter, in a talk in 1997, announcing the discovery of a supernova with expansion velocity z = 0.83, concluded, "We may live in a low mass-density universe." The High-z Supernova Search Team published a definitive paper in 1998,[10] followed almost at once by one from the Supernova Cosmology Project.[11] The results startled the astronomical world. These papers, which quickly became among the most often cited in physics, are likely to prove as revolutionary as Max Planck's late 1899 paper introducing the quantum

Fig. 15.3. Hubble diagram for 42 high-redshift Type Ia supernovae from the Supernova Cosmology Project, and 18 low-redshift Type Ia supernovae from the Calan/Tololo Supernova Supervey, plotted on a linear redshift scale to display details at high redshift. The accelerating expansion of the universe emerges from the high-redshift data. *Courtesy: Carl Pennypacker.*

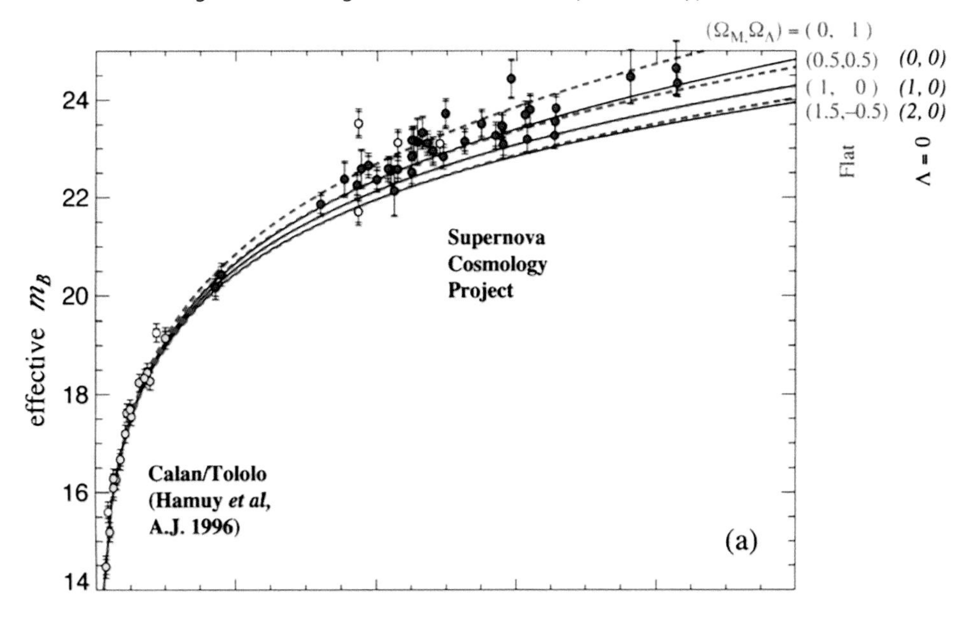

theory. The supernova surveys would earn Riess, Schmidt, and Perlmutter the 2011 Nobel Prize in physics.

It was evident at once that the rival teams, working independently, had obtained corroborative results. From their analysis of cosmological redshifts for about 50 Type Ia supernovae (with expansion velocity z from 0.16 to 0.83), they found something that was—as much as Slipher's discovery of the large redshifts or Zwicky's discovery of "dark matter"—completely unexpected. Once again, it seemed, nature had surprised us.

Though when astronomers first started thinking about using supernovae as probes of the very distant (early) universe in order to determine its fate, they expected to determine the "deceleration" of the universe—in other words, they sought evidence favoring either the "Big Crunch" or the "Big Chill—when the experiment finally came to be done, they found—neither.

The main observational result of the supernovae surveys was that the supernovae discovered by both teams were fainter than what would have been expected for what was then the standard low matter density universe. When analyzed in detail and combined with constraints from measurements of the Cosmic Microwave Background, it was further determined that the universe not only has a low matter density, but a total density equal to the critical density. In other words, Ω still equals 1. Because of this, the space between us and a distant supernova is larger than expected on the basis of the low matter density universe. Hence the shocking conclusion: the universe is not only expanding, its expansion is accelerating. It follows that there must be another source of density than matter. Both supernova search teams interpreted this as a non-zero cosmological constant. By analogy to Zwicky's "dark matter," University of Chicago theoretical cosmologist Michael Turner called this cosmological constant term "dark energy." Dark energy, whatever it is, is opposed to the attractive force of gravity, and involves a repulsive force driving the galaxies farther apart. Its effects (acceleration of the expansion of the universe) have become evident only in the last 5 billion years of the history of the universe, since only then has the expansion of the universe has reached the stage where the repulsive force of Dark Energy exceeds the attractive force of gravity.

Berkeley theoretical cosmologist Eric V. Linder has pointed out that even in a universe without matter or radiation but just the positive cosmological constant (this is called a de Sitter universe; see chapter 11—admittedly it sounds a bit bizarre) one still has acceleration. In that case, there is nothing but a positive uniform energy. But when matter and radiation are added back in, things become more interesting:

> Matter and radiation have the usual gravitational attraction that pulls objects together, fighting against expansion. They act to decelerate the expansion. Depending on the relative contributions then between

> matter etc. and the vacuum, the final result can be either a decelerating or accelerating universe. One of the great paradigm shifts in cosmology was the realization [from the supernova surveys] that we live in a universe that accelerates... This is really a striking development... After Copernicus we have moved beyond thinking that the Earth is the center of the Universe; with the development of astronomy we know that the Milky Way Galaxy is not the center of the Universe; through physical cosmology we know that what we are made of—baryons and leptons—is not typical of the matter in the Universe; and now we even realize that the gravitational attraction we take as commonplace is not the dominant behavior in the Universe.[12]

What is dark energy? We do not yet know, though there are some intriguing surmises. One of the things that astronomers are trying to do now, and will work on for at least the next 20 years, is to formulate some basic questions about dark energy. If it indeed is real, and many observations suggest that something is causing an acceleration in the expansion of the universe, one of the first things we would like to know is whether dark energy is constant or if it evolves over time. Constant dark energy corresponds to the the cosmological constant (which Einstein first introduced in his 1917 cosmological solution of the gravitational field equations, in order to produce a static universe; his greatest blunder!). However, it is also possible that pressure from dark energy may vary with time, so that it increases in relative importance as time goes on. It may also be that dark energy reflects an outward pressure from space itself (or more precisely the vacuum; the vacuum is the lowest energy state of a quantum field) to expand, so that as more and more space is created by the universe's expansion, the greater becomes the corresponding energy of the vacuum and the greater the acceleration. There are many theoretical ideas for how this can happen. Most of these ideas arise from particle theory and some are based on string theory. What we may ultimately find is that the effects we see on the largest structures in the universe and even the universe itself comes down to the smallest particles. This would not be the first time, and probably not the last, that a connection was found between large features of the universe and atomic/particle physics. Dark energy does more, however, than determine the overall structure of the universe. It also profoundly affects its constituents, namely: clusters of galaxies, individual galaxies, even, perhaps, stars and planets. Simply put, if the universe is accelerating due to dark energy, it becomes more difficult for structures to form in the universe, and when they do form, it becomes more difficult for them to form even larger structures. If the universe were dark-matter dominated, then we would expect to find even larger galaxies and galaxy clusters than is actually the case. It might turn out that most of the properties we see in the universe today are due to the interactions between entities that are unseen, namely dark energy and dark matter. The nature of these interactions constitutes one of biggest unsolved problems in astronomy today, and though the answers are still obscure, it seems that the universe is providing us tantalizing shadows into a still unknown truth and reality that seems to grow, in the words from *Alice in Wonderland*, "Curiouser and Curiouser."

* * *

1. Alan Guth, "Was Cosmic Inflation the 'Bang' of the Big Bang?" *The Beamline, 27,* 14 (1997).
2. Guth, *Inflationary Universe,* 26.
3. For a very accessible discussion, see Martin Rees, *Just Six Numbers: the deep forces that shape the universe.* New York: Basic Books, 1999.
4. See: Robert V. Wagoner, "Determining qo from Supernovae," *The Astrophysical Journal,* 214 (1977), L5-7.
5. Ronald N. Kahn, "Desperately seeking supernovae," *Sky & Telescope,* June 1987, 594-597.
6. Ibid., 596.
7. Lee T. Macdonald, cited in: "Meeting of the Royal Astronomical Society, January 13, 2012," *The Observatory, 132,* 1229 (August 2012), 220-223.
8. Ibid.
9. C. Pennypacker, S. Perlmutter, G. Goldhaber, A. Goobar, et al., "Supernova 1992i in anonymous galaxy," *IAU circular no. 5652* (November 12, 1992).
10. Adam G. Riess, Alexei V. Filippenko, et al., "Observational evidence from supernovae for an accelerating universe and a cosmological constant," *Astronomical Journal; arXiv.astro-ph/9805201v1* (May 15, 1998).
11. S. Perlmutter, G. Aldering, G. Goldhaber et al., "Measurements of [omega] and [lambda] from 42 high-shift supernovae," *Astrophysical Journal; arXiv.astro-ph/9812133v1* (Dec. 8, 1998).
12. Eric V. Linder, "Frontiers of Dark Energy," in *Adventures in Cosmology,* David Goodstein, ed. (Singapore: World Scientific, 2012), 357-358.

16. Afterglows

When I heard the learn'd astronomer,
When the proofs, the figures, were ranged in columns before me...
How soon unaccountable I became tired and sick,
Till rising and gliding out I wander'd off by myself,
In the mystical moist night-air, and from time to time,
Look'd up in perfect silence at the stars.

—*Walt Whitman, "When I heard the learn'd astronomer"*

The Milky Way consists of 100,000 million stars. It appears sedate, placid, unchanging as it curves, a river of stars across the bowl of the night-sky. We know, however, that it is a dynamic structure; it must have had a violent youth in which it was cobbled together from various bits and pieces, from blobs of dark matter, streams of gas, and smaller stellar systems.

Majestic winding spiral that it is now, it was a ragged and untidy thing once, like the galaxies we see now when we look into deep space and deep time of the universe, their collisions and rip-tides being visible in the Deep, Ultra Deep, and Extremely Deep fields of the Hubble Space Telescope.

We can see what the Milky Way was once from what the galaxies in those fields are now. Indeed, a telescope is a time-machine, and we look not into the past but into an eternal present. From the edge of our universe looking back, to anyone who happens to be out there, the Milky Way is still unformed, still shaped, or misshaped, as it was 12 billion years ago (there would be, needless to say, be nothing to distinguish it to that distant viewer).

We puzzle for a moment over the resemblance of the stars in the Galaxy to the neurons in a microscopic cross-section of brain. It is easy to imagine that in the microscopic view, we are looking across great rhythmic patterns like the spiral-density waves of galaxies, tides of neurons whose numbers are of the same order as stars in the Milky Way – 100,000 mil-

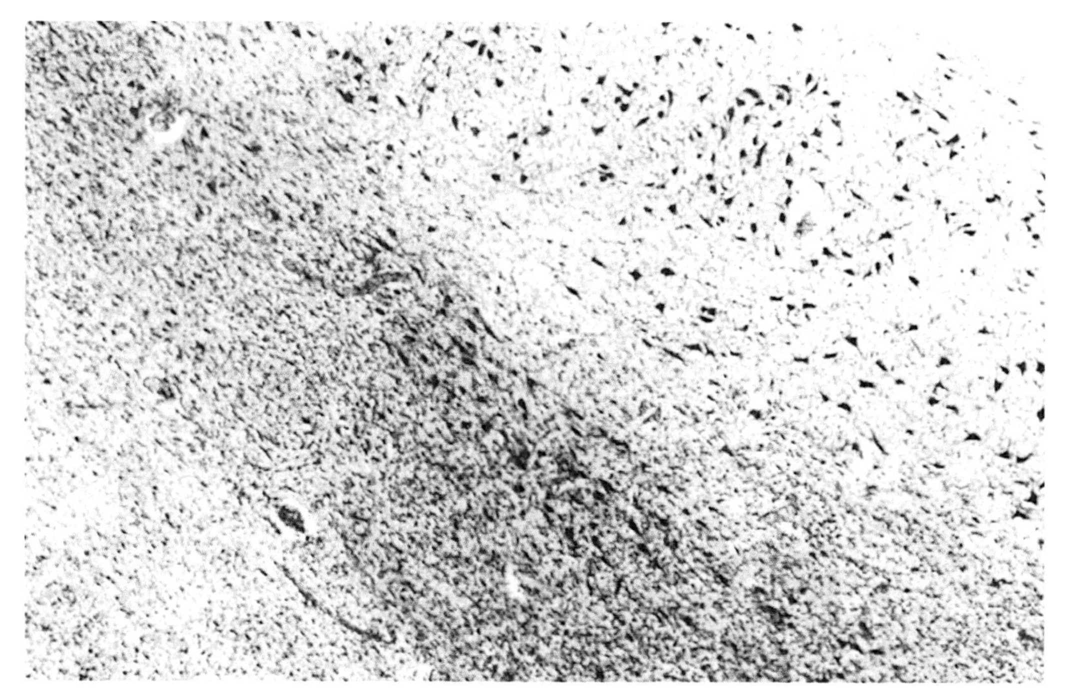

Fig. 16.1. Neurons in the human medulla. Nissl stained micrograph. *Courtesy: William Meller, M.D., University of Minnesota Medical School.*

lion. They follow definite whorls and gyri. They seem in places to consist of dramatic starbursts. There are clusters of neurons no less than there are clusters of stars. We call these neuron-clusters in the brain, nuclei. As we know that in the Milky Way the regions of star-formation, giving rise to the diamond-like clusters that bejewel the Galaxy, occur in the dust and molecular clouds associated with the spiral arms, it is impossible not to imagine that some similar density-wave – associated with chemotactic factors, hormones, and other chemical tracers– pulses through the developing nervous system to orient or commit cells to their eventual specializations and functions.

There have been numerous analogies to explain the human mind, none perhaps grander than the one we just described, and which we illustrate by juxtaposing bursts of neurons in the medulla as shown in microscopic view (Figure 16.1) with the bursts of stars along the dust-lanes of the Andromeda Spiral (Fig. 16.2). There is an astounding symmetry in these structures, and perhaps between the cosmography of inner and outer worlds. In both, we are presented with complexity, in both what we see must be explicable in part in terms of complexity theorists' "ideas of heterogeneous mixtures, of multiple levels and scales, of intricate connectivities, and of nonlinear dynamics."[1]

Fig. 16. 2. Detail of spiral arm structure in the Andromeda Galaxy. The view is dominated by NGC 206, the brightest star cloud in the Andromeda Spiral and similar to M24, the Sagittarius star cloud in the Milky Way, but much larger. *Credit: William Sheehan, University of Minnesota Medical School.*

Obviously, only something as complicated as the brain—which we are told is "among the most complicated natural objects in the known universe"[2]—can understand something as complicated as the universe. To the great Spanish neuroanatomist Santiago Ramón y Cajal is attributed the saying, "the universe is the reflection of the structure of the brain."[3] Perhaps, however, it is the other way around, and the structure of the brain is a reflection of the universe. Perhaps both are true.

The brain is, in a real sense, the ultimate scientific instrument. It was not, of course, "designed" for doing astronomy and cosmology—anymore than the eye, which we considered in the first chapter, was "designed" to be an astronomical detector. The brain—and the eye—are modifications of ancestral structures, organs that go back to primitive mammals, reptiles, fishes. They have been adapted to the exigencies of the here-and-now—the immediate pressures of tracking game, getting enough food to eat, finding mates, and the like. They represent the tinkering of natural selection over thousands of millennia. Our ability to use them to see into the remote corners of the Galaxy, or to solve the equations of General Relativity, are ex-aptations—to use the term evolutionists have coined to describe shifts in the function of a trait over the course of evolution. The traits we rely on were originally subservient to more immediately practical purposes. But it is no accident that survival of the species was well served by the passion for knowledge of the natural and social worlds. Acquisition of knowledge would have conferred advantages in the struggle for survival, as the animal who had the keenest sense of touch or smell or hearing, the most developed ability to determine the color of a surface or the expression of a face, would have the upper hand in the great struggle for survival. Moreover, the one that was best able to acquire that knowledge would have needed in greatest degree what we would call intelligence; for, as the neuroaestheticist Semir Zeki has said, "it takes but a moment's thought to realize that obtaining that knowledge is no easy matter. The brain is only interested in obtaining knowledge about those permanent, essential, or characteristic properties of objects and surfaces that allow it to categorize him." So, because information reaching the brain from objects and surfaces is in continual flux, vision "is an active process that depends as much upon the operations of the brain as upon the external, physical environment; the brain must discount much of the information reaching it, select only what is necessary in order to obtain knowledge about the visual world, and compare the selected information with its stored record of all that is seen. A modern neurobiologist should approve heartily of Matisse's statement that "'*Voir, c'est déjà une operation creatrice, qui exige un effort.*'"[4]

Humans being—well, human—we have always been tempted to consider ourselves the acme of Creation. The mind of man seems to exhibit astonishing perfections that range beyond animal-nature; so much so as to make us believe that man is separate, alone, other—the only creature God created in his image, the only one into which he breathed a soul. Somehow it has always seemed (to us) that our nature comprehends the rest of nature; that we are, in one, what all the other animals are separately, a point of view well attested in Sophocles' *Antigone:*

> Manifold are the marvels of the earth,
> And none more marvelous than man appears.
> Through the grey sea on southern storms,
> Through the swells that raven around him he roams;
> And eldest of gods, earth enduring
> And all unwearied, he works away,
> As yearly he follows the yoke and the plowshare,
> Scarring the sod with the stallion's foal....
> A thing of wisdom this wile of his craft,
> And thus he holds it, all hope surpassing,
> As he wields it now ill, now well, and aspires...[5]

Man's sense of his own uniqueness must be qualified, however, when one turns from the impressive workings of his mind to a consideration of the substance of his brain. That brain has some remarkable attributes, indeed, but they are such as to make it different in degree rather than in kind from those of primates and other mammals. Indeed, the human brain is more like that of a lizard than a circuit-board is to a vacuum tube. Emerson had it nearly right when he said:

> A mollusk is a cheap edition [of man] with a suppression of the costlier illustrations, designed for dingy circulation, for shelving in an oyster-bank or among seaweed.[6]

One of the authors (W.S.) well remembers his first introduction to the brain in medical school. He had seen pickled brains—a bit unimpressive, like jellyfish lying on the beach, desiccated, collapsed, blue-lipped, lifeless. He even dissected them. The first real living brain he saw was during a neurosurgery rotation: an operation was underway, and he looked on, in equal parts fascinated and repulsed. What was this pia-mater and meninges-wrapped, convoluted, three pounds' worth of fatty flesh, this computer made of meat? Well might he have said with Hamlet, "My gorge rises at it!"

It was hard to believe, even as a medical student, that there could be any relation at all between what we know, immediately, as mind—the mysterious realm of thoughts, moods, dreams, actions—and brain. And yet—though we know the brain only from the inside-out—the thoughts, moods, dreams, and actions are what we perceive immediately—only the brain can understand the brain. If indeed the universe is a reflection of the structure of the brain, or vice versa, then, as Cajal again is supposed to have said, "as long as our brain is a mystery, the universe ... will also be a mystery." The brain, in bootstrapping its way to understanding the structure and evolution of the Galaxy, must use the brain to study the brain itself.

It is common to point out that the universe is conscious of itself only through us. In other words, insofar as we are conscious of the universe, it is in some measure conscious also of itself. Obviously, consciousness did not come about for that purpose. It cannot have been the purpose of the universe, which undoubtedly was (as Steven Weinberg implied in the quote at the end of chapter 14) blind and purposeless, to produce conscious beings like ourselves. For that matter, even consciousness is really only the tip of the iceberg of the brain's activity; most of what goes on in the brain is that of which we are not consciously aware. Consciousness is the frosting and the candles of the brain, as the bright young stars and HII regions are of the Galaxy; it is only the froth, but it is what we notice. Underlying consciousness are all sorts of processes of which we are never aware, just as the backbone of the Galaxy consists of the old red stars of the bulge (to say nothing of dark matter).

Equipped with all these caveats, and divested of all the grandiose notions that have had the stuffings knocked out of them by Copernicus, Darwin, Shapley, and Steven Weinberg—we nevertheless return to the small observatory on the North Fork of the Crow River, and open again our dome to the sky. The telescope swivels to the point where we direct it, the CCD imaging session begins—and we behold images that astonish and delight.

To the extent that we are conscious beings—and capable of formulating questions about the nature of the universe—we are also feeling beings. Underlying the conscious mind—the intellectual and decision-making centers of the prefrontal cortex—are deeper, older structures; those in which lie the pacemakers that govern the oscillations and rhythms of life (many of which are tied in with the cosmos, and swing in diurnal, lunar, or circannual phases) and those that connect us with the viscera, anchor brain to body, give rise to a rich palette of emotions, including fear and awe and the thrill of discovery. The impulse to study the stars—the tremblings, trepidations, aspirations, excitements—that drive us to open up the dome to the heavens and to capture the remote galaxies on a CCD chip may seem immediate but they are founded in a far longer and more mysterious ancestral past. The writer Lafcadio Hearn, who followed Percival Lowell to the Far East and once said, "we owe more to our illusions than to our knowledge,"[7] has put this as well as anyone can:

> It is now certain that most of our deeper feelings are superindividual,—both those which we classify as passional, and those which we call sublime. The individuality of the amatory passion is absolutely denied by science; and what is true of love at first sight is also true of hate: both are superindividual. So likewise are those vague impulses to wander which come and go with spring, and those vague depressions experienced in autumn,—survivals, perhaps, from an epoch in which human migration followed the course of the seasons, or even from an era preceding the apparition of man.... The delight, always toned with awe, which the sight of a stupendous landscape evokes; or that speechless admiration, mingled with melancholy inexpressible, which the splendor of a tropical sunset creates,—never can be interpreted by individual experience. Psychological analysis has indeed shown these emotions to be prodigiously complex, and interwoven with personal experiences of many kinds; but in either case the deeper wave of feeling is never individual: it is a surging up from that ancestral sea of life out of which we came.[8]

It is that which keeps us coming back to our little dome on the North Fork of the Crow River—that feeling which is among the most subtle and valuable we can experience, of something oceanic, unfathomable, cosmic. We know that the universe is crushingly vast, fearfully violent, strangely wrought, utterly impersonal; it is indifferent to us and to our fate,

> And we are here as on a darkling plain,
> Swept with confused alarms of struggle and fight,
> Where ignorant armies clash by night.[9]

It cannot care for us; but why should it? If we were not too long steeped in our own arrogance, we would not presume upon the role of being a central actor in the drama of Creation. And yet—it still moves us. We leave the lecture room and wander out into the mystical moist night air. There is intoxication in moonlight we feel still even if in some sense we know that the Moon is a dead world; there are charms in the light of distant suns.

We look up at the stars, ponder the powdery precincts of the Milky Way (whose image, sadly, cannot even be glimpsed by billions on the globe, obscenely defaced by the unnecessary glare of artificial lights erected to keep at bay our primal fear of night; we are lucky here, and can see it, and our spine tingles). The image builds on the screen, as the CCD does its work. Again and again—like Monet painting the haystacks, the water lilies, or the façade of the Cathedral of Rouen—we keep returning time and time again to the objects which have always enchanted; to the Orion Nebula, centered on the hot young massive star θ^1 Orionis, to the Pleiades, tangled like Nerea's hair in the dust of its reflection nebula, to the jeweled splendor of the Double Cluster in Perseus, to the twisty tendrils of the Crab, and on through endless night.

We know, thanks to the journey we have trod in these pages, that these splendors are merely a soufflé, the frosting and candles; the dust and gas and bright young hot stars which dominate our view of the universe making up only 2% of the mass of the Galaxy. We are merely looking at the surface of things. And yet we continue to enjoy our surface, our canvas, and to paint again and again, unwearyingly, the ever-varying light—preferring the romanticism of Monet to the hard and sharp realities of the cubists.

With Marcus Aurelius, we feel resigned: "Whatever is right with you, O Universe, is right with me!"

We suppose—now that planetary systems have been found around thousands of others stars, and no end in sight—there must be many worlds which harbor life, and some which harbor viscoelastic brains out there, pondering as we ponder, conscious of the universe as we are. How many?

We don't know. But the universe is vast, so perhaps billions and billions. How many of them know an equation similar to the following:

$$R_{\mu\nu} - \frac{1}{2} g_{\mu\nu} R + g_{\mu\nu}\Lambda = \frac{8\pi G}{c^4} T_{\mu\nu}$$

How many of our cohorts in the Milky Way realize that the Andromeda Spiral is headed toward us, and that in several billions of years the two will collide, and merge into a giant elliptical system?

How many of those cohorts have worked out the nature of dark matter, or discovered that the universe's expansion is accelerating (thus that the cosmological constant is not zero but corresponds to something we can call, till we understand it better, "dark energy")? How many have worked the expansion of the universe back to the first 10^{-43} sec after the big bang; how many have determined what happened before that, or worked through the intricacies of string theory?

It is something to be able to say that, if there are intelligent creatures elsewhere in the Galaxy, or in the universe, they will come to know some of these things (and perhaps, who can say, others that go far beyond what we know). The human brain is a notoriously unreliable instrument of knowledge; its construction is elaborate and imperfect—in some ways its multimodular structure, consisting of several brains jerry-rigged and strapped together rather like the Space Shuttle which is precariously configured around solid-state and liquid-fuel boosters. It has its makeshift joints, its O-rings; we cannot trust its knowing implicitly. Its unreliability is attested by the numerous illusions, misperceptions, even frank delusions and hallucinations, that its visco-elastic flesh is heir to (it is sobering, for instance, to note that seventy percent of schizophrenics have no insight into their illness, and do not believe they are ill). One of the brain's features is that the virtual reality it produces, a performance we call "mind," consistently mixes up inner and outer worlds. Ever since those internally-visioned ancestors of ours who painted the Great Beasts of the Ice Age on the cave walls of Chauvet, Lascaux, and Altamira—lighting the darkness around them with the emanations of the inner light of dreams—we, as a species, have been only too apt to confuse our inner visions with outer realities, and to regard our conceptual categories, predilections or dispositions, with the word or will of God. It is no doubt true, as Francis Bacon said in the *Novum Organum* (writing of the Idols of the Tribe, fallacies natural to humanity in general):

> For man's sense is falsely asserted to be the standard of things: on the contrary, all the perceptions, both of the senses and the mind, bear reference to man and not to the universe; and the human mind resembles

> those uneven mirrors which impart their own properties to different objects ... and distort and disfigure them.[10]

He spoke also of the Idols of the Cave, the errors peculiar to the individual man. "For every one ... has a cave or den of his own, which refracts and discolors the light of nature."[11]

Every thinking being in the universe must have its peculiar biases, its "Idols of the Tribe," also its "Idols of the Cave." Even so, we would like to think that—distorted and disfigured, refracted and distorted as it may be—the universe we describe is not a mere personal—or even a collective—hallucination. That our science is neither sophistry nor solipsism. That what we know is what other minds, say at Sirius or Aldebaran or M31 or the Coma Cluster, would know and—after suitable translation from their alien language to ours, and mostly looking like mathematics—describe much as we do. To the extent this is true, science becomes the study of the covariant truths, those that hold in all places and all times in our wonderfully homogeneous and isotropic universe, perhaps even in all the universes that can be that can support such beings as we.

The basis of science is the admission of the power of the human mind to perceive and understand the world – and itself – but also the realization that that power is limited, the instrument of knowing flawed and imperfect. It is, as Jacob Bronowski put it in *The Ascent of Man,* reflecting on what happened at Auschwitz:

> This is where people were turned into numbers. Into this pond were flushed the ashes of some four million people. And that was not done by gas. It was done by arrogance. It was done by dogma. It was done by ignorance. When people believe they have absolute knowledge, with no test in reality, this is how they behave. This is what men do when they aspire to the knowledge of gods.
>
> Science is a very human form of knowledge. We are always at the brink of the known, we always feel forward for what is to be hoped. Every judgment in science stands on the edge of error, and is personal. Science is attributed to what we can know although we are fallible. In the end the words were said by Oliver Cromwell: "I beseech you, in the bowels of Christ, think it possible that you may be mistaken."[12]

We don't yet know how pervasive life is in the universe. We do know that there was liquid water once present on Mars, and the Saturnian moon Titan seems to contain hydrocarbon lakes on its surface. The building-blocks of life pervade the Solar System; we can guess that the Galaxy, indeed the universe, must be teeming with life, at least of the single-celled variety. How often do multicellular organisms occur? (On Earth, life remained single-celled for almost three billion years.) The universe may be colonized with microbes, but what we really want to know is: are we alone? We would take small comfort in looking up at the stars and imagining colonies of staphylococci or streptococci, worlds inhabited with smallpox or even sponges and jellyfish. How often, among the myriad planets which exist, are they barren wildernesses, abortive, waste? How often does life arise? Any kind of life? In the Galaxy, is intelligent life common? How do we define intelligent life? Is it life capable of understanding the Pythagorean theorem? Life capable of sending and receiving radio messages? Life capable of grasping the equation we cited above?

The universe itself appears to be highly intelligent. A cosmos. It has mathematical structure –elegance –rationality—beauty. We see it in the logarithmic spirals which unfold in the spiral galaxies, in Pythagoras's harmonics which strike the galactic tuning forks. We can almost imagine in it the workings of mind.

But is it a human mind? A mind with any affinity to ours? Or is it not only the illusion of a mind, governed through the same laws of complexity, seeming intelligent only because our own intelligence has specifically evolved to understand that coherence – those laws – that gave rise to us?

The wish to believe in something greater than ourselves comes from within ourselves. We continue, almost instinctively, to desperately seek some echo of response in the universe at large to the chord struck within. Desperately, because we are unlikely ever to receive it.

In the end, we remain the hopeful and terrified creatures that long ago huddled around the hearth in the Cave of the Night. We remain forever children, small, shivering, afraid of the dark. But also children who are curious and can't resist the challenge of piecing together the parts of a puzzle, and when the pieces fit, are filled with pleasure and a sense of achievement. And eager to tackle another.

Above, where the Beasts of the Ice Age, which we then and still hold in awe, stampede across the sky, the fiery stars maintain their icy silences. We continue to seek wisdom in their mute counsel.

* * *

1. On complexity, see: M. Mitchell Waldrop, *Complexity: the emerging science at the edge of order and chaos* (New York: Simon & Schuster, 1992) and Roger Lewin, *Complexity: life at the edge of chaos* (New York: Macmillan, 1992).
2. Gerald M. Edelman, "Building a Picture of the Brain," in *The Brain, Daedalus* (Journal of the American Academy of Arts and Sciences), Spring 1998, 48.
3. We have tried to discover the source of that remark, without success. It is not, at any rate, in Cajal's autobiography, *Recollections of my life,*

E. Horne Craigie with Juan Cano, trans. (Cambridge, Mass.: The MIT Press, 1989).

4. Semir Zeki, "Art and the Brain," in *The Brain, Daedalus,* 73-74.
5. Translated from the Greek by Michael S. Armstrong, of Hobart and William Smith Colleges; used with permission.
6. Ralph Waldo Emerson, "Power and Laws of Thought."
7. Lafcadio Hearn, *Glimpses of Unfamiliar Japan* (Rutland, Vermont, and Tokyo, Japan: Charles E. Tuttle, 1976 reprinting of 1894 edition), xxv.
8. Hearn, *Kokoro: hints and echoes of Japanese inner life* (New York and Tokyo: ICG Muse, 2001 reprinting of 1896 edition), 197-198.
9. Matthew Arnold, "Dover Beach."
10. Francis Bacon, *Novum Organum,* Aphorism XLI
11. Bacon, *Novum Organum,* Aphorism XLII
12. Jacob Bronowski, *The Ascent of Man* (Boston and Toronto: Little, Brown and Company, 1973), 374.

Author Index

Subject Index

Printed by Publishers' Graphics LLC
MLSI140925.19.35.1